Introduction to Infrastructure:
An Introduction to Civil and Environmental Engineering

Introduction to Infrastructure:
An Introduction to Civil and Environmental Engineering

Michael R. Penn
University of Wisconsin–Platteville

Philip J. Parker
University of Wisconsin–Platteville

About the Cover

Top left: A young engineer standing in the influent sewage pipe during construction of upgrades to the Fort Wayne, Indiana wastewater treatment facility. Source: Courtesy of Donohue & Associates/Stacy Cooke.

Top right: September 2, 2005—A staging area for rescue operations in New Orleans, Louisiana, during the flooding caused by hurricane Katrina. Source: Courtesy of FEMA/Joscelin Augustino.

Bottom: A portion of the Big Dig project in Boston, under construction. Source: Courtesy of Massachusetts Department of Transportation.

VICE PRESIDENT & EXECUTIVE PUBLISHER	Don Fowley
ACQUISITIONS EDITOR	Jennifer Welter
ASSISTANT EDITOR	Alexandra Spicehandler
EDITORIAL ASSISTANT	Samantha Mandel
MARKETING MANAGER	Christopher Ruel
MEDIA EDITOR	Thomas Kulesa
PHOTO RESEARCHER	Sheena Goldstein
SENIOR PRODUCTION MANAGER	Janis Soo
ASSOCIATE PRODUCTION MANAGER	Joyce Poh
ASSISTANT PRODUCTION EDITOR	Yee Lyn Song
COVER DESIGNER	Seng Ping Ngieng

This book was set in 10/12 Palatino-Times Roman by Laserwords Private Limited and printed and bound by Quad/Graphics. Company. The cover was printed by Quad/Graphics.

This book is printed on acid free paper. ∞

Founded in 1807, John Wiley & Sons, Inc. has been a valued source of knowledge and understanding for more than 200 years, helping people around the world meet their needs and fulfill their aspirations. Our company is built on a foundation of principles that include responsibility to the communities we serve and where we live and work. In 2008, we launched a Corporate Citizenship Initiative, a global effort to address the environmental, social, economic, and ethical challenges we face in our business. Among the issues we are addressing are carbon impact, paper specifications and procurement, ethical conduct within our business and among our vendors, and community and charitable support. For more information, please visit our website: www.wiley.com/go/citizenship.

Library of Congress Cataloging-in-Publication Data
Penn, Michael R., 1964-
     Introduction to infrastructure : an introduction to civil and environmental engineering / Michael R. Penn, Philip J. Parker.
          p. cm.
     Includes bibliographical references and index.
     ISBN 978-0-470-41191-9 (pbk.)
     1. Civil engineering. 2. Environmental engineering. I. Parker, Philip J., 1971- II. Title.
     TA147.P46 2012
     624—dc23

                                   2011029037

Printed in the United States of America
V10003940_082418

Dedicated to the memory of Matt Rynish, an outstanding civil engineering alumnus of the University of Wisconsin-Platteville whose life was tragically lost, at age 27, in an occupational accident during the final editing phases of this book. He is shown in the rear of the concrete canoe below, after winning a race at the 2006 Midwest Regional Concrete Canoe Competition.

Foreword

We view our nation's infrastructure—the built environment—from a variety of personal perspectives shaped by locale and experience. A useful perspective, however, is that given by the federal budget, nearly three-fourths of which is annually spent on just five issues—defense, social security, welfare, Medicare, and Medicaid. All but defense are considered social issues, matters that affect us as members of society, and spending on them is intended to maintain or improve our quality of life. Engineers concerned with the built environment recognize that sound infrastructure also contributes to our quality of life—our economic health and our physical well-being. In this respect, spending on infrastructure and spending on social issues benefit society in the same ways. Infrastructure, however, is not viewed in the same light, yet it competes for the same dollars. When we consider infrastructure in the context of social issues, we recognize that spending on infrastructure is a wise investment in our health, our mobility, and our overall quality of life.

To see infrastructure as a major factor in determining our quality of life, we need only to travel back in time 100 years. In the early 1900s, only 14 percent of homes in the United States had indoor plumbing, and fewer than 8 percent had telephones. There were just 8,000 automobiles, and only 144 miles of paved roads. The average wage was 22 cents per hour and fewer than 6 percent of Americans were graduated from high school. The population of Las Vegas was 30 people, and California was the 21st most populous state.

One hundred years ago, water-borne disease was the third leading cause of death and the average life expectancy was just 47 years. Dr. Louis Thomas, who served as the head of the Yale Medical School, wrote in a letter to Senator Kit Bond of Missouri: "... the greatest advances in improving human health were the development of clean drinking water and sewage systems. So we owe our health as much to civil engineering as we do to biology." If we think that sound infrastructure contributes only to our economic wellbeing, Dr. Thomas reminds us that our physical health depends on it as well.

Tragically, our infrastructure is in crisis—it has taken care of us, but we have neglected it. As our infrastructure crumbles, it endangers our prosperity as well as our emotional and physical health. Unfortunately, time is working against us. In the past 100 years, our infrastructure has contributed greatly to increased longevity, to better health, and to vast improvements in our quality of life. Sadly, ongoing deterioration of our infrastructure is threatening to undo those gains.

Recent dramatic examples of crumbling infrastructure have dominated news headlines. Yet after a failure is no longer in the news, our infrastructure is easy to ignore, once again out of sight, out of mind. The northeast blackout of August 2003, which triggered huge economic losses, left 50 million people in the dark across eight states and one Canadian province. A spectacular, escalating failure of infrastructure, it disrupted New York City transit for over 32 hours and left the City of Cleveland without safe drinking water for 3 days. In 2005, Hurricane Katrina devastated Southeast Louisiana and the Mississippi and Alabama coasts. In New Orleans, Katrina overwhelmed improperly designed and poorly maintained infrastructure killing more than 1,100 people, wrecking more than 137,000 homes; leaving more than 190,000 abandoned cars; and piling up more than 50 million cubic yards of debris. In 2007, the tragic collapse of the I-35 W Bridge in Minneapolis took the lives of

13 people. These failures are a sober reminder of just how much we depend on our infrastructure to work for us.

It's not just the headline-grabbing failures that we should heed. Calling to us like the drip-drip-drip of a leaky faucet, regular maintenance is needed to keep our infrastructure at peak performance. Today, one in four bridges is deficient, and nearly half of our urban highways are congested, a waste of time and fuel. We lose more than 7 billion gallons of treated water each day leaking from old, deteriorating pipes—enough water for a thirsty population the size of California. The U.S. Environmental Protection Agency (EPA) estimates that 850 billion gallons of combined sewer overflows enter the nation's waterways each year, and a report from the staff of the House Transportation and Infrastructure Committee concludes that without continued investment, the nation may "wind up with dirtier water than existed prior to the enactment of the 1972 Clean Water Act."

While our infrastructure deteriorates, population growth and rapid urbanization worldwide have placed increasing stress on the planet's resources. Fifty years ago we were using just half of the earth's biocapacity. Today, humanity's demand on the earth's resources—our ecological footprint—exceeds the planet's capacity by 30 percent—that is, we need 1.3 planets to meet our demands and we have just one! This will worsen in the next two decades when the world's population grows by 2 billion people, creating unprecedented demands for food, water, energy, and infrastructure.

Infrastructure strategies to meet this challenge include increased water recycling, reclamation, and conservation; improved energy efficiency using enhanced conservation measures; reduced dependence on fossil fuels, particularly petroleum; and more energy and materials from renewable sources. We build infrastructure for the long haul; it defines our communities and lasts for generations. Our current infrastructure, which has served us well, was designed and constructed using conventional standards and technologies without considering the need for sustainability, now a critical need for the 21st century.

The public assumes that infrastructure will always be there, and that the services it provides will always be adequate. Unfortunately, repairing and replacing infrastructure using conventional technologies will perpetuate a legacy of inefficiency that will last for generations. To achieve sustainability, we cannot let expediency trump efficiency, nor can we let politics trump science. Delivering sustainable infrastructure is the new challenge for the next generation of engineers. Infrastructure components are parts of larger systems whose efficiency and effectiveness depend on smooth integration. It also requires that we define and communicate values that are difficult to describe and even more difficult to quantify.

Two criteria determine if an infrastructure project will contribute to sustainability. The first is: Did we do things right? For example, did we push the boundaries by seeking opportunities to improve sustainable performance? Did we raise the bar in sustainability by going beyond first-cost thinking to total life-cycle costing and to designing for efficiency, resiliency, and reuse? Did we achieve what was reasonable within the realm of technical feasibility and acceptable business risk? The second is: Did we do the right thing? Did we consider how the project aligns with the principles of sustainability, and how it will impact future generations? Is the infrastructure project seen as fair and equitable? Does it align with community goals, account for community issues, and make good of community resources? Does the project create knowledge and advance our understanding of sustainability?

To meet the challenge of the 21st century, we must find the will and the way to focus our attention and increase our investment in infrastructure to enhance our economic health and our physical wellbeing. We must also recognize that our current way of doing things is not sustainable. We build infrastructure to last a long time. If we don't do the right things, and we don't do things right, the infrastructure we build may last, but it will not be sustainable. If we don't build it to be sustainable this time around, it will fall on the next generation to make it right—if they get the chance.

In the past century, engineers conceived and built infrastructure to be a social enabler for our quality of life. Our built infrastructure has improved our health, fueled our nation, driven our economic wellbeing, powered the information age, and provided safety and security. In this century, our infrastructure must do all that and more—it must drive us to a sustainable future. Our resources are too valuable to squander. Engineers met the challenge to create and build the infrastructure that led to vast improvements in our quality of life over the past century. This new century demands that engineers create and build sustainable infrastructure that not only enhances our quality of life, but preserves it for future generations.

I am excited about this new book—it fills a void by introducing both civil and environmental engineering within a framework of infrastructure and sustainability. By including meaningful examples from around the world and by covering non-technical issues that are critical to engineering success, it is both global and holistic. Practicing engineers often find that their greatest challenges are the non-technical considerations, many of which are described in this book, a book that provides a solid foundation for the civil and environmental engineers needed to meet the challenges of the 21st century.

Lawrence H. Roth, P.E., G.E., D. G.E.
Consulting Engineer
Sacramento, California
Former Executive Vice President, American
Society of Civil Engineers, and Chief of Staff, American
Society of Civil Engineers Hurricane Katrina External Review Panel
January 2011

Preface

Intended Audience

The target audience for this textbook is freshmen and sophomore civil and environmental engineering students. It is assumed that the reader has not yet taken courses in Calculus or Statics.

Goals and Motivation

Over the 15-week period of a typical college semester, the population of the United States will increase by approximately 700,000 people. This is roughly equivalent to the population of Charlotte, North Carolina, the 18th largest city in the U.S. During this same period, the world population will increase by approximately 21 million people, which is greater than the population of New York State, the third most populous state in the U.S. These semester-long numbers correspond to an annual growth in U.S. population of 2 million and an annual growth in world population of 90 million people. Imagine the new infrastructure (the "built environment") that must be constructed and the existing infrastructure that must be modified or expanded to meet the demands of these additional people: roads, rails, hospitals, power supplies, drinking water treatment facilities, schools, bridges, etc. And, even if the population was not growing, the need to build new or maintain existing infrastructure would still be necessary, as people move, the infrastructure ages and deteriorates, standards and regulations change, and user demands change (e.g., vehicle use and water use, on a per person basis).

Albert Einstein said, "we can't solve problems by using the same kind of thinking we used when we created them." This belief has motivated us to write a textbook that not only teaches students the *what* of the infrastructure but also the *how* and the *why* of the infrastructure. In particular, students must be able to view the infrastructure as a system of interrelated physical components, and also understand how those components affect, and are affected by, society, politics, economics, and the environment.

We further believe that studying infrastructure allows educators and students to develop a valuable link between fundamental knowledge and the ability to apply that knowledge. In terms of Bloom's Taxonomy (a classification scheme for learning objectives within various levels of cognitive competence), infrastructure offers a bridge between Level 1 (Knowledge) and Level 3 (Application). Too often, we educators jump from the fundamental knowledge gained in lower level classes into the applications of upper level courses, without spending time to ensure that students grasp the meaning of the knowledge (Comprehension, Level 2) or are able to translate the knowledge into new contexts.

Some have made the analogy that in each course of the undergraduate curriculum, students are given a component with which to build a bicycle. Upon graduation, some students carry a box of parts across the stage, whereas others ride a bike across the stage. The latter have seen how the parts fit together. We feel that the infrastructure is an ideal means of introducing students to civil and environmental engineering, as it illustrates the interrelationships of the subdisciplines. We fear that all too often, students do not see how all the parts of civil and environmental engineering are interrelated until their senior year (or later). The topics and themes of this book are intended to help students see the "big picture" in the first or second year of the curriculum. Given this established foundation, subsequent coursework will be more meaningful.

Furthermore, engineers are stereotyped as not being holistic. Too often this stereotype holds true, and often results from a technical education that does not incorporate non-technical considerations. A substantial portion of this text is devoted to integrating these non-technical considerations into the design and analysis of infrastructure systems.

Unique Features

Several features unique to this textbook include:

1. **Introductory case studies**—Most of the chapters open with a brief introduction, followed by an introductory case study. These case studies are revisited throughout the chapter in order to emphasize the learning objectives.

2. **Sidebars**—Material is placed in the margins to enrich the content. Sidebars contain examples, definitions, "factoids," and synopses of current events.

3. **Cases in point**—Case studies, in addition to each chapter's introductory case study, are included throughout the textbook.

4. **Conversational style**—Throughout the text, we have chosen to speak directly to the reader, using "we" to refer to ourselves, the authors, and "you" to refer to the reader. We recognize that this breaks with convention, but believe that it makes the material more accessible.

5. **Outro**—An "outro" is the concluding section of a piece of music, and each chapter of this text concludes with an outro. It is intended to serve as a brief summary of the chapter and to introduce students to the next chapter or to the remainder of the textbook.

6. **Extensive graphics**—This textbook includes more than 450 graphs, schematics, and photographs. We have purposely included photographs of some very commonplace components of our infrastructure, as experience has taught us that many first- and second-year students are unaware of the purpose of these components.

7. **Analysis and design applications**—Four chapters are included that describe examples of how civil and environmental engineers analyze and design infrastructure components. Two of these chapters are presented at the introductory level midway through the textbook and two at a slightly more advanced level at the end of the textbook. These chapters are more calculation-intensive than the other chapters of the textbook.

8. **Icons**—Icons are used in margins to alert students to additional information on the textbook website, including videos, articles, reports, photos, etc. Other icons identify the geographic location of places mentioned in the text.

9. **Facebook Page**—A Facebook page, *www.facebook.com/IntroductionTo Infrastructure*, has been formed. Infrastructure-related news items, at the

appropriate technical level for the intended audience of this textbook, will be frequently posted.

Instructor Resources

The following resources are available to instructors on the book website:

Solutions Manual—Solutions for homework problems are provided.

Image Gallery—All figures, photographs, schematics, and graphs from the text are provided in electronic form.

Student Resources

The following resources are available to students (and instructors) on the textbook website:

Image Gallery—Each photograph from the textbook is provided in color at a high resolution to overcome the limitations of the relatively small one-color photographs found in the textbook. Additional photographs are available, many of which are specifically referenced in sidebars.

Web Resources—These are referenced with icons (see example icon to the right), placed in the margins throughout the textbook.

Video Gallery—A variety of brief videos are available on the textbook website.

Tutorials—Examples include rudimentary spreadsheet functions, use of an engineering scale, etc.

Facebook Page—Students can "Like" (i.e., become a fan of, or join) the textbook-companion Facebook page, *www.facebook.com/IntroductionTo Infrastructure,* and receive frequent news updates related to infrastructure.

Acknowledgments

Much of the inspiration for this textbook has come from activities supported by two National Science Foundation grants (EEC 0530506—"Creating Citizen Engineers through Infrastructure Awareness" and DUE 0837530—"Infrastructure at the Forefront: Development and Assessment of Two Pilot Courses") awarded to the Civil and Environmental Engineering Department at the University of Wisconsin-Platteville. Several of the department faculty members contributed to the textbook, either by adding content, providing ideas, or by reviewing portions of this textbook. Department faculty members who contributed to the grants and textbook development include: Max Anderson, Christina Curras, Kristina Fields, Andy Jacque, Mark Meyers, Tom Nelson, Sam Owusu-Ababio, Matt Roberts, Bob Schmitt, and Keith Thompson.

Two UW-Platteville students provided feedback on the chapters: Ben Heidemann and Steve Pritchett.

Michael Gay provided in-depth reviews of Chapters 11 and 14.

We are indebted to the many photographers who allowed us to use their photographs in this textbook.

Our editor, Jenny Welter, has been very supportive in all phases of the writing process.

We appreciate the insightful contributions of the following reviewers, whose efforts enhanced this book: Kristen Sanford Bernhardt, Lafayette College; Angela R. Bielefeldt, University of Colorado at Boulder; Fred Boadu, Duke University; Paul J. Consentino, Florida Institute of Technology; Anirban De, Manhattan College; David P. Devine, Infrastruct.Net Corp; Charles R. Glagola, University of Florida; Steven D. Hart, United States Military Academy; Rick Lyles, Michigan State University; Emmanuel U. Nzewi, Southern University and A&M College.

We wish to thank the many students over the years that have motivated us to become better teachers.

Finally, we wish to thank Mary, Charlie, Rebekah, Abigail, and Lydia.

Sample Syllabi

Sample syllabi for five possible courses are provided.

One-Credit Introduction to Civil and Environmental Engineering Course

Week	Topic	Readings
1	Introduction to Course	
2	Introduction to Infrastructure	Foreword & Chapter 1
3	History and Heritage	Chapter 8
4 & 5	Construction Engineering	Chapter 6
		Chapter 9 (Construction Engineering Analysis Application)
		Chapter 10 (Construction Engineering Design Application)
6 & 7	Environmental Engineering	Chapter 5
		Chapter 9 (Environmental Engineering Analysis Application)
		Chapter 10 (Environmental Engineering Design Application)
8 & 9	Geotechnical Engineering	Chapter 3 (Foundations)
		Chapter 9 (Geotechnical Engineering Analysis Application)
		Chapter 10 (Geotechnical Engineering Design Application)
10 & 11	Structural Engineering	Chapter 3
		Chapter 9 (Structural Engineering Analysis Application)
		Chapter 10 (Structural Engineering Design Application)
12 & 13	Transportation Engineering	Chapter 4
		Chapter 9 (Transportation Engineering Analysis Application)
		Chapter 10 (Transportation Engineering Design Application)
14	Interrelationships between subdisciplines	Chapter 7
15	Ethics and Sustainability	Chapters 13 & 17

One-Credit Introduction to Infrastructure Course

Week	Topic	Readings
1	Introduction to Course	
2	Introduction to Infrastructure	Foreword & Chapter 1
3	Natural Environment	Chapter 2
4	Structural Infrastructure	Chapter 3
5	Construction Sites	Chapter 6
6	Roads, Mass Transit, and Non-Motorized Transportation	Chapter 4
7	Aviation, Rail, and Waterway Transportation	Chapter 4
8	Energy, Drinking Water, and Wastewater	Chapter 5
9	Stormwater, Solid and Hazardous Waste	Chapter 5
10	Systems View of Infrastructure	Chapter 7
11	Analysis Process	Chapter 9
12	Analysis Applications (Select two subdisciplines)	Chapter 9
13	Design Process	Chapter 10
14	Design Applications (Select two subdisciplines)	Chapter 10
15	Ethics and Sustainability	Chapters 13 & 17

Two-Credit Introduction to Civil and Environmental Engineering Course

Week	Topic	Readings
1	Introduction to Course and Infrastructure	Foreword & Chapter 1
2	Natural Environment	Chapter 2
3	Structural Infrastructure	Chapter 3
4	Transportation Infrastructure	Chapter 4
5	Environmental Infrastructure	Chapter 5
6	Construction Sites	Chapter 6
7	Systems View of Infrastructure	Chapter 7
8	History, Heritage, and Future	Chapter 8
9	Analysis Process and Applications	Chapter 9
10	Analysis Applications and Design Process	Chapters 9 & 10
11	Design Applications	Chapter 10
12	Planning	Chapter 11
13	Sustainability and Environmental Considerations	Chapters 13 & 15
14	Social Considerations	Chapter 16
15	Ethics	Chapter 17

Three-Credit Introduction to Infrastructure Course

Week	Topic	Readings
1	Introduction to Course and Infrastructure	Foreword & Chapter 1
2	Natural Environment and Environmental Infrastructure	Chapters 2 & 5
3	Structural Infrastructure and Construction	Chapters 3 & 6
4	Transportation Infrastructure	Chapter 4
5	History, Heritage, and Future; Systems	Chapters 7 & 8
6	Infrastructure Analysis Fundamentals	Chapter 9
7	Infrastructure Design Fundamentals	Chapter 10
8	Planning and Energy	Chapters 11 & 12
9	Sustainability	Chapter 13
10	Environmental Considerations	Chapter 14
11	Economics Considerations	Chapter 15
12	Ethics; Other Considerations	Chapters 17 & 19
13	Infrastructure Security	Chapter 18
14	Advanced Infrastructure Analysis	Chapter 20
15	Advanced Infrastructure Design	Chapter 21

Three-Credit Capstone Design Course – Supplementary Text

Week	Topic	Readings
1	Introduction to Course	
2	Introduction to Infrastructure	Foreword & Chapter 1
3	Structural Infrastructure and Construction	Chapters 3 & 6
4	Transportation Infrastructure	Chapter 4
5	Environmental Infrastructure	Chapter 5
6	Infrastructure as a System	Chapter 7
7	Planning	Chapter 11
8	Sustainability Considerations	Chapter 13
9	Economics Considerations	Chapter 14
10	Environmental Considerations	Chapter 15
11	Social Considerations	Chapter 16
12	Ethics	Chapter 17
13	Security Considerations	Chapter 18
14	Other Considerations	Chapter 19
15	Open	

About the Authors

Mike and Philip met and quickly became friends at Clarkson University in 1994. Since 1998 they have been teaching courses and sharing ideas at the University of Wisconsin-Platteville. They were drawn to, and have remained at UW-Platteville, due to its focus on undergraduate education and the dedication of their colleagues. The Civil and Environmental Engineering (CEE) Department at UW-Platteville offers degrees in both civil and environmental engineering, with substantial overlap between the curricula. Students and faculty in both programs interact on a daily basis, and in fact it is often difficult to distinguish between students of the two programs. Both programs are heavily application based, with a focus on consulting engineering. The authors, and the vast majority of CEE faculty, are registered Professional Engineers. While Mike and Philip are active in research and consulting, their primary motivation comes from their daily interactions with undergraduates. They have had countless discussions about how students learn, how to improve teaching, and *what* students should learn.

In 2001, Mike began working with the Wisconsin Section of the American Society of Civil Engineers (ASCE) to co-author the first version of Wisconsin's Infrastructure Report Card. This project truly opened his eyes to the vastness, complexity, and importance of infrastructure. Soon thereafter, a vision was developed for an *Introduction to Infrastructure* course. Attending the Infrastructure Solutions Summit hosted by ASCE and touring the damage caused by Hurricane Katrina provided further inspiration. Serving on the local Water and Sewer Commission provided first-hand experience with many of the infrastructure challenges that are commonly faced across the country. A few years later, the UW-Platteville CEE Department was awarded a grant from the National Science Foundation to develop an introductory infrastructure course. The department faculty and practicing engineers developed a framework for the course, which largely serves as the table of contents for this textbook.

As a result of these efforts, an Introduction to Infrastructure course is now offered. Initial offerings of this course have received rave reviews. One sophomore that completed the course told Philip: "To be honest, before I took this course I didn't really have any idea what civil engineering was all about." Another student, caught in what he termed the "death march" of calculus, science, and engineering fundamentals courses, said after completing the course: "At the start of this semester, I was planning on dropping civil engineering and getting a [non-engineering] degree in construction management. But after taking this course, I am motivated to continue my degree." Such comments are especially gratifying as they agree with the authors' predictions about the potential impact of the course.

Philip and Mike have distinct but similar teaching styles, both focusing on student comprehension and student engagement. They are both frustrated by traditional engineering education that seems to focus on how to design components rather than understanding how the components *fit into a system* or, perhaps more importantly, *why* the component is even needed. Over the years, their courses have increasingly incorporated non-technical content including economics, public perception, risk management, and politics. This shift is not based solely on their personal preferences; rather, employers of UW-Platteville CEE graduates, with whom they continuously communicate, have indicated that student awareness of the non-technical aspects of infrastructure is more important now than ever.

They are passionate about teaching as well as assessing what students *actually* learn. Some lessons that they have learned about teaching include:

- Students enjoy learning about the non-technical realities of engineering

- Many students do not have the experience to see how all the pieces fit together

- Helping students develop a broader perspective favors deep learning rather than memorization

- Students respond favorably to practical applications of their coursework

- Just because students are pursuing a degree in civil or environmental engineering does not imply that they understand, or are even aware of, the civil infrastructure that supports their lives

- Repetition is a key to increased student learning

- The true measure of success is the knowledge that students retain after completion of a course.

They have distilled their understanding of, and passion for, infrastructure into a format that will introduce students to the fields of civil and environmental engineering. As a result of using this textbook, they hope that students become much more aware of the infrastructure surrounding them, more appreciative of its complexity, more understanding of its importance to our national and economic security, and inspired to continue their studies in civil and environmental engineering.

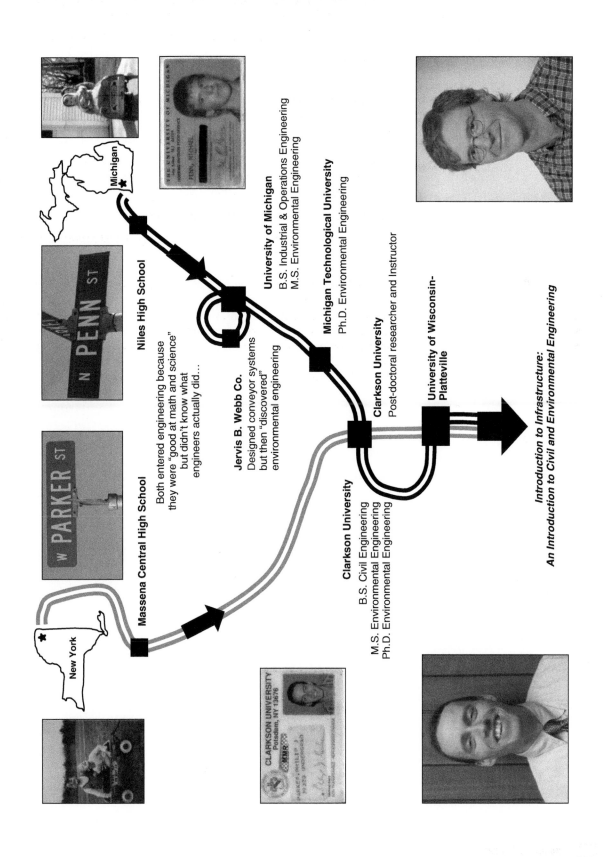

Massena Central High School

Niles High School

Both entered engineering because they were "good at math and science" but didn't know what engineers actually did....

Jervis B. Webb Co.
Designed conveyor systems but then "discovered" environmental engineering

University of Michigan
B.S. Industrial & Operations Engineering
M.S. Environmental Engineering

Michigan Technological University
Ph.D. Environmental Engineering

Clarkson University
Post-doctoral researcher and Instructor

Clarkson University
B.S. Civil Engineering
M.S. Environmental Engineering
Ph.D. Environmental Engineering

University of Wisconsin-Platteville

Introduction to Infrastructure:
An Introduction to Civil and Environmental Engineering

Brief Table of Contents

Detailed Table of Contents

chapter One **Introduction to Infrastructure and Careers in Civil and Environmental Engineering**

Learning Objectives

After reading this chapter, you should be able to:

1. Define infrastructure and explain how it affects nearly all aspects of your life.
2. Describe the "infrastructure crisis."
3. Describe the role of civil and environmental engineers in infrastructure management and design.
4. Describe the role of each of the subdisciplines of civil engineering.
5. Discuss professional issues as they relate to the success of your future career.

Introduction

Infrastructure is defined by Merriam Webster as the underlying foundation or basic framework of an organization or system. As it relates to the work of civil and environmental engineers, infrastructure is the system of **public works** of a country, state, region, or municipality.

Major infrastructure sectors (termed the **built environment**) are provided in the following list, with the first nine sectors comprising the **civil infrastructure**:

1. Transportation systems (roads, highways, railroads, bridges, tunnels, canals, locks, ports, airports, mass transit, and waterways)

2. Structures (including buildings, bridges, dams, and levees)

3. Water supply and treatment systems

4. Wastewater treatment and conveyance systems

5. Solid waste management systems (collection, reuse, recycling, and disposal of wastes)

6. Hazardous waste management systems

7. Stormwater management systems

8. Parks, schools, and other government facilities

9. Energy systems (power production, transmission, and distribution)

10. Communications systems (telephone, computer, etc.)

This book will focus on the first nine sectors listed above. Although civil and environmental engineers are potentially involved with communications systems (and most certainly rely upon them), the design and maintenance are largely carried out by electrical engineers and technicians. Additionally, civil and environmental engineers are also involved with private sector work. The design, maintenance, and analysis of the private infrastructure are very similar to that of the public infrastructure; for example, designing a parking lot for a "big-box" retail center is similar in many respects to the design of a parking lot for a courthouse.

The **natural environment** supports all forms of life and all aspects of our built environment. Human health, recreation opportunities, aesthetics, biodiversity, and sustainability depend on clean soil, air, and water, as well as properly functioning ecosystems. We utilize and alter the natural environment for development as well as for transportation and other uses. Also, the raw materials used to construct the built environment originate from the natural environment.

Introductory Case Study: I-35 W Bridge Collapse

On August 1, 2007, the I-35 W bridge over the Mississippi River in Minneapolis, Minnesota, collapsed carrying rush hour traffic (Figure 1.1). This tragedy resulted in 13 deaths and 145 injuries. The bridge, which carried 140,000 vehicles per day, was replaced and opened for use only 13 months after its collapse, an extraordinarily short length of time for a project of this magnitude.

The events leading up to and following the collapse highlight many aspects of infrastructure engineering and management: design, analysis, maintenance, inspection, emergency response, traffic routing (immediately following the collapse and during reconstruction), planning, and construction, all of which will be discussed in this textbook. The failure of this single infrastructure *component* had a tremendous impact on regional infrastructure *systems* (e.g., road, rail, and river transportation.)

Figure 1.1 The Aftermath of the I-35 W Bridge Collapse in 2007.

Source: U.S. Coast Guard/K. Rofidal.

The Minnesota Department of Transportation estimated that the bridge collapse had an impact of $60 million on the economy.

The collapse made national headlines and soon came to symbolize the poor condition of our nation's infrastructure. It helped spur new funding for bridge inspection and maintenance, which is ironic, given that the collapse of the bridge was not primarily due to insufficient maintenance or inspection. Rather, the collapse was due to a design flaw, as determined by the National Transportation Safety Board (NTSB). The NTSB also cited the fact that the bridge had become heavier over time due to modifications, and this additional weight, along with concentrated loads from construction materials, contributed to the collapse.

NTSB I-35 W
Bridge Report

The Prevalence of Infrastructure

Consider spending an extended period of time without access to any of the major infrastructure systems. Unless you have survived a catastrophic infrastructure failure or spent time in an underdeveloped nation, you may not appreciate how vital infrastructure is to the quality of life. At a significantly lesser level, we have all had to deal with the *inconvenience* associated with a temporary loss of services provided by the infrastructure (e.g., a detour resulting from road resurfacing or a water outage caused by nearby underground utility work).

Infrastructure affects nearly every aspect of your daily life. The alarm clock that wakes you in the morning does so thanks to a power grid that supplies electricity to your home or apartment. The shower that provides pressurized water is possible thanks to the underground network of distribution pipes and valves, water storage tanks, and pumps. The fact that this water, along with the water from your bathroom or kitchen faucets, is safe to drink is due to treatment that has occurred at a drinking water treatment facility. When you flush a toilet, the waste is conveniently flushed away and carried in a series of pipes to a wastewater treatment facility where it is extensively treated before being released to the environment. The fact that, for a matter of pennies, your breakfast consists of a banana from South America and a slice of bread made from wheat that was grown perhaps hundreds of miles away is possible

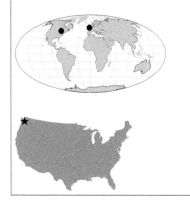

thanks to the widespread transportation network of highways, railways, and waterways. Your trip to class, either by foot on a sidewalk, by bike in a bike lane, by mass transit via a bus or light rail system, or in a vehicle on roadways, is possible because of infrastructure.

Modern society depends upon infrastructure. Residential, commercial, and industrial development can be encouraged or discouraged by a municipality through the availability of infrastructure (e.g., roads and water supply). The dramatic population growth of the arid Southwestern United States was largely enabled by massive water supply projects (e.g., Colorado River diversion) to transport water hundreds of miles from sources to users. China's South-to-North Water Diversion Project, to be completed in 2050 at a cost of over $60 billion, will divert 45 billion m^3 of river water per year to the arid north for domestic, industrial, and agricultural use. This water will allow new development to occur away from currently overpopulated areas. Such massive projects potentially provide economic growth but can also cause significant social and environmental impacts as will be discussed in later chapters.

Much of the infrastructure we use is invisible to the public due to being buried below ground, or is *virtually* invisible given that we are so used to seeing it every day. It has been said that, "our infrastructure is like our stomachs—when working properly, we don't even realize it's there." This is largely true as society often takes the infrastructure for granted. Unfortunately, this often leads to neglect. For example, people *assume* that the bridges they cross daily will support traffic loads and that safe and abundant water will be supplied from their bathroom faucet. Society also tends to assume that the bridge and the faucet will continue to meet their needs indefinitely with minimal care and attention. And day after day, this is indeed the case. However, infrastructure components and systems do fail, sometimes with catastrophic consequences and always with consequences that have far-reaching effects. Failures can occur for many reasons, such as:

- inadequate design (as in the case of the I-35 W bridge collapse)
- improper construction
- lack of critical maintenance
- unforeseen changes over time in factors affecting performance
- complications caused by another component within the system
- natural disasters
- accidents
- terrorism, vandalism, and war

Dewitt Greer, a former Texas Highway Commissioner, said "We do not have great highways because we are a great nation. We are a great nation because we have great highways." Larry Roth, former Deputy Executive Director of the American Society of Civil Engineers (ASCE), has aptly suggested that we substitute the term "infrastructure" for "highways" in this quote.

The reason for such a bold assertion will be explained in many ways in this book, but consider briefly the effect of infrastructure on public health and wealth. Dramatic increases in public health (life expectancy has risen from less than 50 years to nearly 80 years since 1900, a 60 percent increase) and in wealth (**per capita gross domestic product** has increased from approximately $5,000 to $45,000 in **inflation-adjusted** dollars since 1900, an 800 percent increase) can be *directly* linked to our infrastructure.

Increasing Urbanization

The percentage of Americans living in urbanized areas of greater than 50,000 people is steadily increasing and is currently 83 percent. For comparison's sake, the percentage was approximately 40 percent in 1900 and 60 percent in 1950. More than 50 percent of Americans live in large metropolitan areas with populations greater than 1 million. Thus the majority of Americans depend on complex and interdependent infrastructure systems.

Dramatic increases in urban populations are occurring in developing countries around the world as rural residents "flock to the cities" in search of jobs, creating enormous demand for infrastructure that often outpaces its development and often leads to serious public health and social problems. For example, 29 percent of India is urbanized today, but by 2050, this number is expected to grow to 55 percent, corresponding to 900 million people. In comparison, Brazil is currently 86 percent urbanized. The trends for the 50 least developed countries in the world are shown in Figure 1.2.

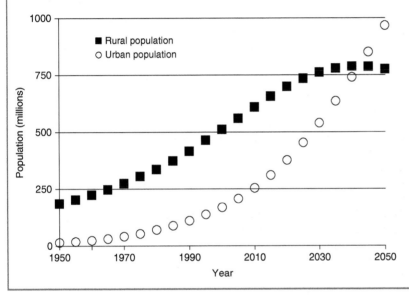

Figure 1.2 Trends in Urbanization for Least Developed Countries.

Data source: United Nations Department of Economic and Social Affairs, 2009.

Continued prosperity requires infrastructure investment not only to meet increased demands from a growing population, but also to maintain and replace our existing infrastructure. *Increased* prosperity requires additional investments above and beyond this level.

Infrastructure Systems

Infrastructure components are integral parts of infrastructure systems. A railway line is part of the rail system, which in turn is part of the larger and more complex transportation system. An individual rail line might meet the needs of the rail system by moving people or goods from point A to point B. However, if an interface with another system, such as an at-grade (i.e., non-elevated) road crossing is not designed appropriately, traffic congestion may result and the road system may fail to meet its objectives.

At a grand scale, the concept of systems is well illustrated by the extensive planning and investment required by a city hosting the Olympics. In 2008, the summer Olympics were hosted by Beijing, China, which received extensive international news coverage of the challenges faced in preparing for the events. The entire infrastructure, with each of its sector systems and subsystems, had to be evaluated to meet a sudden (and importantly, temporary) increase in demand. Lodging, food and supplies, transportation, communications,

The Infrastructure Crisis

Civil Engineering magazine, published by ASCE, released a special report in January 2008 titled "The Infrastructure Crisis."

The Infrastructure Crisis

Table 1.1

Infrastructure Sector Grades

Sector	Grade
Aviation	D
Bridges	C
Dams	D
Drinking water	D–
Energy	D
Hazardous waste	D
Inland waterways	D
Levees	D–
Public parks and recreation	C–
Rail	C–
Roads	D–
Schools	D
Solid waste	C+
Transit	D
Wastewater	D–

Source: ASCE, 2009.

 ASCE's Report Card for America's Infrastructure, 2009

security, waste disposal, energy, water, and wastewater needs had to be met for the demand created by athletes, delegates, media, and spectators. New structures were constructed for many of these needs while considering potential uses *after* the Olympics were completed.

In a similar way, large industries and military bases also must be treated as systems. These are often developed in a manner such that they are nearly or entirely "self-contained," with their own systems of roads, energy, waste handling, water supply, and security.

Funding Infrastructure

Frequently throughout this book, we will refer to ASCE's *Report Card for America's Infrastructure*, an assessment that is completed periodically to inform the public, engineers, regulators, lawmakers, and other decision-makers as to the condition of our national infrastructure. The 2009 grades for major infrastructure components are provided in Table 1.1.

The average grade for the U.S. infrastructure was a D, and ASCE estimated the funding needs over the next 5 years to be $2.2 trillion. Importantly, this is the amount of money required to raise the overall infrastructure grade to an "acceptable" level (a grade of "B"). The $2.2 trillion far exceeds the estimated $1 trillion currently budgeted. This shortfall in funding will, at best, result in infrastructure functioning at its current level. At worst, the level of performance will decrease.

Low grades notwithstanding, the United States has one of the better infrastructure systems in the world. The World Economic Forum publishes a Global Competiveness Report that ranks nations' economic competiveness based on the quality of legal systems, infrastructure systems, economic stability, higher education, public health, technology, and financial institutions. A summary of the 10 highest ranked nations for five infrastructure categories is presented in Table 1.2. For reference, the rank of the United States in each category

Table 1.2

Ranking of World Infrastructure Systems, Top 10 in Each Category and U.S. Rank

Rank	Roads	Rail	Port	Air	Electric
1	Singapore	Switzerland	Singapore	Singapore	Denmark
2	France	Japan	Hong Kong	Hong Kong	Iceland
3	Hong Kong	Hong Kong	Netherlands	United Arab Emirates	Hong Kong
4	Switzerland	France	Finland	Germany	Finland
5	Germany	Germany	Germany	Switzerland	France
6	Austria	Finland	Belgium	Denmark	Switzerland
7	United Arab Emirates	Taiwan	United Arab Emirates	Netherlands	Germany
8	Denmark	Korea	Iceland	Finland	Sweden
9	Portugal	Singapore	Denmark	France	Netherlands
10	Oman	Netherlands	France	Iceland	Austria
U.S. Rank	11	17	13	20	17

Data source: World Economic Forum, 2009.

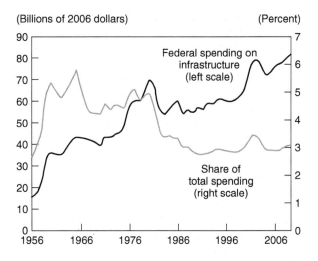

(Billions of 2006 dollars) (Percent)

Figure 1.3 Federal Spending on Infrastructure in the United States.

Source: Congressional Budget Office, 2007.

is also provided. Germany had the highest overall infrastructure ranking, followed by Hong Kong, France, Singapore, and Switzerland. It is important to note that these rankings are based on scores from surveys of business executives, not engineers. Perhaps engineers' rankings would be similar, but to our knowledge, no such assessment has been completed. Note that the United States fails to rank in the top 10 in any category.

Federal government spending on infrastructure is illustrated in Figure 1.3, both in inflation-adjusted dollars and as a percent of total spending. The "good news" from this figure is that the inflation-adjusted expenditures have steadily increased. However, according to Figure 1.3, the amount of spending on infrastructure, expressed as a fraction of total spending, has remained relatively steady for the last 20 years (approximately 3 percent), and is significantly less than it was in the 1960s and 1970s. The rapidly developing nations of India and China are investing two to three times this amount as a percent of total expenditures.

The cost of infrastructure components varies greatly depending on its size and complexity. For perspective, these costs have been summarized in Table 1.3 for various infrastructure components, categorized by order-of-magnitude price ranges.

One very important concept that will be emphasized throughout this book is the need for funds to not only build infrastructure improvements or replacements (i.e., **capital funds**), but also funds to operate and maintain projects (i.e., **O&M** funds). Public funding (i.e., funding by federal, state, and local governments) of these two categories is illustrated in Figure 1.4. Note that since the mid-1970s, O&M expenditures have been greater than capital expenditures. This trend will most likely continue because O&M expenses tend to increase for any individual infrastructure component as it ages and approaches the end of its **design life**.

Failure to Act: Economic Impact Report by ASCE

Costs of Mega-Projects

Every engineered item, be it an automobile shock absorber or a bridge, is designed to function for a certain length of time, termed **design life**.

Sustainability

The demand for resources is greater now than ever before in world history. Not only has the world's population doubled since 1950, the amount of resources consumed *per person* has also increased dramatically. Since resource demand is a function of these two variables (population and per capita consumption), the increase is compounded. For example, world energy consumption has doubled since the mid-1970s. Similarly staggering numbers can be found

Table 1.3

Range of Construction and Design Costs for Various Infrastructure Components

Infrastructure Component	$10 K to $100 K	$100 K to $1 M	$1 M to $1 B	Greater than $1 B
Roads	Local street repair	Local road reconstruction (several city blocks)	Major highway construction	Mega-urban project (e.g., The "Big Dig" in Boston)
Water Resources	Residential subdivision stormwater detention pond	Replace water and sewer lines for several city blocks	Wastewater treatment facility for a small- to medium-sized city	Large hydroelectric dam, large desalination facility
Bridges	Biking/Walking trail bridge	Standard bridge spanning 20 to 100 feet	I-35 W replacement bridge (Saint Anthony Falls Bridge)	Major long-span bridge (e.g., San Francisco–Oakland Bay Bridge)
Buildings	Equipment storage shed, picnic shelter	One-story office building	City Hall, civic theater, sports arena, school, prison	Mega-skyscraper (e.g., One World Trade Center), sports stadium complex (e.g., Cowboys Stadium)
Parking	Small, paved surface lot (less than 30 stalls)	Large commercial lot for retail center or sports complex	Multi-level parking structure	Space Shuttle docking station at the International Space Station

Figure 1.4 **Capital and O&M Spending on Infrastructure in the United States.**

Source: Congressional Budget Office, 2007.

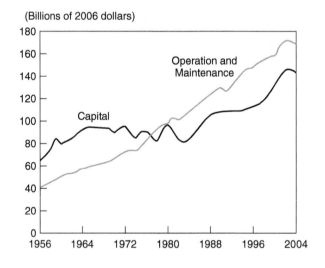

that quantify recent increases in water consumption, mining activity, waste production, vehicle miles travelled, cropland converted to development, etc. In light of these increases, the need to ensure that sufficient resources are available for future generations is greater than ever. In an increasingly global political and economic environment, we must all be concerned not only with sustainability within the United States, but throughout the world. Motivation to do so may come from a sense of ethical duty or humanitarian desire, or more pragmatically, from security concerns; a lack of resources and political instability in some countries, such as Somalia and Yemen, are contributing factors to the rise of terrorism.

The Civil and Environmental Engineering Professions

In the following pages, we will introduce the civil and environmental engineering professions. Once considered a subdiscipline of civil engineering, environmental engineering is now a distinct engineering discipline. For the purposes of this text, we have chosen to combine the two fields as civil *and* environmental engineering.

Civil and environmental engineers identify, solve, manage, and prevent problems that arise within the construction, environmental, geotechnical, structural, transportation, and water resources subdisciplines of the fields. The interrelationships between these subdisciplines are shown in Figure 1.5, with infrastructure being the focal region of overlap. A primary objective of this textbook is to introduce infrastructure and demonstrate the interrelationships of civil and environmental engineering. Note, however, that within each subdiscipline in Figure 1.5, there is a non-infrastructure component (not shown in the figure). This distinction is important in that civil and environmental engineers do not solely work on public infrastructure projects. As an example, structural and environmental engineers may design separate portions of a *privately-owned* car washing facility. Likewise, structural engineers may design airplane wings and environmental engineers may design industrial treatment processes.

Collection of More Than 200 Profiles of a Variety of Civil and Environmental Engineering Projects

CONSTRUCTION

Construction engineers design, plan, and manage construction projects. Major focus areas include site layout and earthwork, temporary structures, material quantity and cost estimating, scheduling, quality assurance and control, contracting, bidding, safety, inspection, and personnel management. Note that many engineers with expertise in the other areas of civil and environmental

Construction Photo Gallery

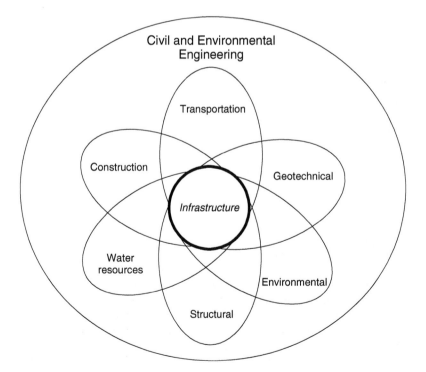

Figure 1.5 Subdisciplines of Civil and Environmental Engineering, with Infrastructure as the Region of Overlap.

engineering also manage the construction of projects that they and their team design.

ENVIRONMENTAL

Environmental Photo Gallery

Environmental engineers prevent, minimize, or remediate the contamination of air, soil, water, and the associated human and environmental health aspects of such contamination. Major focus areas include drinking water treatment, water supply, wastewater treatment, surface water quality, air quality, contaminated sites, wetland mitigation, solid and hazardous waste management, and stormwater management. Today, many universities offer degrees in environmental engineering. It is still common, however, to get a degree in civil engineering with an emphasis in environmental engineering.

GEOTECHNICAL

Geotechnical Photo Gallery

Geotechnical engineering deals with the behavior of earth materials. Major focus areas include: subsurface sampling and classification; engineering properties (e.g., compressibility, strength) of soils and rock; groundwater seepage; foundations for buildings, bridges, and other structures; earthen structures (e.g., levees, dams); slope stability; retaining walls; and geosynthetics.

STRUCTURAL

Structural Photo Gallery

Structural engineers design and analyze structures to meet safety constraints while considering architectural requirements and aesthetics. Structural engineers must determine the loads that a structure must withstand, the engineering properties of the materials used for construction, the mechanics (i.e., reaction to a load) of the materials, and interaction of structural members (e.g., beams, connections). Major focus areas include buildings, bridges, tunnels, and towers. Structural engineers often work closely with architects to turn the architect's vision into reality.

TRANSPORTATION

Transportation Photo Gallery

Transportation engineering deals with the safe and efficient movement of people and goods. Transportation engineers analyze current "traffic" (e.g., pedestrians, automobiles, buses, and ships) and predict future traffic in order to design appropriate systems. Major focus areas include parking, roads, highways, mass transit, airports, shipping ports, and rail.

WATER RESOURCES

Water Resources Photo Gallery

Water resources engineering deals with the quantity, quality, and transport of water. Major focus areas include surface water hydrology, groundwater hydrology, and hydraulics (flow of water in natural or built open channels, and in closed conduits).

The above listed subdisciplines are the major areas of civil and environmental engineering as defined by the National Council of Examiners for Engineering and Surveying (NCEES), which develops professional licensure examinations. ABET, Inc., formerly the Accreditation Board for Engineering and Technology, is the organization responsible for accrediting university programs in engineering and technology, and requires that civil engineering programs provide instruction in at least four of the major subdiscipline areas.

Municipal engineering is not a distinct subdiscipline, but rather a combination of transportation, water resources, and environmental engineering. This term is often used synonymously for civil engineering in a broad sense.

Note that in most of the descriptions of the subdisciplines, the focus was on design (creating a new component or portion thereof) and analysis (studying an existing component). However, civil and environmental engineers are also increasingly responsible, solely or collaboratively, for the research, planning, financing, inspection, construction, start-up, operation, maintenance, security, rehabilitation, inspection, monitoring, and demolition of infrastructure systems and components. Additionally, engineers are responsible for integrating new components into existing systems while meeting economic, social, political, environmental, and ethical constraints. Security is of increasing concern in light of terrorist activity (e.g., the events of 9/11) and natural disasters (e.g., Hurricane Katrina, Japan tsunami/Fukushima nuclear reactor disaster). Additional focus is required to bolster the security and resilience (the ability to "rebound" after a disaster) of our infrastructure systems.

Thus, civil and environmental engineers bear a substantial responsibility for public health and prosperity. The following excerpt is from *The Vision for Civil Engineering in 2025* published by ASCE in 2007:

The Vision for Civil Engineering in 2025

> *Entrusted by society to create a sustainable world and enhance the global quality of life, civil engineers serve competently, collaboratively, and ethically as master:*
>
> - *planners, designers, constructors, and operators of society's economic and social engine—the built environment;*
> - *stewards of the natural environment and its resources;*
> - *innovators and integrators of ideas and technology across the public, private, and academic sectors;*
> - *managers of risk and uncertainty caused by natural events, accidents, and other threats; and*
> - *leaders in discussions and decisions shaping public environmental and infrastructure policy.*

To meet these responsibilities, civil and environmental engineers must be highly trusted professionals.

Integration with Other Professions

Several types of infrastructure can be viewed in the urban setting depicted in Figure 1.6: buildings, lighting, parking, pedestrian walkways, a retaining wall, a bridge, and a waterway. As such, the work of engineers with expertise in several different subdiscipline areas is evident (e.g., construction, structural, transportation, municipal).

No single component of the infrastructure is the domain of a single subdiscipline area, neither is it the domain of civil and environmental engineers only. Rather, various subdiscipline areas of civil and environmental engineering

Figure 1.6 The San Antonio, Texas, Riverwalk.

Source: M. Penn.

must work together in the design and construction of various components and work with experts from other areas. For example, consider the walkways (one in the foreground and two along the river) in Figure 1.6. Given the setting, the design and construction of the walkways were integrated into other component designs involving structural engineers (bridge), geotechnical engineers (retaining wall), transportation engineers (road), environmental/water resources engineers (river), and construction engineers (installation). Additionally, experts (perhaps engineers, perhaps not) in lighting, safety, aesthetics, urban planning, horticulture, architecture, and other areas were also likely involved in the planning and design. Additional professions with which civil and environmental engineers often work include construction contractors, surveyors, engineering technicians, equipment and material suppliers, maintenance personnel, regulators, public safety personnel (e.g., police, fire), database specialists, and utility representatives.

The most readily apparent challenge in working with different professions is that of different perspectives and priorities. An additional challenge of working with such a diversity of professions is that of vocabulary. Different terms may be used to describe the same thing amongst professions, and in rare occasions, the same term may have two different meanings.

Integration with the Public

More so than those of any other field of engineering, civil and environmental engineering projects involve the public. Citizens are the end users and funders of the vast majority of infrastructure. Also, for many projects, citizens will shape the direction and scope of projects through public meetings. In many cases, citizens will be members of committees that are responsible for approving development plans or reviewing proposals and selecting companies for the design of infrastructure projects. For many engineers, the ability to communicate technical information to non-technical audiences is a formidable challenge. Citizens become extremely frustrated with engineers that cannot speak "their" language. The old advice "know your audience" rings especially true here.

The title of this Case in Point is a quotation attributed to President Truman. Although an ethical engineer would never purposely confuse the public, often their use of technical language does just that. Consider this excerpt from the official NTSB press release on the results of the study into the cause of the failure of the I-35 W bridge:

"Shortly after 6 p.m. a lateral instability at the upper end of the L9/U10W diagonal member led to the subsequent failure of the U10 node gusset plates on the center portion of the deck truss. Because the deck truss portion of the I-35 W bridge was considered non-load-path-redundant, the total collapse of the deck truss was unavoidable once the gusset plates at the U10 nodes failed." (NTSB, 2008b)

News releases are intended for the general public, and apparently the NTSB felt that the general public would understand the terms "lateral instability" and "non-load-path-redundant" as well as understand the significance of the U10 nodes (as compared to the U11 nodes or the P10 nodes, if such nodes even exist).

The Burden and Glory of Engineering Responsibility

"The great liability of the engineer compared to men[1] of other professions is that his works are out in the open where all can see them. His acts, step by step, are in hard substance. He cannot bury his mistakes in the grave like the doctors. He cannot argue them into thin air or blame the judge like the lawyers. He cannot, like the architects, cover his failures with trees and vines. He cannot, like the politicians, screen his shortcomings by blaming his opponents and hope the people will forget. The engineer simply cannot deny he did it. If his works do not work, he is damned . . . . No doubt as years go by the people forget which engineer did it, even if they ever knew. Or some politician puts his name on it. Or they credit it to some promoter who used other people's money . . . . But the engineer himself looks back at the unending stream of goodness which flows from his successes with satisfactions that few professions may know. And the verdict of his fellow professionals is all the accolade he wants."

—Herbert Hoover, mining/civil engineer and 31st President, 1929–1933.

Source: Herbert Hoover Presidential Library Association.

[1] Note that when President Hoover said this, very few engineers were women. Indeed, the 1930 U.S. Census enumerates 256,078 engineers, 18 of whom were women.

Professional Issues

LICENSURE

Each state, or jurisdiction such as Washington, DC, has its own requirements for licensure as a professional engineer (PE). A PE license is a requirement for many responsibilities of civil and environmental engineers; thus, obtaining (and maintaining) licensure is encouraged or required by many employers. Typical requirements for becoming a PE include:

1. Obtain an ABET-accredited baccalaureate degree in engineering

2. Pass the Fundamentals of Engineering (FE) examination—a day-long exam, typically taken near the time of graduation, covering basic mathematics and science, engineering fundamentals (statics, mechanics of materials, thermodynamics, engineering economics, etc.), as well as civil and environmental engineering fundamentals

3. Complete 4 years of engineering experience under the supervision of a licensed engineer

4. Pass the Principles and Practices examination—a day-long exam, taken after meeting items #1 through #3, covering civil and environmental engineering fundamentals and details of a chosen subdiscipline.

We encourage you to pursue licensure as it provides professional stature and creates additional employment opportunities.

Other Resources

The Year in Infrastructure

Infrastructure 2010: Investment Imperative Report

Infrastructure Blog: "The Infrastructurist"

GRADUATE DEGREES

As the challenges that civil and environmental engineers face become more complex, many entry-level positions now require a master's degree as the minimum level of education. Master's degrees can be solely based on additional course work (typically 30 credits), or on a combination of course work and research. Research assistantships often provide stipends (in the range of $15,000 to $20,000 per year and higher) for living expenses and also cover tuition costs.

In 2006, the NCEES voted to extend the minimum education requirements for licensure to 30 credits of acceptable upper-level undergraduate or graduate coursework beyond the bachelor's degree. This requirement may take effect as early as 2015. The rationale for the requirement includes the following:

1. Bachelor's Civil and Environmental Engineering (CEE) degrees today typically require 120 to 130 credits, whereas the requirement several decades ago was 150 credits,

2. Other professions (e.g., law and medicine) once required only a bachelor's degree, but now require advanced education, and

3. The demands of the engineering profession are increasing such that a bachelor's degree may be inadequate education for professional practice.

The impact of this "requirement" on the engineering profession and its rationale remain a source of debate. It remains the responsibility of the Board of Professional Engineers in each state or jurisdiction to adopt these additional requirements. Notwithstanding, the NCEES vote and its support by ASCE demonstrates significant support for advanced education.

LIFE-LONG LEARNING

Life-long learning (or continuing education as it was once commonly termed) is tremendously important for career development. A bachelor's degree can *at best* provide the foundation for an engineering career. Life-long learning provides opportunities for growth in specialized technical fields as well as in business and management skills necessary for advancement. It may come in many forms, through employer-directed learning, professional workshops, Internet seminars ("webinars"), and "short courses" (2- or 3-day intensive training). When examining employment opportunities upon graduation, we recommend that you determine the degree to which a potential employer supports life-long learning.

Professional Societies

Just as staying updated on local, national, and international news is important, so is keeping up to date in your professional field. To help you keep updated, each subdiscipline area of civil and environmental engineering has one or more professional societies that act as a forum for discussion, debate, advocacy, and technology transfer. There may be student chapters of several of these organizations at your university, and you are encouraged to consider membership in one or more of them.

Throughout this book we will refer to ASCE (American Society of Civil Engineers), which is considered the "umbrella" organization for civil and environmental engineers and has more than 140,000 members. ASCE also publishes a daily "Smart Brief" on infrastructure. We encourage you to sign up for daily infrastructure e-mail news updates and current events at: http://www.smartbrief.com/asce/.

TEAMWORK AND LEADERSHIP

The letters of the word "TEAM" are commonly expanded to the phrase Together Everyone Achieves More. Indeed, the nature of most engineering work is team oriented. ABET mandates that CEE curricula require students to work in teams for course projects. Successful teamwork requires that team members understand:

- team formation and evolution
- problem solving
- personality profiles
- time management
- team dynamics

Moreover, team members must have the ability to foster and integrate a diversity of perspectives, knowledge, and experiences (ASCE, 2008).

Additional skills are necessary for working in multi- or cross-disciplinary teams. Such teams are becoming more commonplace, and require collaboration among diverse disciplines (e.g., engineering, economics, and public policy). In such teams, you will need to be able to explain your recommendations and your reasoning to non-engineers, and must understand their viewpoint and appreciate the skills that they bring to the team.

Without appropriate leadership, a team, even one composed of effective team members, is unlikely to succeed. Leadership is the ability to accomplish tasks and reach goals through the efforts of other people. Effective companies, organizations, and government agencies are typically led by effective leaders. The converse is equally true, where ineffective leadership leads to only moderate success or leads to the failure of an organization.

There are many definitions of effective leadership, but most definitions include the following traits of leaders:

- honest
- dependable
- competent
- cooperative
- inspiring
- supportive
- imaginative/innovative
- determined

Without leadership, a more suitable expansion of "TEAM" may result: Together Everyone Always Meanders. Certainly, some people are "born leaders," but by no means does this imply that leadership skills cannot be learned. Gaining leadership skills often requires deliberate intention. These skills will dramatically enhance opportunities for career advancement, and will also provide you with abilities to lead other groups outside of your profession (e.g., volunteer, community, church, and sports.)

The Situational Leader

"Effective leaders make things happen. They don't sit around watching other people and waiting to react to whatever situations occur. They know what ought to happen, plan a way to make it happen, and take steps to see that it does!"

—Paul Hersey, author of
The Situational Leader

Abraham Lincoln's Leadership Skills

"Among Lincoln's greatest qualities were empowering and trusting his gifted subordinates; communicating brilliantly and understandably to the public; maintaining a determined core of beliefs; running his political party like a master puppeteer; and practicing patience as the nation's leader. He liked to say he was slow to take an important step, but once he did, he never retreated. Lincoln also had sense of humor: He loved to laugh, and his laughter helped America survive its most tragic period."

—Harold Holzer, historian and author of several books on Lincoln

Reprinted courtesy of
American Profile.

Entry-Level Job Responsibilities

Typical entry-level civil engineering jobs have the following responsibilities:

- preparing detailed calculations, specifications, and design drawings
- ensuring conformance with applicable codes and standards
- visiting project sites to inspect construction progress
- compiling data for project reports
- collecting and analyzing field data
- preparing operations and maintenance manuals
- assisting with proposal writing and cost estimating.

Figure 1.7 Percentage of Civil Engineering B.S. Degrees Awarded to Women.

Source: National Science Foundation, 2008.

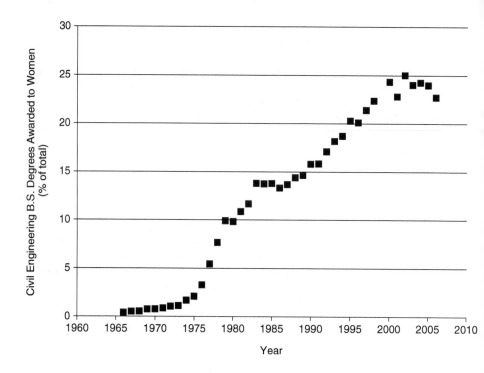

Senior-Level Job Responsibilities

Typical senior-level (10 to 20 years of experience) positions have the following responsibilities:

- managing project teams
- establishing and managing project priorities
- writing project proposals
- developing and maintaining client relations
- maintaining project quality assurance
- managing regulatory permitting and construction documents
- overseeing project design and specifications
- conducting strategic planning

The Future of Civil and Environmental Engineering

The stereotype of an engineer from previous generations was that of a white male working at a drafting table, wearing thick-rimmed glasses, a white short-sleeved shirt, and a dark, skinny tie. Indeed, the engineering workforce has largely been the domain of white males, which explains the use of the male pronoun solely in the quote by President Hoover in a previous sidebar. However, the demographics of the engineering workforce are changing. For example, the percentage of civil engineering degrees awarded to women has increased from virtually none in the 1960s to more than 20 percent in the past decade (Figure 1.7). The fraction of women in the civil engineering workforce is expected to increase as these new engineers will replace retiring, predominantly male engineers. Similar increases have occurred, and are expected to occur in the future for minorities, who now constitute approximately 9 percent of engineering graduates.

Along with becoming more diverse with respect to race and gender, the engineering field has become more diverse with respect to professional responsibilities. A generation ago, newly hired engineers were typically restricted to design and analysis functions. Today as a young engineer, you can expect to take on many additional tasks such as meeting with clients, presenting at public meetings, writing proposals, and managing construction projects. The responsibilities of future young engineers can be expected to increase even further.

Employment Opportunities

According to the Bureau of Labor Statistics, civil engineering is the largest field of engineering in the United States (Table 1.4). Environmental engineering ranks eighth but is among the fastest growing fields. Employment opportunities in both civil engineering and environmental engineering in the future are favorable, and growth rates are above the average growth rates of many other occupations.

About This Book

Traditional engineering education was such that engineering courses dealt exclusively with technical issues. Engineering students were expected to get a holistic education from non-engineering courses in humanities, fine arts, and social sciences. This model proved to be inadequate as the complete separation of the two areas (i.e., engineering from non-engineering) did not provide an opportunity for engineers to place their designs in context. Increasingly, engineering courses include "non-engineering" content. This textbook aims to introduce the first- or second-year engineering student to the "bigger picture" within the context of a potential future career in civil and environmental engineering.

This book is written in a conversational style that we hope you will find very accessible. As such, it breaks with many conventions of engineering communication. For example, the use of "you" (to refer to the reader) and "we" (to refer to the authors) is forbidden in many forms of engineering communication, but we find writing in this style to be more direct and less "academic."

Also, we have intentionally addressed certain critical topics and applications in more than one chapter. We feel that repeatedly *revisiting* core topics and demonstrating interconnections will lead to improved comprehension. Conventional textbooks are structured so that each chapter deals with an individual and self-contained set of concepts. While this conventional approach may promote concentration on those concepts, once you have moved on to the next chapter the previous material is often "out of context." Indeed, the traditional approach to textbooks is analogous to traditional engineering education discussed above, and suffers from similar shortcomings.

Outro

Civil and environmental engineers have made and will continue to make a tremendous impact on public health and prosperity. As students, you are being trained to be future leaders and stewards of the built and natural environments. Future challenges will be great in light of social, economic, environmental, and technological changes. Through collaborative professional, ethical, and sustainable engineering practice, you can indeed make the world a better place.

Table 1.4

Distribution of Engineers in the United States by Field, in 2006

Civil Engineers	**256,000**
Mechanical engineers	227,000
Industrial engineers	201,000
Electrical engineers	153,000
Electronics engineers, except computer	138,000
Aerospace engineers	90,000
Computer hardware engineers	79,000
Environmental engineers	**54,000**
Chemical engineers	30,000
Health and safety engineers, except mining safety engineers and inspectors	25,000
Materials engineers	22,000
Petroleum engineers	17,000
Nuclear engineers	15,000
Biomedical engineers	14,000
Marine engineers and naval architects	9,200
Mining and geological engineers, including mining safety engineers	7,100
Agricultural engineers	3,100
All other engineers	170,000

Source: Bureau of Labor Statistics, 2009.

chapter One Homework Problems

Calculation

1.1 Using the information in Table 1.4, calculate the percentage of engineers that are:

a. civil engineers
b. environmental engineers

1.2 Recreate Figure 1.2 to show the percent of people living in urban areas on the y axis. *Pg 5*

Short Answer

1.3 What is per capita GDP and why is it potentially a more meaningful value than the GDP of a nation?

1.4 In your own words, summarize ASCE's Vision for Civil Engineering in 2025.

1.5 Why is the United States experiencing an "infrastructure crisis"?

Discussion/Open-Ended

1.6 List the advantages and disadvantages of pursuing a master's degree in civil and/or environmental engineering.

1.7 Does ASCE's Vision for Civil Engineering in 2025 apply to the field of environmental engineering? Discuss.

1.8 Describe how the responsibilities of entry-level engineers have changed.

1.9 Why hasn't the "infrastructure crisis" been resolved?

1.10 Which area of emphasis within civil and environmental engineering appeals to you most? Why?

1.11 Which area of emphasis within civil and environmental engineering appeals to you least? Why?

1.12 Do you think that a master's degree should be required to practice as a professional engineer? Why or why not?

1.13 Identify the student chapters of professional civil and environmental engineering societies/organizations at your campus. How would membership in one or more of these organizations benefit students?

Research

1.14 Read the "Findings" section of the NTSB Accident Report for the I-35 W bridge collapse (available at www.wiley.com/college/penn). Was the failure of the bridge due to the factors that had caused the bridge to be classified as "structurally deficient?"

1.15 Read the "Findings" section of the NTSB Accident Report for the I-35 W bridge collapse (available at www.wiley.com/college/penn). What are some lessons that young engineers can learn from the findings?

1.16 Explore a subdiscipline area of civil and environmental engineering. Answer these questions:

a. What are some typical careers in this area?
b. How essential is a master's degree in this area? Note that in some subdisciplines, a master's degree is more important than in others.
c. What is the outlook for jobs in this subdiscipline area?
d. What are some specializations within this subdiscipline?
e. What are some potential future projects in this area that you find exciting?
f. What are some *non-infrastructure* aspects of this subdiscipline?

1.17 Choose a professional journal from ASCE (a list is available at www.wiley.com/college/penn) in an area of interest and prepare a one-page summary of an article published in the last 2 years.

1.18 Research a labor strike associated with some sector of the infrastructure, and summarize the effect on society. The example you choose is not to be one of the strikes mentioned in this chapter.

Key Terms

- built environment
- capital funds
- civil infrastructure
- design life
- inflation-adjusted dollars
- infrastructure
- municipal engineering
- natural environment
- operation and maintenance (O&M) funds
- per capita gross domestic product
- public works

References

American Society of Civil Engineers (ASCE). 2007. *The Vision for Civil Engineering in 2025*. Reston, VA: American Society of Civil Engineers.

American Society of Civil Engineers (ASCE). 2008. *Civil Engineering Body of Knowledge for the 21st Century (2nd Edition)*. Reston, VA: American Society of Civil Engineers.

American Society of Civil Engineers (ASCE). 2009. *Report Card for America's Infrastructure*. Reston, VA: American Society of Civil Engineers.

Bureau of Labor Statistics. 2009. *Occupational Outlook Handbook, 2010–11 edition, Engineers*. http://www.bls.gov/oco/ocos027.htm, accessed August 6, 2011.

Congressional Budget Office. 2007. *Trends in Public Spending on Transportation and Water Infrastructure, 1956 to 2004*. Publication No. 2880.

Herbert Hoover Presidential Library Association. West Branch, IA. http://www.hooverassociation.org/hoover/quotes.php, accessed August 6, 2011.

Hersey, P. 1997. *The Situational Leader*. Escondido, CA: Center for Leadership Studies.

Holzer, Harold. "Abraham Lincoln in Retrospect." *American Profile*. http://www.americanprofile.com/articles/abraham-lincoln-in-retrospect/, accessed August 6, 2011.

National Science Foundation. 2008. *S&E Degrees 1966–2006*. NSF 08–321.

National Transportation Safety Board. 2008a. *Collapse of I-35 W Highway Bridge Minneapolis, Minnesota August 1, 2007*. NTSB/HAR-08/03 PB2008-916203.

National Transportation Safety Board. 2008b. *NTSB Determines Inadequate Load Capacity Due to Design Errors of Gusset Plates Caused I-35 W Bridge to Collapse*. http://www.ntsb.gov/news/2008/081114.html, accessed August 6, 2011.

United Nations, Department of Economic and Social Affairs. *World Urbanization Prospects: The 2009 Revision*. http://esa.un.org/unpd/wup/, accessed August 6, 2011.

World Economic Forum. 2009. *The Global Competiveness Report, 2009–2010*. https://members.weforum.org/pdf/GCR09/GCR20092010fullreport.pdf, accessed August 6, 2011.

chapter Two The Natural Environment

Learning Objectives

After reading this chapter, you should be able to:

1. Describe the hydrologic cycle, and explain the function of wetlands, rivers, and groundwater.
2. Explain the hydrologic budget and define recurrence intervals.
3. Delineate a watershed.
4. Describe the fundamentals of geologic formations.
5. Summarize the interrelationships between infrastructure and land, water, and air.
6. List the potential impacts of climate change on infrastructure.

Introduction

Our infrastructure (the built environment) is inextricably linked to the natural environment. Not only is the natural environment affected by the built environment, but the natural environment is in many ways a critical component of the infrastructure. The materials that we use to build infrastructure come from the natural environment. We build on (or below) the earth's surface. We utilize rivers for transportation and drinking water. We tunnel through soil and rock.

The purpose of this chapter is to introduce you to the aspects of the natural environment such that you can understand the interrelationship between the environment and the infrastructure. As such, this chapter provides context for the following chapters.

Hydrologic Cycle and Hydrologic Budget

A substantial portion of civil and environmental engineering work is related to water. In addition to the more "obvious" applications such as drinking water and wastewater, applications include road drainage, dewatering during construction excavation, construction site storm erosion control, **rip-rap** armoring of bridge abutments, and stormwater outfalls. The **hydrologic cycle** is the basis for understanding many of these applications.

A generalized schematic of the hydrologic cycle is shown in Figure 2.1. Water, in various forms, is temporarily stored in several hydrologic components of the natural environment, including the atmosphere (as clouds and water vapor), oceans and other surface water, ice and snow, and in the subsurface (e.g., groundwater in geologic formations). Water continually moves, or cycles, between these hydrologic components via evaporation, precipitation, **transpiration** (the release of water vapor from leaf surfaces), **infiltration** (soaking of rainwater into the soil), groundwater flow, and **runoff** (the flow of non-infiltrated and non-ponded rainwater or snowmelt across the ground surface).

The mass of all the water in the natural environment remains virtually constant. However, over time, the mass of water stored in one form or hydrologic component does vary. This change can be analyzed with a **hydrologic budget**, and is expressed in equation form in Equation 2.1.

$$\text{Change in Volume of Storage} = (\text{Volume Entering}) - (\text{Volume Exiting}) \quad (2.1)$$

For example, a hydrologic budget for a lake would tabulate the inputs (e.g., runoff, groundwater inflow, and direct precipitation) and outputs (e.g., evaporation, outflow to river); the difference between the inputs and the outputs is the change in volume in the lake, which varies seasonally and annually as evidenced by changes in lake levels.

Consider that when we speak of water "shortages," we mean that the mass of water stored in groundwater, for example, has decreased at one location or

Rip-rap refers to irregularly shaped rocks (size from 4 inches to greater than 36 inches, and weighing from 3 pounds to more than a ton) that are used for erosion protection and slope stabilization. Larger sizes are needed at sites with greater water velocities or steeper slopes.

The **hydrologic cycle** refers to the circulation of water between the atmosphere, surface water, and the subsurface.

A financial budget tracks revenues and expenses, with the difference between the two being an increase or decrease in savings. Similarly, a **hydrologic budget** tracks all of the hydrologic inputs and outputs for a hydrologic component (e.g., the atmosphere or groundwater). The difference between these inputs and outputs is the change in the volume of water stored in the hydrologic component.

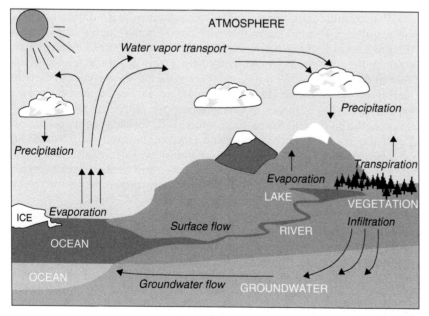

Figure 2.1 Simplified Regional Hydrologic Cycle Including Water Storage (Hydrologic Components in Capital Letters) and Exchange/Transport Processes (Items in Italics).

Source: Adapted, with permission, from Trenberth et al., 2007.

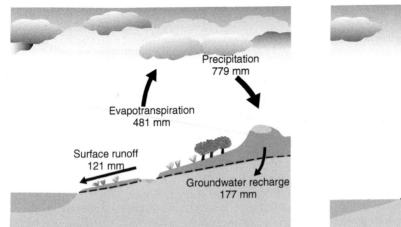
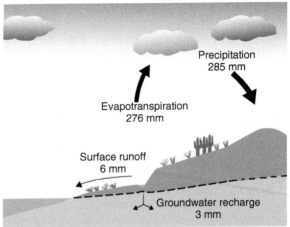

Figure 2.2 Global Variation in Hydrologic Exchanges in Germany (left) and Namibia (right). Note that values are given in millimeters of water depth. Evapotranspiration is the sum of evaporation and transpiration. When water depth is multiplied by the land area of the nation, a volume of annual water exchange is obtained. Also note that the numeric values alone may be misleading; evapotranspiration is greater in Germany than Namibia (481 mm compared to 276 mm), but the relative fraction of evapotranspiration to precipitation is much greater in Namibia (276 mm/285 mm = 0.97 compared to 481 mm/779 mm = 0.62 for Germany). Note that arrow width is proportional to magnitude of transfer.

Source: Redrawn from Struckmeier et al., 2005.

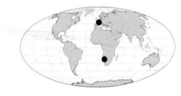

Precipitation intensity is the rate of precipitation, measured as in/hr or cm/hr.

region. That is, the water added to the groundwater over a certain time period is less than the water extracted from the groundwater. Where has the "lost" water gone? It must be in surface water, the atmosphere, the oceans, snow or ice, or in groundwater somewhere else on the planet.

The quantity and timing of water movement (exchange between components) is highly variable spatially at global (Figure 2.2), national, and regional scales. Moreover, variability exists temporally also at these same scales. For example, according to Figure 2.3, there is a large amount of variability for the Great Lakes region over the past century, with annual rainfall depths ranging from less than 25 inches to more than 40 inches.

The predicted depth and **intensity of precipitation** affects the design of every hydraulic component of the infrastructure (e.g., dams, levees, culverts, gutters). Rainfall storm intensities and annual depths of precipitation (Figure 2.4) vary widely across the United States. The annual average precipitation in the contiguous United States ranges from as little as a few inches per year in the arid Southwest to greater than 160 inches per year in the Pacific Northwest. Most of the United States located east of the Mississippi River experiences 30 to 50 inches of precipitation per year. This depth of precipitation is the sum of all forms (rain, snow, hail) in water equivalents, where 1 inch of rain equals 1 inch of water and 1 inch of snow may equal only 0.1 inch of water, depending on the density of the snow. In colder climates, spring snowmelt of accumulated winter precipitation can cause dramatic increases in river flows. In the arid west, reservoirs capture mountain snowmelt and store the water for use during the dry summers.

Hydraulic components of the infrastructure are *not* designed based on annual *average* precipitation. If they were, they would be undersized 50percent of the time. Rather, hydraulic components are designed based on a selected **recurrence interval**. Common terminology, for example, is a "100-year storm" or "100-year drought." This does **not** mean that the event will

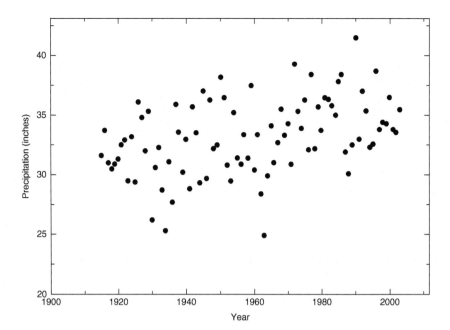

Figure 2.3 Long-Term Regional Variation in Annual Precipitation in the Great Lakes Region of the Midwest.

Source: Redrawn from Hodgkins et al., 2007.

occur once every 100 years; rather, it means that the event has a 1 in 100 (1 percent) chance of occurring in any given year. Although unlikely, the 100-year storm could occur twice in a very short time; for example, Charlotte, North Carolina experienced 100-year storms in both 1995 and 1997.

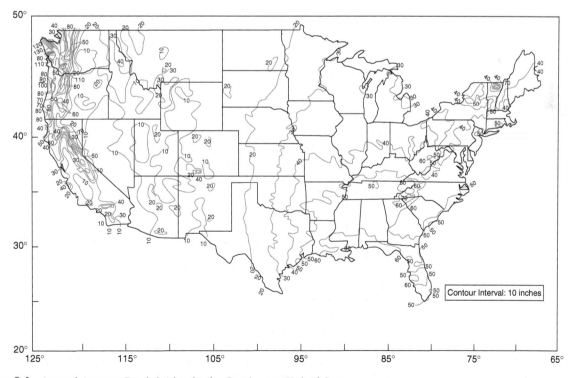

Figure 2.4 **Annual Average Precipitation in the Contiguous United States.**

Source: National Oceanic and Atmospheric Administration, National Climatic Data Center/Owenby et al.

Recurrence Intervals

The **recurrence interval** (sometimes called a "return period") is equal to the inverse of the probability of an event happening in any given year. For example, a 10-year storm has a 1/10, or 10 percent, probability of occurring in any given year. This has many applications in infrastructure engineering, as infrastructure components need to be designed to not fail in response to a given floodwater depth, rainfall intensity, stormwater runoff, etc. A floodwall built to withstand a 100-year flood will be taller than one designed to withstand a 50-year flood. A culvert designed to convey the runoff from a 10-year storm will have a smaller diameter than the culvert designed to pass the 50-year storm. In each of these cases, the longer return period affords a lower risk and a greater sense of security but also costs more. Consequently, the infrastructure component is not expected to "handle" events that are larger than that for which it was designed. That is, a culvert designed to pass the 10-year storm is expected to "fail" (i.e., not be able to convey the runoff) for any storm larger than the 10-year storm. The failure of the culvert will lead to a flooded roadway and perhaps soil erosion and related damage to the roadway. However, if this roadway conveys minimal traffic, the selection of a 10-year storm may be appropriate given that only a few residents will be affected. (Of course, for the affected residents, this flooding may be totally unacceptable.) The recurrence interval chosen as the basis for design increases as the number of affected people and the economic risk of failure increases; for example, culverts under freeways are typically designed at the 50-year (2 percent) level.

The relationship between the storm intensity and its recurrence interval is illustrated by intensity-duration-frequency (IDF) curves (e.g., Figure 2.5). Local IDF curves are available for many regions of the country. An IDF curve is utilized when designing infrastructure components to meet desired levels of risk protection from flooding. According to an IDF curve, for a given storm duration the depth of precipitation increases as the recurrence interval increases. For example, according to the IDF curve in Figure 2.5, the 2-year intensity for a 60-minute storm duration is approximately 1.1 in/hr while the 100-year intensity for the same duration storm is approximately 2.4 in/hr.

Figure 2.5 Intensity-Duration-Frequency (IDF) Curves for Pierre, South Dakota. Note that both axes are logarithmic scale.

Source: Redrawn from South Dakota Department of Transportation.

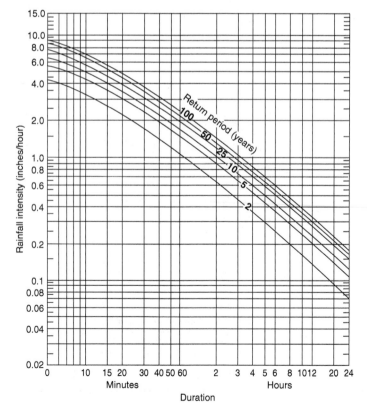

The total precipitation depth of a storm is the product of the intensity and the duration. Also as shown in Figure 2.5, for a given recurrence interval, the intensity increases as the duration decreases.

Watersheds

A **watershed** is the area draining (i.e., contributing runoff) to an outlet point. The outlet point may be the mouth of a river or stream, a culvert, or a stormwater inlet. When designing hydraulic structures, engineers must **delineate** (i.e., draw the outline or boundaries of) contributing watershed areas.

 How to Delineate a Watershed

Watershed characteristics of interest include the type of land cover (e.g., crops, forest, pavement), the soil types (e.g., sand or clay), and topography (flat, hilly, mountainous).

Columbia River Watershed

The watershed for the Columbia River is shown in Figure 2.6 (the watershed outlet being the mouth of the Columbia where it enters the Pacific Ocean). The boundaries of the watershed are mountain ridges; for example, precipitation falling west of the watershed boundary in Montana will travel to the Columbia River, whereas precipitation east of the watershed boundary (in the adjacent watershed of the Missouri River) will travel to the Gulf of Mexico. Notice that the Columbia River watershed is contained within seven states and one Canadian province of British Columbia, making management of the river quite challenging. The Columbia River has many tributaries (rivers feeding into it). Each of them has its own watershed, which is considered a sub-watershed of the Columbia River. The largest sub-watershed, the Snake River watershed, is delineated with a dashed line in Figure 2.6; the outlet is the confluence of the Snake and Columbia Rivers. The Snake River also has many sub-watersheds for each of its tributaries.

Figure 2.6 The Columbia River Watershed (Shaded Area) and Its Largest Sub-Watershed (Snake River, Delineated by Dashed Line).

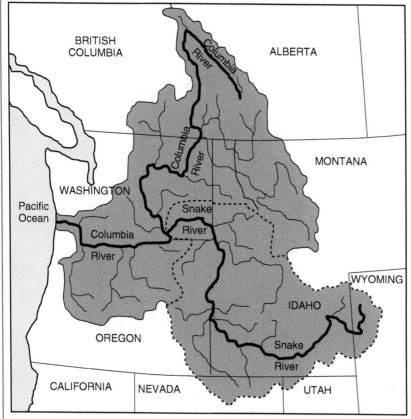

Figure 2.7 Map of Major U.S. Watersheds. The Mississippi River basin is delineated in dark blue.

Source: U.S. Geological Survey.

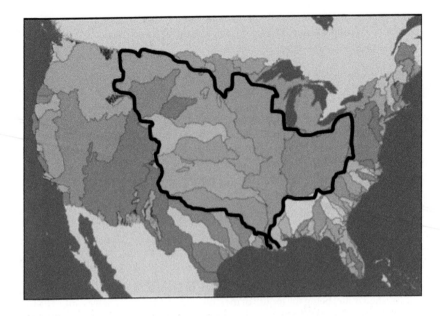

Major watersheds in the United States are mapped in Figure 2.7. The Mississippi River watershed, one of the largest in the world, is outlined with a dark solid line. This **basin** (large watershed) includes the watersheds of the Missouri River, the Ohio River, the Arkansas River, and many other major rivers. Knowing the boundaries of this watershed aids in managing the quality and quantity of water in the Mississippi River. For example, knowing the contributing area is necessary in predicting floods for various reaches of the river. Also, predicting the mass of pollutants, such as sediment and nutrients, entering the Gulf of Mexico can be accomplished by **mathematical modeling** of the river and its watershed and sub-watersheds.

As used in this text, the term "**mathematical model**," or often simply "model," refers to a collection of mathematical expressions or formulae that predict how systems, both natural and built, will behave.

Rivers

Because of complexities such as soil infiltration rates, snow melt, and the varying travel times of storm runoff to rivers, a 100-year *storm* does not necessarily result in a 100-year *flood*. Rather, flood probabilities are based on long-term measurements of actual river **stage** (water elevation), rather than precipitation. An example of such historic data is provided in Figure 2.8.

Figure 2.8 Maximum Annual River Stage (Elevation) for the Susquehanna River at Williamsport, Pennsylvania. Note that data has been collected at this site for more than 100 years. The record stage of 35 feet occurred in 1972 when Hurricane Agnes produced 18 inches of rain over a 5-day period.

Source: U.S. Geological Survey.

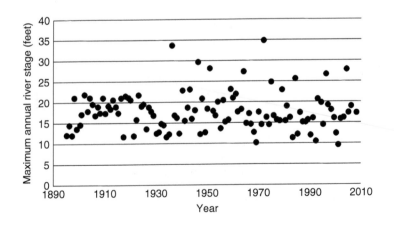

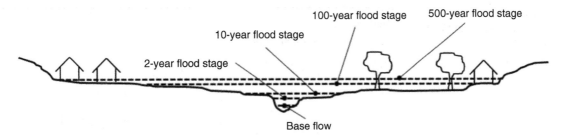

Figure 2.9 Cross-Section View of a River and Floodplain Valley.

A cross-section of a river and **floodplain** can be seen in Figure 2.9, with several flood stages noted. In Figure 2.9, **base flow** is identified, which corresponds to the "normal" water depth for the river at times between substantial runoff events. For runoff events with recurrence intervals greater than 2 years, the depth of water will be higher than the top of the riverbank, and given the topography of the land, this water will spread laterally across the floodplain to a land elevation equal to the water depth. Figure 2.10 is a plan (aerial) view of this same floodplain and the extents of the 100-year and 500-year floods.

The 100-year floodplain is especially noteworthy, given that most modern building codes prohibit building permanent structures in the 100-year floodplain. The delineation of the floodplain is provided by the National Flood Insurance Program of the Federal Emergency Management Agency (FEMA). Many older structures, including homes, are located in the 100-year floodplain. This may be because the homes are very old and were built before floodplain maps were available. Also, consider a home that was built along a river for which flooding is minimized by an upstream dam. If the dam is removed, the boundary of the 100-year floodplain will change substantially.

Alternatively, as a watershed becomes increasingly developed with more **impervious** surfaces (paved parking lots, roads, and rooftops that limit infiltration), more runoff occurs for any given storm event. Thus, the flood stage for any recurrence interval tends to increase over time. For flood insurance and development planning reasons, FEMA is updating floodplain

> A **floodplain** is the land along a stream or river that is inundated when the stream or river overtops its banks.

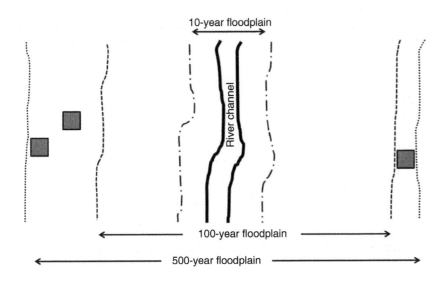

Figure 2.10 Aerial (Plan) View of River and Valley from Figure 2.9. Note that the homes (squares) are located outside of the 100-year floodplain as required by building codes.

maps to more accurately delineate current flood zones. This is an important, challenging, and expensive endeavor, costing more than $1 billion, and will utilize mathematical models and data collected by engineers and scientists.

The Fickle Power of Nature

Rivers and streams create their own routes, which move over time. Site investigations have shown that some river channels have moved laterally by as much as 100 feet in the last two centuries. Many civil engineering projects have tried (often successfully, sometimes not) to "contain" rivers to a desired path. Consideration of the tendency for river channels to migrate is important when infrastructure projects are located near rivers (e.g., choosing bridge locations; see Figure 2.11).

Figure 2.11 Armored Bridge Abutments. Note the armoring of bridge abutments with rip-rap (6 inches to 18 inches in size) used to prevent scour during flood stages. Also note that construction of this particular bridge required the filling of wetlands in the floodplain.

Source: M. Penn.

Habitat/Native Species

Development inherently eliminates wildlife habitat and otherwise affects the natural environment; however, it is possible to minimize habitat loss by controlling the location or type of development. Many community and regional planning organizations identify rare or important natural resources and prevent or discourage development in those areas. Federal and state regulations also limit development in areas documented as habitat for endangered species. Some types of development can include green space that can be designed to provide wildlife habitat. Designing a functioning habitat depends on the species desired, and is not as simple as planting a few trees and shrubs. Professionals, most likely without engineering backgrounds, that have training and experience in habitat design should be consulted.

Wetlands

Wetlands are often viewed as "stumbling blocks" to many development and transportation projects due to their highly regulated nature. Such regulations are necessary, given that in the past 150 years, more than half of U.S. wetlands have been drained or filled for urban, suburban, and agricultural uses. Wetlands provide many potential benefits ranging from recreational activities to floodwater storage. Some wetlands provide habitat for rare or endangered plants and wildlife.

The Use of Native Species

The use of native species of plants has gained popularity from an aesthetic and ecological landscaping perspective. In some cases, local or state regulations recommend or require the use of native plant species. For example, prairies, with very diverse grasses and flowers, once ranged from the Great Plains to the Midwest prior to settlement and subsequent conversion to agriculture. Some states have begun planting thousands of acres of native prairie plants along highway roadsides (Figure 2.12). Native plants offer the advantages of being best suited (i.e., naturally selected) for the climate at the site location, reducing erosion, increasing water infiltration, and offering natural microhabitats. In addition, during the summer and fall bloom period, the wildflowers greatly enhance the beauty of the roadside.

Roadsides for Wildlife brochure, Minnesota Department of Natural Resources

Figure 2.12 **A Roadside Prairie Sign.**

Source: Minnesota Department of Natural Resources.

Wetlands, by legal definition, are not necessarily always "wet" (i.e., containing ponded water). When you think of a wetland, you may envision a swampy area or a marsh with ducks and cattails. Indeed, these examples are classified as wetlands. But many types of wetlands do not fit this stereotypical view of a wetland (Figure 2.13).

Proper identification and location of wetlands is critical at development sites, and is accomplished using **wetland delineation**. Failure to identify a wetland may lead to costly fines and potential project delays. Definitions vary at the local, state, and federal level, but in general, wetlands are areas that have, for a significant portion of the plant growing season, groundwater within a few feet of the ground surface or ponded water at the ground surface. This definition stems from the fact that certain plants and the wildlife that depend upon them, flourish in wet soils, whereas other plants do not.

Wetlands will be discussed in more detail in Chapter 15, Environmental Considerations.

Trained scientists and engineers evaluate the hydrology, soils, and plants at a site to determine whether or not a wetland exists and to determine its spatial extent (termed **wetland delineation**).

Geologic Formations

Having an understanding of geologic formations is necessary to understand groundwater flow, infiltration, and runoff. Such understanding is also

Figure 2.13 **Students Learning to Identify Wetland Plants and Soils in a "Wet Prairie."** There is no "standing" water typically in this type of wetland.

Source: M. Penn.

necessary for designing rock cuts for roads and foundations for buildings. Additionally, geologic formations are sources of building materials for roads, levees, and other infrastructure components.

Soil is the uppermost (surface) layer of **unconsolidated material** that lies above the uppermost layer of rock (**consolidated material**), termed **bedrock**. Soil is the biologically active upper layer of unconsolidated material and may range from approximately 2 to 20 feet in depth below the surface. The biologically inactive unconsolidated material below soil is termed **subsoil**. The depth of unconsolidated material varies based on how it was deposited (e.g., by wind or by glacier) and on the shape of the bedrock surface below; this surface may be "uneven" due to prior geologic occurrences (e.g., an ancient river gorge). The depth of unconsolidated material can vary dramatically over small distances or in some cases can be very uniform over large distances. Figure 2.14 is a depth to bedrock map for Stearns County, Minnesota. Based on our definitions of bedrock and unconsolidated material, we could also describe Figure 2.14 as a map of the thickness of the unconsolidated material. Note that in the southwest corner of the county, over a distance of only approximately 15 miles, the depth of unconsolidated material ranges from less than 100 feet to nearly 500 feet, which demonstrates the need for proper geotechnical site investigations for any project.

For example, at the University of Wisconsin-Platteville (the authors' campus), a new building was planned for construction. The initial site investigation included several **borings** to locate the bedrock depth for designing the building's foundation (Figure 2.15). Based on the results of the borings, engineers predicted a bedrock surface, and used the depth of this predicted surface when designing the foundation and estimating the cost of excavation. However, when excavation commenced, bedrock was encountered at shallower depths than expected in some areas between the locations of the borings (Figure 2.16). The project was delayed and additional costs were incurred because of the additional unanticipated blasting and excavating.

In some locations across the United States, the bedrock represents a massive formation which may extend *vertically* for thousands of feet. In other areas, many layers of rock exist, as is evident for the cross-section of Kansas provided in Figure 2.17. Each layer has different properties that can affect engineering decisions. These properties include bedrock strength, which will affect the design of foundations, and the **porosity** and extent of fractures, which will affect groundwater flow.

Unconsolidated material refers to loosely arranged organic and mineral particles that are not cemented together.

Borings are vertically drilled holes used to identify geologic materials present at various depths.

Porosity is a measure of the void spaces in a geologic formation. Porosity is equal to the volume of void spaces in the sample divided by the total sample volume. For example, some types of sandstone have porosities of 0.1; thus, a 1 cubic foot of sample of sandstone would contain 0.1 cubic feet of void space and 0.9 cubic feet of sand.

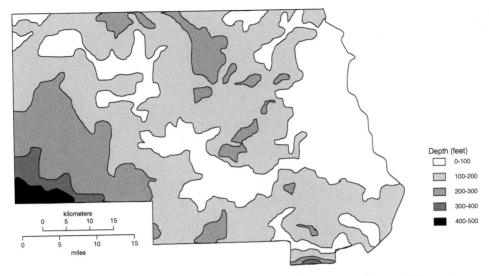

Figure 2.14 **Depth to Bedrock (or Thickness of Unconsolidated Material) in Stearns County, Minnesota.**

Source: Redrawn from Minnesota Geological Survey.

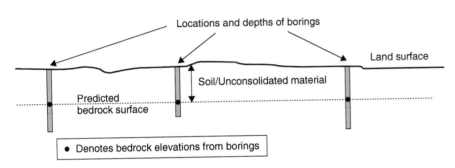

Figure 2.15 Location of Borings and Predicted Bedrock Elevations (Dotted Line) for a Building Foundation.

Groundwater flow, like many natural processes, is typically hidden from view. One case in which groundwater flow can be observed is depicted in Figure 2.18. Groundwater is seeping out between layers of rock. In this instance, the groundwater preferentially flows outward in the horizontal direction as compared to downward in the vertical direction. In general, large fractures or highly porous formations favor groundwater flow, while impermeable formations hinder groundwater flow. The "icicles" in the photo form at the openings of vertical fractures.

While you may think of the earth's crust (the uppermost rock layers, tens of thousands of feet in thickness) as a stable body, it is not. Movement of geologic formations creates stress that can cause large planar fractures, or faults. Sudden movement along faults is the cause of seismic activity

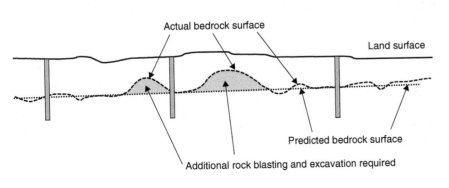

Figure 2.16 Predicted and Actual Bedrock Surface, and Additional Rock Removal Required.

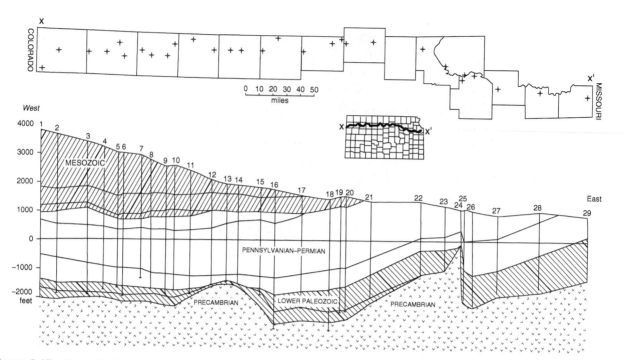

Figure 2.17 **Geologic Cross-Section of the State of Kansas Illustrating Sedimentary Rock Layers.** Note the exaggerated vertical scale. Names refer to geologic periods of deposition. Numbers and vertical lines in cross-section correspond to borings at locations marked (+) in plan view.

Source: Redrawn from Merriam and Hambleton.

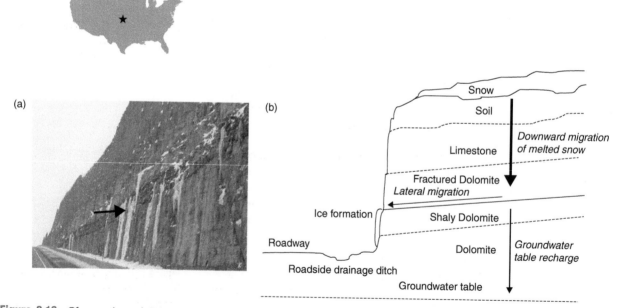

Figure 2.18 **Observation of Groundwater Flow.** (a) Icicles forming on the face of a roadcut from horizontal groundwater seepage along the top of a rock layer (darker formation shown with arrow) that is more restrictive to vertical flow. (b) Cross-sectional view of the geologic formations shown in the photo. The shaly dolomite layer is more restrictive to vertical groundwater migration than the other rock types, but some water does pass through to recharge the groundwater table. Arrows are scaled by width (not length) to represent the relative amount of groundwater flow. Note that the groundwater table (defined in the next section) is below the elevation of the roadway.

Source: M. Penn.

Figure 2.19 Damage Caused by the 1989 San Francisco Earthquake.

Source: U.S. Geological Survey.

(earthquakes), which varies in magnitude from undetectable to catastrophic. Earthquakes occur continuously, literally hundreds across the globe in any given week.

The "Bay Area" of California (San Francisco, Oakland, and San Jose), one of the 10 largest metropolitan areas in the country, is located on and near more than one dozen major faults. In 1906, an earthquake here killed more than 3,000 people in what is considered one of the worst natural disasters in our nation's history. In 1989, another earthquake caused an estimated $6 billion in damages (Figure 2.19) and claimed more than 60 lives. The decreased number of fatalities can be in part attributed to the lesser magnitude of the 1989 earthquake (6.9 as compared to 7.9 on the Richter scale); however, credit must also be given to the work of civil and environmental engineers—both in the design of earthquake-resistant structures and in the development of infrastructure that allows speedy emergency response. An effective infrastructure also allows for a relatively rapid recovery and less disruption to daily lifestyles, while, as witnessed by the recent Haiti earthquake (Figure 2.20), an infrastructure-poor country takes much longer to recover.

While California and Alaska experience the most earthquakes, damaging earthquakes have occurred across the United States. Many geotechnical and structural engineers specialize in designing foundations and structures that can withstand seismic activity.

Many civil and environmental engineering programs require a course in geology or **geomorphology**. Understanding geologic formations and processes is vital to infrastructure engineering.

Map of Latest Earthquakes in the World—Past 7 Days, U.S. Geological Society.

Geomorphology is the study of landforms and the processes that shape them.

Groundwater

Water is found in the subsurface in pores of unconsolidated material and in pores and fractures in rock. Below the **water table**, or groundwater table, all pores and fractures are filled with water. The depth to the water table from the ground surface may range from zero in a wetland to hundreds of feet (Figure 2.21).

An **aquifer** is a geologic formation that stores water. Aquifers may be shallow (less than 100 feet below ground surface) or deep (sometimes greater than 1,000 feet below ground surface) and are composed of granular unconsolidated material, of fractured bedrock, or of porous rock (e.g., sandstone). Water is extracted from aquifers by drilling a well (a vertical borehole in the ground) and pumping the water out of the well. Groundwater availability is

The **water table** occurs at the depth at which the geologic formation is saturated.

Figure 2.20 **Aftermath of Haiti Earthquake.**

Source: Copyright © Claudia Dewald/iStockphoto.

Figure 2.21 **An Aerial View of an Operating Limestone Quarry.** Quarry excavation typically continues downward until the water table is reached, which has yet to occur in this quarry. In this case, the water table is greater than 150 feet below ground surface. The dark area in the center of the photo is a stormwater pond at the lowest elevation to collect stormwater runoff. Note the minimal setback distance (approximately 30 feet) from open excavation to the industrial building in the background; such a short setback is allowed because of the stability of the geologic formation.

Source: Copyright © Yvan Dubé/istockphoto.

highly variable across the United States. Where it is plentiful enough to meet large water demands, groundwater is used for public drinking water systems and industries. In regions where it is not economically feasible to extract groundwater, surface waters are used for water supply. Most rural residences have individual private drinking water wells as the small demands required can readily be obtained.

One important characteristic of groundwater can be understood by applying a hydrologic budget to an aquifer. As development occurs, previously pervious areas are made impervious by paving driveways and roads and by constructing rooftops. As a result, more runoff occurs and there is a corresponding decrease in infiltration to groundwater (termed **recharge**). Consequently, the mass of water stored in aquifers decreases and groundwater levels are lowered. This may lead to decreased stream flows in streams that are fed by groundwater, and to the unwanted conversion of wetlands to "drylands." Similarly, pumping groundwater out of an aquifer at a rate that is greater than the natural rate of recharge will lower groundwater levels.

Many public water supplies that rely on groundwater are finding that groundwater levels are dropping, sometimes at the rate of several feet per year. This is not only occurring in the arid southwestern states, but throughout the country. As a result, groundwater must be pumped up from lower depths, resulting in more costly pumping and possibly the installation of deeper wells if deeper aquifers exist. If no other aquifers are present, alternative water supplies must be identified, often at great expense (e.g., by damming a river).

Atmosphere

The atmosphere is the major transport mechanism for water in the hydrologic cycle, and has complex interrelationships with the land surface and oceans with respect to global climate change (discussed in the following section). **Air quality** primarily deals with the air that we breathe in the lower atmosphere (stratosphere) and the potential adverse health effects of pollution (Figure 2.22). However, air quality also deals with aesthetics (visibility) and degradation of structures (e.g., acid rain). Furthermore, many pollutants cycle between land, water, and air. Some types of air pollution are linked to degraded surface water quality, and thus human and wildlife health via contaminated fish consumption. Interactions between infrastructure and air quality will be discussed throughout this book.

Figure 2.22 Air Quality in Widnes, England, in the Late 1800s was Typical of Industrial Cities in the United States. Due to passage of the Clean Air Act of 1970 and the 1990 Amendment, air quality is much better today. However, air quality remains a public health concern in many urban areas.

Source: Wikipedia: http://commons.wikimedia.org/wiki/File:Widnes_Smoke.jpg.

Climate Change

Human-induced climate change (i.e., global warming) has been debated for decades by scientists, policymakers, and the public. Greenhouse gases (carbon dioxide, methane, nitrous oxide, ozone, and others) have been rapidly accumulating in our atmosphere (Figure 2.23) and trap heat radiating outward from the earth. A common analogy for the greenhouse effect is a car with the windows shut on a sunny day—light (short wavelength radiation) enters through the glass and is converted into heat (long wavelength radiation), which is trapped by the glass.

In 1990, the Intergovernmental Panel on Climate Change (IPCC) released its first report on climate change which, along with subsequent reports, led to the Kyoto Agreement for greenhouse gas emission reductions. Nearly 200 nations have signed the protocol; the United States, the second largest CO_2 emitter (recently surpassed by China), has yet to agree to the requirements at the time of writing of this textbook. The IPCC's claim that the major cause of recent (last 50 years) global warming is very likely to be anthropogenic (human induced) is supported by the national science academies of nearly all industrialized nations and is now considered to be scientific consensus. There is, however, debate as to the extent of this relationship relative to naturally occurring warming processes.

The potential negative consequences of climate change typically intensify with increasing temperature rise. The consequences are vast (Parry et al., 2007), and often regional, including:

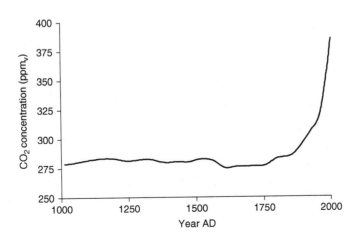

Figure 2.23 Increases in CO_2 Gas Concentrations in the Atmosphere Over the Past 1,000 Years. Although not shown, the trends for nitrous oxide and methane are similar. Data prior to 1950 is based on ice cores while data since 1955 is based on atmospheric measurements.

Data source: Carbon Dioxide Information Analysis Center.

- Increased drought, leading to decreased water availability for hundreds of millions of people
- Increased risk of species extinction
- Loss of coral reefs
- Changes in global carbon cycling
- Shifts in suitable habitat for species
- Increased wildfire risk
- Changes in suitability for crop production
- Increased flooding
- Loss of coastal wetlands
- Increased malnutrition and disease

Complex mathematical climate models not only estimate averaged global temperature, but also regional temperature and weather pattern changes. Some models predict more intense rainstorms, tornadoes, and hurricanes, all of which potentially have disastrous effects on the built and natural environments.

Because of melting glaciers and polar ice caps, sea levels are expected to rise. Estimates range from 8 inches to 20 inches over the next century above current levels (Figure 2.24). To the untrained engineer, this might not seem significant, but it is potentially very significant. If flood protection systems are overtopped, flooding will occur. And given that there have been many cases where floodwaters have come within inches of overtopping levees (e.g., Figure 2.25), a few additional inches of sea level increase could have catastrophic results.

In 2009, the U.S. Army Corps of Engineers adopted a policy whereby future sea level predictions *must* be incorporated in to the planning and design of all of its projects. "To ignore rising sea level in the design of civil works would be like ignoring the health effects of smoking a cigarette . . . we've gotten to that point," Jim Titus, a U.S. Environmental Protection Agency researcher, said, in response to the release of the policy (Luntz, 2009). **Adaptive management** is an increasingly common and effective approach to deal with future uncertainties of all kinds. For example, without adaptive management, two options exist for a new levee: (1) design the levee for the expected future floodwater elevation,

U.S. Army Corps of Engineers Circular on Incorporating Sea Level Change in Civil Works

Adaptive management is a process whereby future uncertainties are incorporated into a plan or design so that if changes do occur, a strategy *is already in place* to deal with them.

Figure 2.24 Historical and Projected Range of Sea Level Rise. Note that scale is in millimeters (100 mm ≈ 4 inches). Actual instrumental measurements have been made since 1870. Prior to this date, estimates are made by a variety of scientific techniques.

Source: U.S. Environmental Protection Agency.

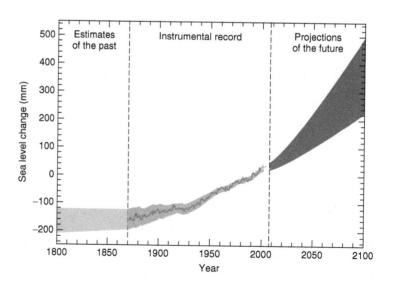

Figure 2.25 **One of the Authors Holds a 1996 Newspaper Published the Day After a Flood Crest Nearly Overtopped the Levees (Over Which the Bridge in the Photo Passes) in Wilkes-Barre, Pennsylvania.** The Susquehanna River level rose 30 feet in a 2-day period following a 2-inch rainstorm and warm weather causing excessive snowmelt in January.

Source: M. Penn.

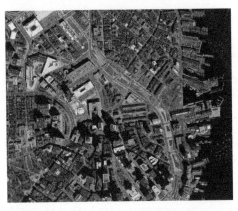

Remember that full-sized versions of textbook photos can be viewed in color at www.wiley.com/college/penn.

Figure 2.26 **Downtown Boston (left) and 100-Year Flood Zones (right).** The existing flood zone is depicted by the solid line. The predicted future flood zones corresponding to a "high level" of greenhouse gas emissions scenario, is depicted by the dashed line.

Source: U.S. Geological Survey, and floodplain boundaries adapted from Frumhoff et al., 2007.

or (2) design an even taller levee for protection against potentially higher floodwaters, which may or may not occur. In the first case, if future floodwaters are indeed higher than expected, the cost to raise the levee will be significant. In the second case, the taller levee would be more expensive to design and build, and would be unnecessary if higher floodwaters do *not* occur. With adaptive management, a "middle ground" can be achieved. A levee design can specify a levee that will contain expected floodwater elevations and that can be readily modified as warranted by future circumstances. In this case, since the initial design accounted for the future possibility of raising the levee, the cost to raise the levee would be less than retrofitting a levee that was not designed to accommodate modification.

Sea level rise is of particular concern for coastal communities such as Boston, Massachusetts (Figure 2.26), many of which have experienced rapid growth in recent years.

Outro

The natural environment supports all life and, in many different ways, all infrastructure. Human activity and infrastructure development has many potential adverse effects on the land, water and air. The *impact of infrastructure on the environment* and *the impact of the environment on infrastructure* must be fully understood in order to plan and design future growth that is sustainable.

chapter Two Homework Problems

Calculation

2.1   Use Figure 2.24 to estimate low-range and high-range projected rates of sea level rise (mm/yr) from 2010 to 2100. Convert these rates to inches per year. Show all work.

2.2   For Pierre, South Dakota (Figure 2.5), what is the precipitation intensity of a storm that:

a. lasts 10 minutes, given a recurrence interval of 100 years?
b. lasts 30 minutes, given a recurrence interval of 10 years?
c. lasts 30 minutes, given a recurrence interval of 2 years?
d. lasts 2 hours, given a recurrence interval of 10 years?

2.3   Using Figure 2.5, calculate precipitation depth for a 3-hour duration event with the following probabilities of occurrence in a given year:

a. 1%
b. 10%
c. 50%

2.4   1.00 million gallons per day (MGD) infiltrates into an aquifer.

a. If 1.05 MGD are extracted from the aquifer, by how many gallons will the aquifer be depleted in one year?
b. If 1.05 MGD are extracted from the aquifer, but development has caused the infiltration rate to drop to 0.95 MGD, how does this change in infiltration change your answer from part "a"?

2.5   A pond has a surface area of 1.0 acre. A stream flows into the pond at a steady flowrate of 1.20 cfs (cubic feet per second) for the entire year. The evaporation rate, also steady for the entire year, is 1.0 inch/month. Assume that the pond is circular and has straight sides (that is, it has the shape of a cylinder). The pond has no outflow stream but a farmer withdraws water for 4 months of the year at a rate of 3.50 cfs.

a. What annual change in pond volume (ft^3) corresponds to the evaporation rate of 1.0 inch/month? That is, what volume of water evaporates in one year?
b. What is the annual change in storage in the pond, in ft^3, due to all inflows and outflows? Use the relationship that volume is the product of flowrate and time, and convert all values to the same units.
c. What is the change in water depth for one year?

2.6   Given that Germany has a land area of 349,520 km^2 and Namibia has a land area of 824,290 km^2, the volume associated with the hydrologic exchanges in Figure 2.2 can be calculated as the product of land area (km^2) and hydrologic exchange (mm).

a. Create a table that compares the volume of surface water runoff, precipitation, evapotranspiration, and groundwater recharge for Namibia and Germany. Present your answers with units of m^3.
b. Apply Equation 2.1 to the land surface of each of these countries and comment on the change in storage that you calculate.

Short Answer

2.7   What is the difference between soil and unconsolidated material?

2.8   What are the components and transfer mechanisms of the hydrologic cycle?

2.9   Why are many existing homes located in floodplains?

2.10   What are some advantages to using native species in plantings?

2.11   What are some benefits that wetlands provide?

2.12   How does the development of land affect aquifers?

Discussion/Open Ended

2.13   What are some possible outcomes of the failure to incorporate climate change in long-term infrastructure planning?

2.14   What are the possible outcomes if infrastructure planning and design accounts for climate change but scientific predictions are incorrect and temperatures do not increase as expected?

Research

2.15   Find information on the depth of water supply wells and the associated geologic formations for a community of your choice. Your state's Department of Natural Resources (sometimes named Environmental Protection Agency or Department of Environmental Conservation) website might be a good place to start.

2.16   Delineate the watershed in which your university is located. Is this a sub-watershed of a larger basin? A topographic map may be downloaded from a variety of websites such as msrmaps.com.

Key Terms

- adaptive management
- air quality
- aquifer
- base flow
- basin
- bedrock
- borings
- consolidated material
- delineate
- evaporation
- floodplain
- geomorphology
- hydrologic budget
- hydrologic cycle
- impervious
- infiltration
- intensity of precipitation
- mathematical modeling
- porosity
- recharge
- recurrence interval
- rip-rap
- runoff
- soil
- stage
- subsoil
- transpiration
- unconsolidated material
- water table
- watershed
- wetland delineation

References

Carbon Dioxide Information Analysis Center. http://cdiac.ornl.gov/ftp/trends/co2/lawdome.combined.dat, accessed August 6, 2011.

Frumhoff, P. C., J. J. McCarthy, J. M. Melillo, S. C. Moser, and D. J. Wuebbles. 2007. *Confronting Climate Change in the U.S. Northeast: Science, Impacts, and Solutions*. Cambridge, MA: Union of Concerned Scientists.

Hardie, D. W. F. 1950. *A History of the Chemical Industry in Widnes*. Imperial Chemical Industries Limited.

Hodgkins, G. A., R. W. Dudley, and S. S. Aichele. 2007. *Historical Changes in Precipitation and Streamflow in the U.S. Great Lakes Basin, 1915–2004*. USGS Scientific Investigations Report 2007–5118.

Luntz, T. 2009. "New Army Corps Policy Forces Project Designers to Consider Rising Seas." *New York Times*. November 11, 2009.

Merriam, D. F., and W. W. Hambleton. 1959. "Relation of Magnetic and Aeromagnetic Profiles to Geology in Kansas"; *in*, Symposium on geophysics in Kansas: *Kansas Geological Survey*, Bull. 137, pp. 153–173.

Minnesota Geological Survey. *Geologic Atlas of Stearns County Minnesota*, C-10, Part A, Plate 5, Depth to Bedrock.

Owenby, J., R. Heim, Jr., M. Burgin and D. Ezell. *Climatography of the U.S. No. 81—Supplement # 3, Maps of Annual 1961–1990 Normal Temperature, Precipitation and Degree Days*.

http://www.ncdc.noaa.gov/oa/documentlibrary/clim81supp3/clim81.html, accessed August 6, 2011.

Parry, M. L., O. F. Canziani, J. P. Palutikof, P. J. van der Linden, and C. E. Hanson, Eds. 2007. *Climate Change 2007: Impacts, Adaptation and Vulnerability*. Cambridge, UK: Cambridge University Press.

South Dakota Department of Transportation. *Road Design Manual*. http://www.sddot.com/pe/roaddesign/plans_rdmanual.asp, accessed August 6, 2011.

Struckmeier, W., Y. Rubin, and J. A. A. Jones. 2005. *Groundwater: Reservoir for a Thirsty Planet?* Earth Sciences for Society Foundation, Leiden, The Netherlands. http://www.yearofplanetearth.org/content/downloads/Groundwater.pdf, accessed August 6, 2011.

Trenberth, K. E., L. Smith, T. Qian, A. Dai and J. Fasullo. 2007. "Estimates of the Global Water Budget and its Annual Cycle Using Observational and Model Data." *Journal of Hydrometeorology*, 8: 758–769.

United States Environmental Protection Agency. *Seal Level Rise Projections to 2100*. http://www.epa.gov/climatechange/science/futureslc_fig1.html, accessed August 6, 2011.

United States Geological Survey, USGS Gauging station 01551500 (WB Susquehanna River at Williamsport).

chapter Three Structural Infrastructure

Learning Objectives

After reading this chapter, you should be able to:

1. List the main subsectors and components of the structural sector of the infrastructure.
2. Explain the purpose of and list some design criteria for various components of the structural sector.
3. Identify various components of the structural sector.

Prelude to Chapters 3 through 5

Chapters 3 through 5 introduce you to the physical portion of the built environment. The broadest categories are termed infrastructure **sectors**, such as transportation, structural, or environmental. Some sectors are further broken down into **subsectors**; for example, the transportation sector includes aviation and mass transit. We will also introduce various **components**, the most fundamental "building blocks" of the sectors. Examples of mass transit components include roadways (with sub-components of pavement, curbs, etc.), buses, rail lines, and trains. Knowledge of these sectors, subsectors, and components is necessary to understand topics presented throughout the remainder of this textbook. In addition to being able to *identify* components, you will also begin to understand the *interrelationships* between components, sectors, society, and the environment.

Note that the sectors, subsectors, and components introduced by these chapters are not inclusive (due to the vastness of our infrastructure), nor is our treatment of each exhaustive. In writing these three chapters, we decided to include all sectors or subsectors from the *ASCE Report Card for America's Infrastructure*, which was introduced in Chapter 1; additional sectors and subsectors are included to demonstrate the variety of infrastructure that exists.

Additionally, some of the sectors, subsectors, and components are covered in more detail than others. Greater level of detail will be provided for the most common sectors, subsectors, and components (e.g., roads and stormwater management), whereas the less ubiquitous examples (e.g., ports, skyscrapers) are introduced with less detail. This varying level of coverage mirrors that of typical undergraduate civil and environmental engineering programs (e.g., skyscraper and port design are *not* commonly taught at the undergraduate level). Also, in an attempt to begin to "paint the big picture," we will briefly discuss design parameters, planning, maintenance, etc., which are topics of more detailed coverage in later chapters.

While reading Chapters 3 through 5, note the interrelationships between sectors and components from each of the chapters; for example, the bridges of this chapter and the roads of Chapter 4, Transportation Infrastructure. Given these interrelationships, categorization of the built environment into three distinct groupings (i.e., structural, transportation, and environmental) is inherently difficult. For example, dams and levees are indeed structures, but have distinct environmental aspects to their design and planning. Likewise the treatment processes at a wastewater treatment facility are designed to protect the environment, but are contained within tanks, reactors, and buildings designed by structural engineers.

Multi-Disciplinary Aspects of Large Infrastructure Projects

Hoover Dam Bypass Project

Videos and Animations Demonstrating Construction of Improvements to the Greater New Orleans Hurricane and Storm Damage Risk Reduction System

Introduction

In this chapter, we introduce many types of structures that are part of the built environment. Skyscrapers have long captured the imagination of many young people, inspiring them to become civil engineers. Many of these structures are national or international landmarks (Figure 3.1 and Figure 3.2). However, many additional types of structures exist with the purposes of housing

Figure 3.1 The Empire State Building (Center) Dominates the Manhattan Skyline. Completed in 1931, this 102-story building was the tallest in the world for four decades. It was named one of the Seven Wonders of the Modern World by the American Society of Civil Engineers (ASCE).

Source: Copyright © oversnap/iStockphoto.

Figure 3.2 Dubai Skyline. The Burj Khalifa, completed in 2010 at a cost of over $1 billion, is the tallest structure in the world at 2,717 feet.

Source: Copyright © Uwe Merkel/iStockphoto.

people; storing goods; holding back water; providing places of employment, education, or worship; allowing for the movement of goods and people across water bodies and other obstructions, etc.

Foundations

Virtually any structure, be it a building or a bridge, requires a foundation. The foundation may be as simple as a concrete slab, or much more complicated. Foundations are necessary to distribute the weight of the structure onto the soil or rock below it. Without a proper foundation, a structure may sink (settle), tip, crack, or collapse. The strength of the soil or rock below a structure is a critical factor in foundation design. "Weak" soil or rock cannot support much weight. Consequently, the structure's weight must be distributed over a larger area in contact with the soil or rock; this is the fundamental principle behind **footings** (Figure 3.3).

Figure 3.3 Concrete Footing with Reinforcing Steel (Rebar) for Walls. The footing is approximately three times the width of the wall that will be cast in place later.

Source: M. Penn.

Figure 3.4 **Steel Piles for a Bridge Foundation.** These steel "H" piles (so named due to the cross-sectional shape) were driven vertically into holes bored into underlying shallow bedrock. Workers are building wooden forms (explained in Chapter 6) around the piles in which concrete will be placed to complete the foundation.

Source: M. Penn.

Very large structures require deep foundations to reach subsoils or rock with greater load bearing capacities. Steel or concrete **piles** are placed (Figure 3.4), often at great depth, to which the structure is connected. For underwater foundation applications (e.g., bridge piers), temporary watertight structures such as **caissons** or **cofferdams** (Figure 3.5) are constructed and pumped dry for subsequent foundation construction work. Accounting for seismic activity from earthquakes is a highly advanced area of study in foundations and structures.

A **pile** (or piling) is a column placed vertically, deep in the subsurface, as a foundation.

A **caisson** is a complete enclosure (with a roof) that is pressurized to keep water from entering.

A **cofferdam** is a walled enclosure (with no "roof") from which water is removed for construction work within.

Skyscrapers

High-rise structures, while typically privately owned, are of particular importance to infrastructure systems due to the demands created by a large number of people occupying a relatively small footprint; a high-rise condominium may "concentrate" the population equivalent of a small city into the area of a single city block. Needs for water, wastewater, parking, and traffic must be carefully considered when designing these structures and their integration into existing infrastructure systems. The structural design of skyscrapers and the design of the foundations are extremely complex, requiring education beyond a bachelor's degree.

Figure 3.5 **Construction Workers Working Inside a Cofferdam.**

Source: Courtesy of the Missouri Department of Transportation.

Schools (ASCE 2009 Grade: D)

Schools present challenges beyond structural design because of the many associated infrastructure demands (e.g., transportation, water and wastewater, energy). Collectively, school construction, operation, and maintenance are typically the largest expenses in a community's budget. Schools are often viewed as integral components of neighborhood identity. Rapidly growing regions face great challenges in providing educational facilities to meet growing populations, especially due to the fact that required expenditures are large and long term. Managing this growth can be particularly challenging and requires predictions of future enrollments. Such predictions are made using population growth estimates and the age distribution of the population (i.e., the number of children). If growth is over-predicted, the community will be faced with substantial debt and underutilized facilities. If growth is under-predicted, or if decisions to expand are delayed, facilities will be undersized, resulting in overcrowded and/or substandard temporary classrooms.

Also challenging are schools in large cities where populations are declining (even as neighboring suburbs grow). The existing educational infrastructure was built for and financed by a larger population. Now underutilized, these facilities lead to high costs per student and often to facility closures and mergers. This problem is also common in many rural areas that are experiencing population declines.

Bridges (ASCE 2009 Grade: C)

Bridges are used to span gorges and valleys or to cross any number of obstructions including waterways and other roads. At its simplest, a bridge consists of a deck and its supports. The deck may be used to convey trains, automobiles, pedestrians, or bicyclists. The supports can be as simple as concrete **abutments** at either end of a span, or as complicated as the towers and cables that support the deck on the Golden Gate Bridge and other suspension bridges.

An **abutment** is a structure, often of concrete, that supports the ends of a bridge.

As you will learn in Chapter 9, Analysis Fundamentals, the two most fundamental forces on structural members are compression and tension forces. Compression force may be thought of as a "pushing force," and is the force you feel in your arm as you push on a door to keep the door from opening toward you. Conversely, if you are trying to pull a stuck door open, you are experiencing a pulling or tension force in your arm. The loads acting on a bridge will cause its structural members to experience compression and tension. Considering these forces is a rudimentary first step in analyzing and designing bridges.

The loads on bridges include the traffic (e.g., rail or auto) that will be carried by the bridge. Additionally, the bridge must be able to support its own weight. Other loads may be added by wind or snow. In certain parts of the world, the bridge must also be able to withstand the movements induced by earthquakes.

In designing bridges, engineers have many options at their disposal. That is, there are many ways to transfer the loads from the deck to the earth. Said another way, there are many methods engineers can use to ensure the bridge carries the loads in structural members via tension and compression.

Bridges come in all shapes and sizes: suspension, cable-stayed, steel or concrete arch, trusses, steel girder, or concrete girder (see Figure 3.6). The choice of which type of bridge to use for a given situation depends on:

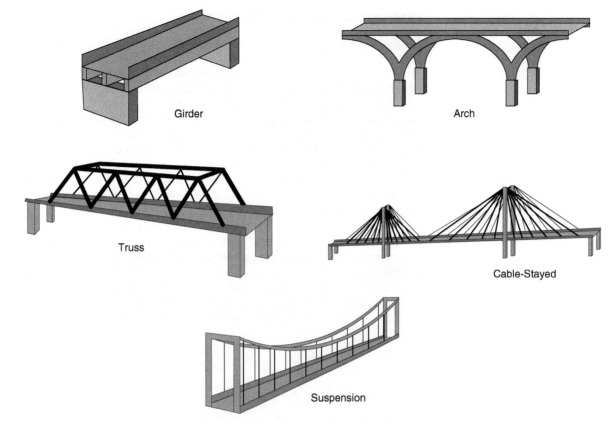

Figure 3.6   Common Bridge Types.

- The distance to be crossed—As the distance increases, additional spans (and therefore additional supports) will be needed; or, if additional supports are not feasible, only certain types of bridges will work (e.g., a suspension bridge).
- The site geology—All loads are transferred from the bridge to the subsurface. The design of the foundations will depend on the strength of the underlying soil or rock.
- The ease by which the entire span length can be constructed—For example, a long water crossing requires much different construction techniques than a short span across a gully.
- The type of traffic, if any, that will pass under the bridge—For example, trucks on freeways, or ships with tall masts on waterways will both affect bridge selection and design.
- The amount of money that project owners are willing to spend—The owners may be willing to spend more money for a bridge that is aesthetically pleasing, or perhaps even more for one that will become a true "symbol" of a city (e.g., Golden Gate Bridge).

The simplest and most widely used type of bridge is the simple girder or beam bridge (Figure 3.7). Placing a plank or log across a ditch would be the simplest example of a beam bridge. Loads to the parallel beams (Figure 3.8) are transferred to the ends of the beams that rest on abutments. Beam bridges are relatively inexpensive, and their design is relatively straightforward.

Figure 3.7 Examples of Single-Span Bridges. The freeway bridge in the background and the relatively short multi-use trail bridge in the foreground are both simple beam bridges.

Source: P. Parker.

Remember that full-sized versions of textbook photos can be viewed in color at www.wiley.com/college/penn.

Figure 3.8 Steel I-Beam Girders and Lateral (left to right in photo) Bracing Supporting the Concrete Road Deck of a Girder Bridge.

Source: M. Penn.

Time-Lapse Video of Colorado River Bridge Construction

"How Stuff Works" Video on Bridge Design

As the span to cross becomes longer or as the loads become greater, beams and girders become impractical. Early engineers solved this problem with the use of arch bridges; arch bridges are still used worldwide (e.g., see Figure 3.9). In arch bridges, the arches are under compression and transfer loads to the ends of the arches.

The introduction of steel to bridge building allowed the use of trusses to become widespread. Truss bridges consist of a network of steel members arrayed in triangular shapes. The design of truss bridges is relatively straight-forward and therefore can be efficiently and rapidly designed by engineers. Furthermore, they offer exceptional rigidity and relative ease of construction, as compared to the masonry arch bridges that they historically succeeded. They require significant maintenance, as members must be painted regularly to inhibit rust. Although not used widely for highway and railroad use any more, they are regularly found in pedestrian bridges (Figure 3.10). Tension and compression forces for truss bridges will be discussed in Chapter 10, Design Fundamentals.

As spans become longer, piers may be needed (Figure 3.11). However, for crossings with deep or fast-moving water, construction of such supports is impractical. In other cases, additional piers are not desirable as they impede ship traffic. In such cases, a suspension bridge may be most feasible. In a suspension bridge, the loads from the deck are transferred to vertical

Figure 3.9 Double Arch Bridge on Natchez Trace Parkway (Tennessee).

Source: Courtesy of B. Moore.

Figure 3.10 Truss Bridge for Pedestrian Use.

Source: Courtesy of A. McNeill.

suspender cables (also termed "hangers") that in turn are attached to the suspension cables. All cables are under tension, and transfer their loads to the columns (which are under compression) and also to the anchorages on the end. Anchorages are large masses of concrete; in the case of the Golden Gate Bridge, the anchorages each weigh 60,000 tons.

Cable-stayed bridges are among the most visually striking bridges. Like suspension bridges, the load from the deck is transferred to cables. Unlike

Figure 3.11 Lake Pontchartrain Causeway. At 24 miles long, it is the world's longest bridge over water. The twin, parallel spans are supported by more than 9,500 pilings.

Source: Copyright © Gary Fowler/iStockphoto.

suspension bridges, cable-stayed bridges do not have anchorages. That is, all the loads are transferred directly to the towers. Cable-stayed bridges are often stiffer than suspension bridges, and whereas suspension bridges are typically limited to two towers, any number of towers can be incorporated into cable-stayed bridges.

Case in Point: Sunshine Skyway Bridge

In 1980, one of the piers of the Sunshine Skyway Bridge over Tampa Bay was struck by a freighter. A portion of the bridge collapsed, and six vehicles and a bus fell into the water, killing 35 people.

The replacement bridge (Figure 3.12) is a cable-stayed bridge, and is considered to be one of the most beautiful bridges in the United States. It has also become a symbol of Tampa Bay, similar to the way the Gateway Arch has become a symbol of St. Louis or the Golden Gate Bridge a symbol of San Francisco. Notably, the cable-stayed construction for the replacement bridge allows a span length

(1,200 feet) that is twice as long as the span opening on the old bridge, which makes it much safer for ship traffic to pass. It also has a larger vertical clearance that can allow taller ships to pass.

The demolition of the old bridge was carried out in dramatic fashion. A video is available on the textbook website.

Video of Demolition of Sunshine Skyway Bridge

Figure 3.12 **The Cable-Stayed Sunshine Skyway Bridge.**

Source: Copyright © Tinik/iStockphoto.

Dams (ASCE 2009 Grade: D)

There are over 80,000 dams throughout the United States for purposes of navigation (in conjunction with locks), flood control, hydroelectric power, water supply (storage reservoirs), and recreation (boating and fishing).

Dams are primarily earthen (Figure 3.13) or concrete (Figure 3.14). Concrete dams must be designed to resist overturning. For slope stability reasons (i.e.,

Figure 3.13 **The Fort Peck Dam on the Missouri River in Montana.**
One of the largest in the United States, this dam is 4 miles long and 250 feet high.

Source: U.S. Army Corps of Engineers/R. Etzel.

Figure 3.14 The Hoover Dam. Perhaps the most famous dam in the United States, the Hoover Dam was constructed from 1931 to 1936 on the Colorado River at the border of Nevada and Arizona. It is 1,244 feet long and 726 feet tall. In the foreground is the Colorado River Bridge under construction, which was completed in 2010.

Source: Federal Highway Administration.

to keep the earthen embankment from moving downward as a mass and thus resulting in dam failure), earthen dams require gentle slopes that necessitate a large footprint (e.g., see Figure 3.13). With a 4:1 (horizontal-to-vertical) slope on both upstream and downstream sides, a 100-foot-high dam would be 800 feet wide at the base.

Major dams are defined as being at least 50 feet in height with a storage capacity of at least 5,000 **acre-feet**, or of any height with a storage capacity of 25,000 acre-feet or more. As the height of a dam increases, so does the water storage capacity, the potential energy for hydroelectric power, and the extent of property and lives to be affected in the event of a failure.

River flows vary greatly with rainfall intensity and snowmelt. Earthen dams are protected from overtopping (and potential scour and failure) by emergency **spillways**. Spillways convey flow through an engineered channel, and act as a bypass of the dam during exceedingly high water elevations (Figure 3.15).

> One **acre-foot** is the volume of water that occupies an area of 1 acre at a depth of 1 foot and is equivalent to 325,851 gallons or 43,560 ft^3.

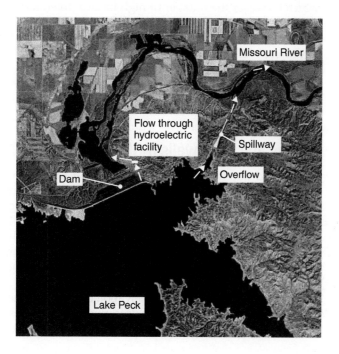

Figure 3.15 Emergency Spillway at the Fort Peck Dam. Routes for normal flow through the hydroelectric facility and emergency flow via the spillway are shown.

Source: Copyright © 2010 TerraMetrics, Inc.

Dams for drinking water supply require water intakes. Newer designs incorporate intakes at several water depths to allow water supply managers to select the withdrawal depth based on water quality that varies seasonally and with depth in reservoirs.

The many potential benefits of dams are countered by adverse environmental and social effects that will be discussed later in this book. When fish migration in a river is necessary for ecological reasons, **fish ladders**, or in extreme cases, **fish elevators**, are used to allow passage around a dam (Figure 3.16). Many dams are being removed to restore rivers to their free-flowing nature.

Many dams in the United States are in deficient condition. The Association of State Dam Safety Officials estimates that there are over 1,700 **high-hazard potential dams** (where loss of human life is probable as a result of dam failure) that need repair—more than 1 out of every 10 (ASDSO, 2007). This number has steadily increased (from approximately 400 needing repair in 2001) as a result of insufficient funds and increased development downstream of dams, which increases the number of high-hazard potential dams. The 2009 *Report Card for America's Infrastructure* states: "In order to make significant improvements in the nation's dams—a matter of critical importance to public health, safety and welfare—Congress, the administration, state dam safety programs, and dam owners will have to develop an effective inspection, enforcement and funding strategy to reverse the trend of increasingly deteriorating dam infrastructure."

Levees and Floodwalls (ASCE 2009 Grade: D−)

Levees and floodwalls are a means of flood protection that are now in the national spotlight after the Hurricane Katrina disaster in New Orleans (which will be discussed in great detail throughout this book). Levees (earthen embankments sometimes referred to as dikes, Figure 3.17) and floodwalls (typically made of concrete, Figure 3.18) are commonly aligned parallel to rivers. When access through a floodwall is needed during normal (non-flood) river stages, floodgates are installed (Figure 3.19). These gates are closed during high water events.

When designing a levee, the elevation of the top of the levee is determined based on probabilities of **stage** (for river flooding) or storm surge water elevation (in the case of hurricanes). As the levee height increases, the risk of

Stage refers to the water surface elevation.

Figure 3.17 **An Earthen Levee Along the Mississippi River in Dubuque, Iowa.** Note that the levee serves as an attractive pedestrian walkway. The top of the levee is approximately 20 feet above the river level in the photo. The parking lot and ground floor of the hotel in the background are approximately 15 feet below the top of the levee. Note that the river side of the levee is armored with large rip-rap to prevent erosion.

Source: M. Penn.

Figure 3.18 **A Newly Constructed Portion of the Breached Floodwall during Hurricane Katrina in the Devastated Lower 9th Ward of New Orleans.**

Source: M. Penn.

Figure 3.19 **Floodgate in a Floodwall Protecting Historic Downtown Galena, Illinois.** In the absence of the floodwall and gate, when 14 inches of rain fell in three days in August 2010, the city would have been under 7 feet of water from the adjacent Galena River.

Source: M. Penn.

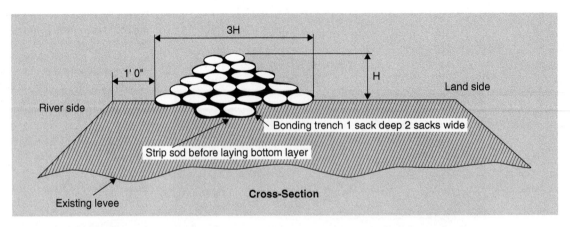

Figure 3.20 Sandbag Placement Atop a Levee.

Source: City of Anderson, Indiana Emergency Management.

overtopping and flooding decreases. In emergency situations, elevations may be temporarily raised by the addition of sandbags (Figure 3.20). Permanent increase in elevations can be accomplished by placing a floodwall on top of an existing levee (Figure 3.21). This increased protection from hazard requires additional expenditures since the design, construction, and maintenance costs increase with levee elevation.

The elevation of the top of a levee is typically based on "100-year" protection, which corresponds to a water elevation with a 1 in 100 (1 percent) chance of occurring in any given year. Currently, there is national debate as to what level of protection should be provided. In the Netherlands, for example, protection is provided at the 2,500-year level (0.04 percent probability) or greater.

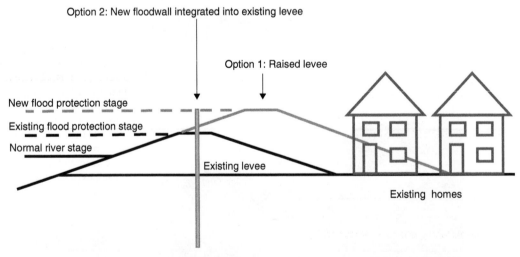

Figure 3.21 A Comparison of the Footprint of a Replacement Levee of Higher Elevation (Option 1) Versus a Floodwall Extension to an Existing Levee (Option 2). A larger footprint (i.e., a wider base) is required for a taller levee because the side slope is a fixed design parameter. To accommodate this larger footprint, existing homes would need to be removed or relocated. Adding a floodwall to an existing levee can raise the flood protection elevation without increasing the required footprint.

Source: Figure modified with permission from ASCE, 2007.

Levees are owned and maintained by various levels of government, from local to national. Largely for this reason, there is no accurate estimate of the total length of levees in the United States. Many levees were designed and constructed several decades ago, and are considered deficient as a result of poor maintenance, differential settling, outdated designs, or improper construction.

Levees and floodwalls should be a *primary*, but not the sole, means of protection. A **multiple lines of defense** strategy toward flood risk minimization has been adopted by the U.S. Army Corps of Engineers for coastal Louisiana:

"No single measure or approach for achieving risk reduction will be sufficient for achieving the multiple risk reduction objectives established for coastal Louisiana. No alternatives can be formulated that will provide total protection to the entire planning area against all potential storms. The reason that total protection is not possible is a matter of practicality, technical inability and construction challenges, and extremely high costs. Therefore, the best strategy is to rely on multiple lines of defense." (USACE, 2009)

This strategy may include structural components (e.g., levees/floodwalls, pumping stations, and elevated structures) and non-structural components (e.g., removal of homes and businesses in low-lying areas, restricting new development in low-lying areas, and well-planned evacuation procedures). This strategy provides redundancy to the system, in that some level of protection will be provided if any one component fails.

> ### Levee Risk
>
> The following predictions have been provided by the U.S. Army Corps of Engineers—Louisville District regarding the deficient levee that protects the city of Anderson, Indiana:
>
> - Within next 10 years—67 percent chance of levee failure
>
> - Within next 20 years—94 percent chance of levee failure
>
> - Within next 50 years—100 percent chance of levee failure
>
> In this case, note that "failure" is not referring to "failure to provide flood protection" (e.g., being overtopped by flood waters), but to *true* structural failure of the levee (e.g., breaching).

Retaining Walls

Retaining walls are structures utilized to prevent the downward movement of soil or rock to protect other structures, parking lots, etc. As shown in Figure 3.22, the use of a retaining wall can provide additional space for development as compared to grading a 3:1 slope or installing benches (steps). The cost of designing, building, and maintaining the retaining wall is often offset by the value of the additional land that is available for development.

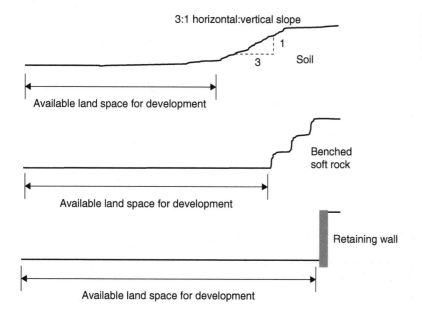

Figure 3.22 Available Land Space for Development with Sloped Soil, Benched Soft Rock, and a Retaining Wall.

Figure 3.23 A Curved Retaining Wall Made of Block.
Note the safety fence above the wall.

Source: M. Penn.

Figure 3.24 An Overturned Concrete Retaining Wall.
The face of the wall was vertical when originally constructed.

Source: M. Penn.

Lateral earth pressure is the pressure (force per unit area) that soil or rock exerts in the horizontal direction.

Hydrostatic pressure is the pressure exerted by water perpendicular to a surface such as a retaining wall.

Retaining walls are made of many materials depending on the height required, type of geologic material being retained, aesthetics, etc. (Figure 3.23). Structurally, retaining walls must support the **lateral earth pressure** caused by the downward movement of the geologic formation and the **hydrostatic pressure** of groundwater (if not properly drained from behind the wall). Improper design or construction may lead to collapse, sliding, or overturning (Figure 3.24).

Outro

The poor grades assigned by the ASCE *Infrastructure Report Card* to the sectors pertaining to structures (schools, bridges, dams, and levees and floodwalls) paint a dismal picture of the structural infrastructure. The next two chapters will introduce you to the transportation and the environmental infrastructure, and these sectors also have received very low grades. Keep in mind however, as you read the next two chapters, that the low grades can be raised with adequate funding.

Additional Resources

The following books are highly recommended for further understanding of the infrastructure components presented in this chapter and the following two chapters:

E. Sobey. *A Field Guide to Roadside Technology*. Chicago: Chicago Review Press, 2006.

K. Ascher. *The Works: Anatomy of a City*. New York: Penguin Press, 2005

B. Hayes. *Infrastructure: The Book of Everything for the Industrial Landscape*. New York: W.W. Norton, 2005.

G. Rainer. *Understanding Infrastructure*. Hoboken, NJ: John Wiley & Sons, 1990.

S. Hule. *On the Grid: A Plot of Land, an Average Neighborhood, and the Systems That Make Our World Work*. Emmaus, PA: Rodale, 2010.

D. L. Schodek. *Landmarks in American Civil Engineering*. Boston: MIT Press, 1997.

chapter **Three** Homework Problems

Calculation

3.1 An existing levee has a top width of 10 feet, a top elevation of 10 feet above the base, and 3:1 (horizontal:vertical) slopes on both sides. If the levee is to be raised to an elevation of 15 feet and must meet new standards requiring a 12-foot top width, how much additional land space (in ft^2) will be needed for the base of the levee per mile of levee length? Assume side slopes remain 3:1. Show all work, and support your answer with appropriate drawings and schematics.

Short Answer

3.2 When designing a bridge, what considerations must be accounted for in addition to cost?

3.3 Why are fish ladders sometimes necessary?

3.4 Explain how, for a certain site, the additional cost of a retaining wall might offset the cost of purchasing additional land.

3.5 What is the difference between a caisson and a cofferdam?

3.6 Why are spillways used?

Discussion/Open Ended

3.7 Sketch the bridge shown in Figure 3.12. On your sketch, show how the loads on the bridge deck are transferred to the subsurface. Also, note which portions of the bridge are in compression and which are in tension.

3.8 A new bridge was planned to cross Lake Champlain, which lies between New York State and Vermont. Preliminary designs of several types of bridges were provided (available at www.wiley.com/college/penn). Which one would you select if you were in the position to do so? Why?

3.9 Currently, levees in the United States have not been catalogued, nor is there a uniform method for assessing them. Why might this be?

3.10 Do you think it is practical to require that *all* levees should provide protection from greater than 100-year floods? Why or why not?

3.11 Photograph a retaining wall in your community from several different angles, including close-up and distance pictures. Describe the wall (height, length, etc.) using a tape measure. Assess the condition of the wall.

Research

Note: For Homework Problems 3.12 through 3.14, 3.16, and 3.18, the *ASCE Report Card* is available at www.wiley.com/college/penn.

3.12 Read the most recent chapter on dams in the ASCE *Report Card for America's Infrastructure*. Based on the information in this chapter, write a brief (approximately 500 words) letter to your local newspaper in which you voice your concerns over dam safety. Refer to information in the report card chapter to justify your claims.

3.13 From the data in the 2009 ASCE *Report Card* chapter on dams, create a graph of the number of high-hazard dams needing repair for the years 2001 to 2007.

3.14 Read the most recent chapter on levees in the ASCE *Report Card for America's Infrastructure*. Summarize the reasons why levees and floodwalls obtained the assigned grade.

3.15 Read the document "So, You Live Behind a Levee!" (available at www.wiley.com/college/penn). In approximately 500 words, explain what you learned and describe what you found most interesting from this article. Also, prepare a list of new (to you) vocabulary presented in the document.

3.16 Read the most recent chapter on bridges in the *ASCE Report Card for America's Infrastructure*. What are some reasons that bridges earned their relatively high grade?

3.17 Define "structurally deficient" and "structurally obsolete." Which one is "worse"? Give some examples of bridge characteristics that would cause that bridge to be classified as: structurally deficient; structurally obsolete.

3.18 From the fact sheet on bridges supplied by the 2009 *ASCE Report Card* chapter on bridges, create a graph that plots the fraction (as a percent) of bridges in the United States that are (a) structurally deficient and that are (b) structurally obsolete for the years 1998 to 2007.

3.19 In what types of situations are pilings required for foundations? Do pilings always rest on bedrock?

3.20 Visit www.floodsmart.gov. Enter location information (specified by your instructor) into the "One Step Flood Risk Profile" and print out the results.

3.21 Use a program such as Microsoft PowerPoint to create a photo gallery that includes photos of each of the bridge types listed in this chapter. Make sure to provide references for photos obtained online.

Key Terms

- abutments
- acre-feet
- caissons
- cofferdams
- fish elevators
- fish ladders
- footings
- high-hazard potential dams
- hydrostatic pressure
- lateral earth pressure
- major dams
- multiple lines of defense
- piles
- sectors
- spillways
- stage
- subsectors

References

American Society of Civil Engineers (ASCE). 2007. *The New Orleans Hurricane Protection System: What Went Wrong and Why*. Reston, VA: American Society of Civil Engineers.

Association of State Dam Safety Officials (ASDSO). 2007. *Statistics on Dams and State Safety Regulation*. Lexington, KY.

City of Anderson, Indiana. *Emergency Management*. http://www.cityofanderson.com/emergency/fws.aspx, accessed July 23, 2010.

City of Anderson, Indiana. *Flood Warning System*. http://www.cityofanderson.com/beta/directory_department_page_view.aspx?id=21&deptid=13, accessed July 23, 2010.

U.S. Army Corps of Engineers (USACE). 2009. Louisiana Coastal Restoration and Protection Final Technical Report and Comment Addendum. http://lacpr.usace.army.mil/default.aspx?p=LACPR_Final_Technical_Report, accessed August 6, 2011.

chapter Four Transportation Infrastructure

Learning Objectives

After reading this chapter, you should be able to:

1. List the main sectors and subsectors of the transportation infrastructure.
2. Explain the function of each transportation sector.
3. Identify transportation infrastructure components.
4. Explain the function of transportation components.
5. List some of the most important design parameters for transportation components.
6. Describe some of the fundamental relationships between transportation and other infrastructure sectors, subsectors, and components.

Introduction

Transportation systems and components allow the movement of people and goods. Nearly 12 billion tons of goods are transported domestically while an additional 2 billion tons are traded to and from the United States annually. Designing, building, and maintaining transportation systems have a significant effect on the economic activity of the United States. Clearly, a strong economy is reliant on efficient and effective transportation systems.

Level of service, or **LOS**, is a qualitative measure of the roadway's operating conditions and the perception of those conditions by motorists and/or passengers. It will be discussed in further detail in Chapter 20.

Design speed is the maximum safe speed that an automobile can operate on a roadway in perfect conditions.

Sight distance is the length of road in front of the vehicle that is visible to the driver.

Cross-section refers to a vertical slice taken of a street, perpendicular to the line of travel.

Roads (ASCE 2009 Grade: D−)

The road subsector includes streets, alleys, and service roads. Streets may be further subdivided into four categories as summarized in Table 4.1.

The planning and design of streets takes into account the following design parameters (Kutz, 2003):

- Traffic volume—Higher traffic volume (i.e., number of vehicles per hour) requires more lanes in order to provide a higher **level of service**.

- Range of driver types—Drivers vary in their ability to see, hear, read, comprehend, and react to stimulus (e.g., consider designing a road for a retirement community).

- Vehicle type—Vehicles vary in terms of size (height, width, length, and weight) and ability to accelerate, brake, turn, and climb hills.

- Design speed—Designers must select a **design speed** based on intended use, desired traffic volume, adjacent land use, and other factors. For example, the design speed for freeways is 70 miles per hour (mph). Selection of the design speed has a significant impact on safety and comfort.

- Sight distance—Road designers must consider the stopping **sight distance**, such that drivers operating vehicles at the design speed are able to stop the vehicle in a controlled manner if an obstacle is blocking the lane of travel. Road designers should also consider passing sight distance, whereby drivers are able to see far enough into the distance to safely pass.

- Topography—Topography affects many aspects of street design, including the geometry and design speed as well as sight distances.

- Land use characteristics—The type of land use (e.g., commercial, residential) will determine the number of vehicles using the roadway.

- Critical need of use—local streets might be allowed to periodically flood, whereas freeways are not; local streets are lowest priority for snow removal.

The design of streets includes the **cross-section** design, profile design, and plan design.

Table 4.1

Street Categories Based on Traffic Volume, Speed Limit and Linkage

Street Category	Traffic Volume (vehicles/day)	Speed Limit (miles/hour)	Linkage
Local	Low (<2,000)	Low (10–30)	Neighborhoods
Collector	Moderately low (2,000–8,000)	Low (25–35)	Local areas
Arterial	Moderately high (8,000–20,000)	Moderately high (25–45)	Regions of moderately high traffic generation
Freeway	High, free-flowing (may exceed 100,000)	High (≥55)	Major traffic generating areas

The cross-section details the width and side slopes of the street. Streets are often **crowned**; that is, they are constructed with a side slope, typically 2 percent (2 inches per 100 inches), to prevent the accumulation of water and ice. A cross-section includes driving lanes as well as shoulders, curbs, medians, and ditches.

- Shoulders allow drivers to regain control once leaving the pavement, provide a space for emergency parking, and provide structural support for the edge of the pavement.

- Curbs, often formed of concrete, keep vehicles off lawns, and in conjunction with a gutter, provide a conveyance channel for stormwater.

- Medians provide separation between opposing lanes of traffic, thereby reducing the probability of deadly head-on collisions. Also, in cold climates, medians provide space for snow storage.

- Ditches convey stormwater, and unlike impervious curbs and gutters, allow for some infiltration and treatment of stormwater.

A sample cross-section for a freeway with center median is provided in Figure 4.1. Elevation is plotted on the vertical axis, while the horizontal axis provides distances to the left or right (in feet) from the centerline of the cross-section. The existing ground is drawn with a dashed line, and you can readily see how the proposed cross-section will require excavation. For this example cross-section, construction of the roadway will require that soil is removed, or **cut**, from the right side and an embankment will need to be graded such that the road side slope matches in with the existing ground. During the design phase, cross-sections such as these are created at every 100-foot **station** (i.e., at stations 0+00, 1+00, 2+00, and so on) and at every change in cross-section design (e.g., entrance/exit ramps or a change in the number of traffic lanes).

Roadway designers also need to consider the materials to be used in the pavement structure. The pavement is typically concrete or asphalt. Concrete pavements are termed **rigid pavements** while asphalt pavements are termed **flexible pavements**. Pavements are supported by a granular material base that transfers traffic loads to the native material (termed **subgrade**) and provides water drainage.

The **vertical alignment**, or **profile view**, is a "side view" of the roadway along a planar vertical surface running along the roadway centerline. The term "profile" is similarly used in the design of other infrastructure components (e.g., storm sewers, water mains, retaining walls). When designing the vertical

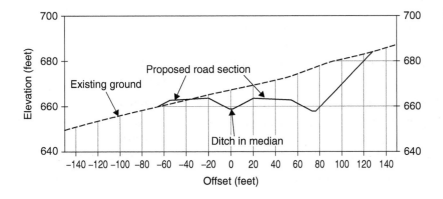

Figure 4.1 Roadway Cross-Section with Offsets.

Figure 4.2 Profile View of the Vertical Alignment of a Roadway.

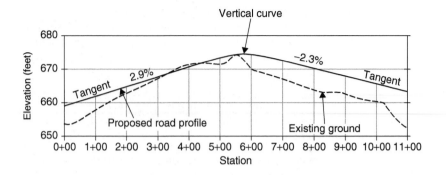

alignment, designers define the **grades** (uphill or downhill slopes) of the roadway, in essence setting the steepness (or "gentleness") of the roadway. Moreover, the vertical alignment must provide transitions between grades using **vertical curves**. The proposed road profile in Figure 4.2 consists of two approaching grades, or tangents, and a vertical curve that transitions between the two tangents. Note that for this section of roadway, a large amount of **fill** material will be needed as nearly the entire proposed roadway is above the existing ground, by nearly 10 feet at station 11+00 (1,100 feet from the beginning of the profile). A small amount of excavation (cut) will be required where the existing land surface is above the design elevation in the vicinity of station 4+00.

The **horizontal alignment**, or **plan view**, is the "bird's-eye" view of the roadway. The term "plan view" is also applied to virtually all infrastructure component designs. The horizontal alignment, like the vertical alignment, is a two-dimensional view of the roadway. At its simplest, a horizontal alignment consists of a series of straight lines (also called **tangents**), with **horizontal curves** providing the transition between the tangents. Two tangents and two horizontal curves are depicted in Figure 4.3. Note that the use of stationing allows you to compare the plan and profile view, as stationing is consistent between Figure 4.2 and Figure 4.3. This figure also denotes the **right of way**.

The **right of way** is the limits of the land which the public owner of the road may use for locating utilities.

Figure 4.3 Plan View of the Horizontal Alignment of a Roadway Curve.

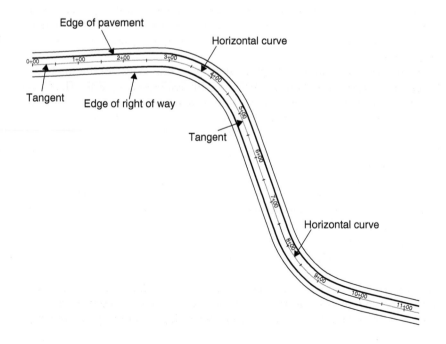

Figure 4.4 Modern Alley. This alley services the garages that are in the back yards of residences and allows for trash and recycling pickup to occur at the back of the homes.

Source: Courtesy of K. Neundorf.

Figure 4.5 A Service (Frontage) Road Serving Businesses Adjacent to a Six-Lane Highway (Left). Note the lower traffic volume on the service road.

Source: M. Penn.

Design of the horizontal curves must hold driver safety paramount. As horizontal curves become longer (i.e., as they have a larger radius), vehicles may traverse the curve at higher speeds; however, the tradeoff is the additional land required.

Alleys are low volume, single-lane roads for access behind commercial buildings or residences (Figure 4.4). They were commonly used in urban land development during the early to mid-1900s, and allowed residents a private and low-volume drive with which to access their garages while providing a place to leave garbage for collection. With larger lots and higher reliance on the automobiles, these fell out of favor but are being used in some subdivision designs today.

Service roads, or **frontage roads**, run parallel to major limited-access arterials or highways (especially in commercial districts) to isolate local traffic from through traffic on the higher traffic volume roadway, and to limit the number of intersections (Figure 4.5).

INTERSECTIONS AND INTERCHANGES

Intersections are nodes in a transportation system and occur when two or more road alignments intersect. Figure 4.6 is an aerial photograph in which a variety of intersections can be seen, including a highway interchange,

Design Speeds and Speed Limits

You may be interested to know that the posted speed limit is not necessarily the design speed of the roadway. Rather, the speed limit may be set 5 to 10 mph below the design speed. Doing so builds in an additional safety factor into the roadway's design. Alternatively, speed limits may be set based on speeds measured after a road is built and is in use. In such instances, speed studies are conducted on vehicles, and the speed limit is set equal to the 85th percentile speed (the speed which 15 percent of the cars are exceeding).

Figure 4.6 An Aerial View of
Several Types of Intersections.

Source: U.S. Geological Survey.

T-intersections, four-way intersections, and even a five-way intersection. The design of intersections considers the type of vehicles using the intersection and the throughput needed (volume of traffic and the speed limit).

A **conflict point** is the point where the paths of two through or turning vehicles diverge, merge, or cross.

Design of intersections takes into account **conflict points**. A typical intersection where two roads cross has 32 conflict points (see Figure 4.7 for an example of a conflict point diagram). Rear-end and sideswipe collisions are possible at the "merging and diverging" conflict points, while angle collisions can occur for "crossing" conflict points. The geometry of the intersection will affect the

Figure 4.7 A Conflict Point Diagram for a Conventional Four-Way Intersection.

Source: U.S. Federal Highway Administration, 2004.

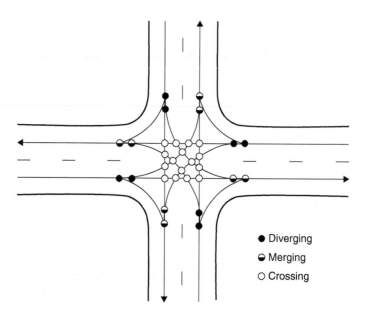

● Diverging
◐ Merging
○ Crossing

Figure 4.8 A Roundabout, an Increasingly Popular Alternative to a Traditional Four-Way Intersection.

Source: M. Penn.

number of conflict points. By designing an intersection with adequate sight distances and incorporating effective traffic signals and/or signs, the severity and frequency of accidents can be minimized.

Roundabouts (Figure 4.8) are increasingly popular in some parts of the country, and have only eight conflict points. Depending on the type of vehicles that will be using the roundabout and the desired travel speeds, diameters are on the order of 200 to 300 feet. As such, they can require a large parcel of land. Because of driver unfamiliarity with roundabouts, many transportation managers provide information to educate drivers and ensure successful implementation.

Intersections for freeways are termed **interchanges**, and you can view an interchange at the bottom of Figure 4.6. A properly designed interchange will allow for the free and safe flow of traffic. Many of the design considerations associated with non-freeway intersections also apply to freeway interchanges. An additional consideration is the length of entrance and exit ramps necessary for acceleration and deceleration.

Roundabout Educational Material

PARKING

Parking is a major consideration in urban planning and suburban development. The location and size of parking facilities has direct impact on the convenience of users. Parking demand is reduced by mass-transit use, car-pooling, and non-motorized modes of travel. Large surface lots require large land areas (approximately 20 acres for the five parking lots in Figure 4.9) and

Figure 4.9 Stadiums and Adjacent Parking Lots for the Pittsburgh Steelers and Pirates Professional Sports Teams.

Source: GeoEye Satellite Image.

Creative Parking Solutions

While engineers are often stereotyped as being non-creative, this stereotype is certainly untrue of all, if not most, engineers. For example, automated parking structures are being developed around the world in a variety of configurations. One of the most famous is actually a new car  "warehouse" at a Volkswagen factory and visitor's center (the Autostadt) in Germany. Two 200-feet tall glass "silos" are used to store vehicles. An automated central mechanism places and retrieves cars. Also in Germany, a luxury condominium is being developed with automated parking on the exterior of the building on balconies adjacent to the condos, dubbed the "car loft."

Autostadt Car Towers Website

While more common in Europe and Asia, some automated parking does exist in the United States. The use of automated lifts allows for a smaller structural footprint because no traffic lanes are needed. In very high-density urban settings where land value is excessive, it is possible for such seemingly extravagant designs to be cost-competitive. Also, it is important to consider that some owners will be willing to pay added expense for the "glamour" associated with these innovative designs.

Figure 4.10 Three Adjacent Parking Structures and an Adjacent Surface Lot.

Source: M. Penn.

when land value is sufficiently high, it must be weighed against the design, construction, operating, and maintenance expenses of **parking structures**. Such structures can be free standing (Figure 4.10) or integrated into the design of a building (Figure 4.11).

To design a parking lot or structure, an estimate of the number of users is needed. Like many other infrastructure projects, parking lots are often designed for peak demands. For example, parking lots serving large retail centers are often sized to hold the "Black Friday" (the Friday after Thanksgiving) crowds. This means that for much of the year, the lots are grossly oversized. Other design decisions include the type of parking stalls (angled or perpendicular), the size of those stalls (length and width), and means for the safe and efficient circulation of vehicles around the lot. In areas of high snowfall, additional space is required for the accumulation of plowed snow.

Figure 4.11 Multi-Level Parking for Cars Below Twin Residential High-Rise Towers in Chicago, Illinois. Each tower has 13 levels of parking below 40 levels of apartments. Note also the boat "parking" in the marina along the Chicago River.

Source: P. Parker.

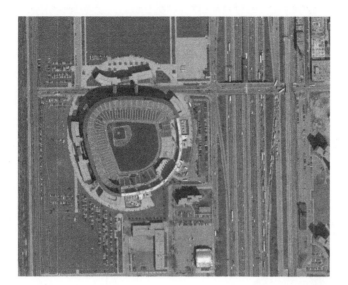

Figure 4.12 The Dan Ryan Expressway in Chicago to the right of US Cellular Field, home of the Chicago White Sox.

Source: U.S. Geological Survey.

TRAFFIC CONGESTION RELIEF

Traffic planners can use many tools to help relieve traffic congestion. For example, bus lanes ensure uncongested bus travel, which in turn helps maintain route schedules and thereby encourages ridership.

HOV lanes (also referred to as carpool lanes) are used in many metropolitan areas to promote carpooling (thus reducing congestion). Driving in a carpool lane in a single occupied vehicle is a traffic offense subject to fines of several hundred dollars.

Automated tolls are rapidly replacing traditional toll booths; traditional toll booths require all vehicles to stop and pay tolls. In contrast, pre-paid accounts and signal transponders placed on front windshields or dashboards allow vehicles to proceed (often at highway speeds) through automatic toll zones, significantly reducing congestion. Drivers that are not subscribed to the local automatic toll service quickly realize the benefits when waiting in long lines of stopped vehicles in adjacent manual pay lanes.

Express lanes are used on freeways to limit congestion from entrance and exit traffic. For example, the Dan Ryan Expressway (Interstate 90/94, Figure 4.12) in Chicago has seven lanes in each direction. Four center lanes are "express" with limited interchanges to the local lanes. Three outer lanes are "local" with frequent entrance and exit ramps to arterials.

Mass Transit (ASCE 2009 Grade: D)

Approximately 5 percent of Americans use mass transit (public transportation) to get to work (Figure 4.13), a percentage much lower than that of most other industrialized nations. Mass transit includes buses, trolleys, and rail.

Buses (Figure 4.14) are the most prevalent form of mass transit, with over 160,000 miles of routes in the United States. However, many communities struggle to manage bus systems, and as ridership decreases, so does revenue, creating further problems for sustaining or expanding the systems.

Ferries are used to transport passengers or vehicles over water. They are a minor mode of transport in the United States, but are critical *local* components in some regions. For example, consider Puget Sound in Washington

Slugging

The benefits associated with HOV lanes has led to the development of a unique solution, known as casual carpooling or **slugging**, (also known as "dynamic ridesharing") in Washington, DC and other

metropolitan areas, whereby professionals hitchhike. Car commuters will "pick up" slugs (hitchhikers), often total strangers, in order to meet the two or three passenger minimum requirement to drive in the HOV lanes and substantially decrease commuting times. Slugging has evolved over the last few decades into a highly organized, citizen-run activity with its own rules of etiquette.

Sluglines.com

Figure 4.13 **Modes of Travel to Work in the United States.**

Source: U.S. Department of Transportation, 2009.

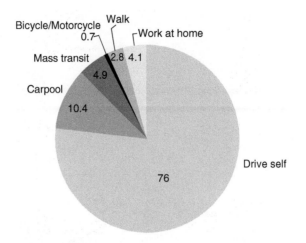

Figure 4.14 **The Famous "Double-Decker" Buses (circled) of London.**

Source: Courtesy of M. Gay.

(see Figure 4.15), where large population, commercial, and industrial centers exist on opposite sides of a large body of water.

Mass transit rail is typically divided into three categories:

- light rail—low-speed streetcars and trams with right-of-ways that can be crossed by vehicles and pedestrians (Figure 4.16);

- heavy rail—higher speed, higher volume rail with dedicated right-of-ways that are not crossed by vehicles or pedestrians, for example, subways; and

- commuter (regional) rail—high-speed, high-volume rail between adjacent cities, or between city centers and suburbs.

Many cities throughout the world have extensive rail systems, either underground (e.g., the "Tube" in London, the "Subway" in New York City) or elevated (the "El," or "L," in Chicago (Figure 4.17); the "People Mover" in Detroit), or a combination of below-ground, above-ground, and at-grade. The trend in the use of mass transit rail in the United States is shown in Figure 4.18. Note the steady increase since the mid-1990s.

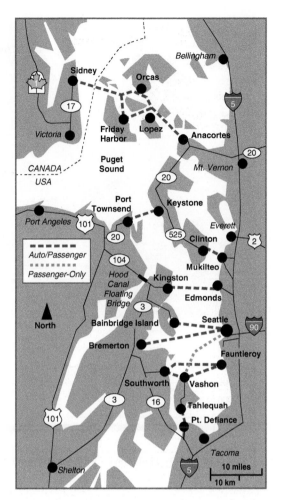

Figure 4.15 Seattle-Area Ferry Routes. Many ferry routes are utilized to link population and business centers on both sides of the Puget Sound.

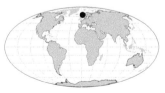

Figure 4.16 A Jointed, Multi-Car Tram in Olso, Norway. Note the interactions between tram, car, and pedestrian.

Source: M. Penn.

Figure 4.17 The "El" (Elevated Heavy Rail Above Street Level) in Chicago. Note parked cars (left) on street approximately 20 feet below rails.

Source: P. Parker.

Figure 4.18 Sum of Rail Car-Miles for Light Rail, Heavy Rail and Commuter Train in the United States. Rail car-miles is the product of miles travelled and number of rail cars.

Source: U.S. Department of Transportation, 2009.

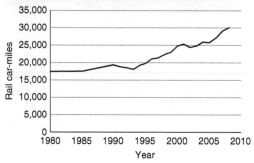

Denver Mass Transit

Dramatic population growth in some areas often necessitates mass transit. For example, Denver, Colorado was the fastest-growing region in the United States during the 1990s, and has seen its population double since 1970 (Figure 4.19). A multi-billion dollar rail system was recently completed for southern suburbs and is expanding over the next ten years in all directions (Figure 4.20). "Our rail investments have been more successful than we ever imagined, and the public's ability to embrace a vision for more in the future has exceeded any of our expectations," states Clarence Marsala, General Manager Denver Regional Transportation District. He continues, "We have billions of dollars in development happening along our rail corridors, especially in the vicinity of our stations." Home values are also increasing, as demand is high for residences close to the stations.

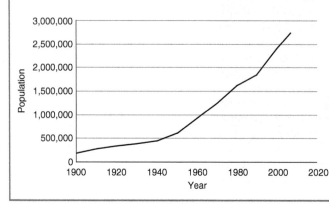

Figure 4.19 Population of the Denver, Colorado Metropolitan Area.

Data source: Metro Denver Economic Development Corporation.

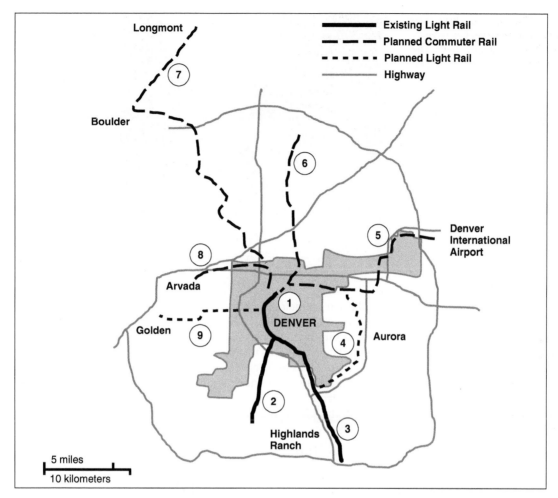

Figure 4.20 Planned Expansion of the Public Rail System in the Denver Region. Over 100 miles of rail lines will be added. Denver city limits shown as shaded region. Numbers refer to rail lines. Existing: (1) Central Corridor, (2) Southwest Corridor, (3) Southeast Corridor. Planned: (4) I-225 Corridor, (5) East Corridor, (6) Northwest Metro Corridor, (7) Northwest Rail Corridor, (8) Gold Line, (9) West Corridor.

Source: Redrawn from Denver Regional Transportation District.

Non-Motorized Transportation

Walking and biking are becoming increasing popular means of commuting and recreation in some parts of the United States. Effectively planned and designed pedestrian/bike routes can provide healthy and pleasant alternatives to mass-transit or automobiles. More pragmatically, every person that commutes by bicycle potentially equates to one less automobile on the road during times of peak congestion, thus making a contribution toward decreasing travel times for all roadway users and decreasing air pollution.

When seeking funding to expand cycling infrastructure, advocates often find themselves in a vicious cycle (pun intended) due to low demand. Funding is often based on demand, so money is not invested in cycling infrastructure. As a result, roads continue to be bike unfriendly, which makes them even less attractive to riding. The number of users remains low, which appears to planners as a lack of demand for bicycling infrastructure, and thus the pattern continues.

Figure 4.21 Example of a Bike Lane.
Shown are the bike lane, bus lane (right), and standard lanes (left). The far left lane is closed for construction. Note the "jaywalker" opting against use of the pedestrian bridge only a few yards away (background).

Source: M. Penn.

Bicycle infrastructure includes off-road trails and bicycle lanes within roadways (Figure 4.21). Signage and pavement markings are key design considerations to increase the safety of bike lanes. Many of the planning and design considerations for off-road bike trails are analogous to those for roads (e.g., route selection, grade, width, turning radius, intersection geometry). Interfaces between pedestrian/bike routes and roadways are of particular concern from a safety perspective, especially in highly developed areas where pedestrian/bike routes did not pre-exist. In such areas, education and signage are very important to alert automobile drivers.

Aviation (ASCE 2009 Grade: D)

"Flying Blind" (PBS Video on the U.S. Air Transportation System)

Airports are fundamental components of population centers for business and personal travel, as well as for package and freight transport. Air traffic has steadily increased, *doubling* since 1980. Airports range in scale from local to regional to international.

Airport design must consider many aspects. The runway length is largely dependent on the type of aircraft the airport intends to serve. An analysis of wind patterns and speeds must be performed, along with an analysis of the available airspace and existing flight patterns. Also, the airport must have appropriate links to other transportation and infrastructure systems. A viable site for an airport or airport expansion must consider the existing topography; a flat site will minimize earthwork. However, many areas with level topography include wetlands (because precipitation collects in depressions), which complicate regulatory requirements for construction. The location of a major airport will also have profound impacts on road traffic, surrounding land use (commercial and industrial development, long-term airport parking, required buffer zones), and property values. Like all infrastructure components, growth will create increased demand and the need for expansion. As with all infrastructure projects, effective design must be preceded by effective planning (discussed in more detail in Chapter 11, Planning Considerations). Without proper planning, the expansion of an airport can lead to great controversy as surrounding areas of development (businesses or residential) will need to be purchased and demolished or moved.

Figure 4.22 Atlanta Hartsfield International Airport.

Source: U.S. Geological Survey.

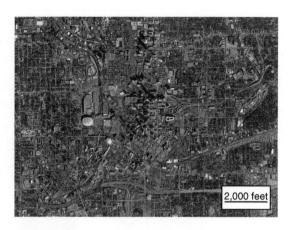

Figure 4.23 **Downtown Atlanta.** Note the scale is the same as in Figure 4.22.

Source: U.S. Geological Survey.

Design elements include a plan design of the runways, taxiways, terminal, parking, and other features. Runways are typically laid out such that the prevailing wind directions do not act as crosswinds for as much of the year as possible. A pavement design must be created that will withstand the landing and taxiing of planes. An additional design element is a stormwater management plan. The large amounts of impervious surfaces (runways, roads, roofs) create stormwater management challenges. These challenges are compounded in colder climates, where chemicals used to de-ice planes end up in the stormwater runoff.

Four U.S. airports rank in the top 10 in the world in terms of passenger arrivals and departures. The world's busiest airport, Atlanta's Hartsfield-Jackson International Airport, has a footprint of approximately 6 square miles (approximately equal to 4,000 soccer fields, or the size of a city of 20,000 residents). It is larger than Atlanta's central business district as shown by comparing Figure 4.22 and Figure 4.23. Moreover, it has tremendous infrastructure demands; for example, it is the third largest water user in the state of Georgia. Its combined terminal and concourse space is nearly 6 million square feet.

Remember that full-sized versions of textbook photos can be viewed in color at www.wiley.com/college/penn.

Waterways (ASCE 2009 Grade: D−)

Cities often originated in coastal areas and along major rivers because historically ships were a primary means of transportation. While roads and railways are the major modes of transportation today, waterways remain a significant mode of transportation for goods. Approximately 12 percent of domestic goods and 80 percent of international goods transport occurs via waterways.

Rivers such as the Mississippi, the Missouri, and the Ohio are often called "working rivers" because they are as much a means of transportation as they are aspects of the natural landscape. There are approximately 30,000 miles of navigable waterways in the United States. Managing these rivers for transportation requires controlling flows and dredging bottom sediment to keep navigation channels open at necessary water depths. These engineered activities are often at odds with environmental requirements, some of which will be discussed later in this book.

Figure 4.24 **The Port of Los Angeles.**
The port encompasses 7,500 acres with over 40 miles of waterfront and 27 cargo terminals. For scale, the opening in the seawall (foreground) for ship passage is approximately one-half mile wide. It is estimated that the port creates 900,000 regional jobs and generates $40 billion in annual wages and tax revenues.

Source: Los Angeles Harbor Department.

PBS Video on Ports

A Commentary on Ports

Infrastructure Needed to Support a Port

You are encouraged to use an Internet mapping service with satellite imagery such as maps.bing.com or maps.google.com to scan the array of infrastructure associated with the Port of Los Angeles and the adjacent Port of Long Beach.

Draft is the depth of a ship's bottom below the water surface.

Inner Harbor Navigation Canal Lock Replacement Project Video

Ports

Ports are integral in the transportation of goods. Ports allow for the **intermodal** transfer of goods, between water (ships) and land (trains and trucks). Design, construction, and maintenance of ports is typically carried out by "specialty" firms. This is because there are approximately 150 ports in the United States, a number dwarfed by the number of other infrastructure components such as roads, bridges, parks, and water and wastewater facilities. Thus, the vast majority of civil and environmental engineers are not involved in work on ports. Nonetheless, there is great potential for interested engineers in this line of work.

Seven of the world's 20 busiest ports (and three of the top five) are located in China; this is perhaps not surprising given the tremendous quantity of exports from the country. Three U.S. ports rank in the top 20 (rankings in parentheses): Los Angeles (10, seen in Figure 4.24), Long Beach (12), and New York/New Jersey (18). As noted previously with respect to airports, shipping ports have dramatic impacts on surrounding land use and must be appropriately linked to other infrastructure systems, especially major freight modes such as trucking and rail. For intermodal freight transport, standardized containers are used commonly with ships, rail, or trucks to significantly increase transfer efficiency between modes of transport via crane (Figure 4.25 and Figure 4.26). The containers come in several standardized sizes with the most common being approximately 8 feet wide by 10 feet tall by 50 feet long.

As with roadways, ship traffic volume is an important design consideration for ports. Other design considerations include: the size of the vessels (width, length, and **draft**); the type of cargo (for handling and storage needs); types of vessels (e.g., passenger, commercial, or recreational); breakwater design (to dissipate wave energy); and the structural design of piers and wharfs.

Locks

There are approximately 250 locks in the United States, 29 of which are on the Mississippi River. **Locks** are utilized to permit ship travel around a dam (Figure 4.27) or through a shallow stretch of water (Figure 4.28). Locks function by opening or closing gates to allow water to flow in to raise a ship, or to flow out to lower a ship. Their design depends on the type of vessel (barges on the upper Mississippi versus oceangoing vessels in the St. Lawrence Seaway),

Figure 4.25 A Cargo Ship Carrying More than 500 Intermodal Containers About to Be Unloaded.   Note the crane and container storage area in the background. Several million containers are handled annually at the Port of Los Angeles.

Source: Los Angeles Harbor Department.

Figure 4.26 **Standardized Intermodal Containers Being Loaded Onto Trucks from a Ship on the Rhine River in Germany.**

Source: Copyright © Joerg Reimann/iStockphoto.

Figure 4.27 **A Lock for Passage of Ships, and Dam on the Mississippi River in Dubuque, Iowa.**   The drop in water surface elevation across this lock is approximately 15 feet.

Source: P. Parker.

the water surface elevations upstream and downstream of the locks, and the speed at which locks are desired to fill and empty.

Tunnels

Tunnels are a costly but potentially very effective means of "shortcutting" transportation routes for rail or roadways (e.g., routing *through* a mountain, rather than around it). In urban areas, tunnels also provide a means of moving surface traffic underground when space is not available aboveground. In addition to transportation uses, tunnels are also used for water supply conveyance, hydroelectric power (connecting low elevation turbines to high elevation water reservoirs), utilities, and stormwater/wastewater storage. Large diameter tunneling machines, while expensive, dramatically reduce construction times for tunnels.

Article on "How Tunnels Work"

Figure 4.28 The "Soo Locks" at Sault Ste. Marie, Connecting Lake Superior and Lake Huron. Notice the shallow water (light color) to the left of the locks.

Source: U.S. Army Corps of Engineers.

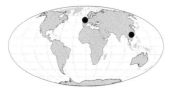

The World's Longest Transportation Tunnel

In October 2010, rock boring of a 35-mile long tunnel through the Swiss Alps was completed. The $10 billion rail tunnel is expected to open in 2017.

The Channel Tunnel (or "Chunnel"), between Britain and France, is a multi-billion dollar project completed in 1994. It was declared one of the Seven Wonders of the Modern World by the American Society of Civil Engineers (ASCE). It consists of three borings (two 25-foot diameter rail tunnels and one 16-foot diameter service tunnel) over 30 miles long, which at the lowest point are 250 feet below sea level. In Shanghai, China, twin tunnels were constructed under the Yangtze River, each to hold three lanes of traffic and one lane of rail. The world's largest tunnel boring machines (50 feet in diameter) were used to create the tunnels (Figure 4.29). The most complex tunneling project in U.S. history, the "Big Dig" in Boston, will be discussed in later chapters of this book.

The most important design consideration for tunnels is the geologic formation through which the tunnel is being built. The design of a tunnel must specify its diameter, length, and longitudinal profile. If the geologic formation is not self-supporting, the type of tunnel lining (e.g., concrete as shown in Figure 4.30) must also be specified. The design will vary depending on whether it is to be used for rail or auto. Ventilation must be provided in longer tunnels to

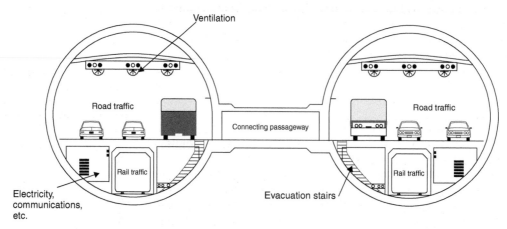

Figure 4.29 A Cross-Section of the Twin Tunnels in Shanghai, China, Under the Yangtze River.

Source: Redrawn from Wang et al., 2008.

Ground Freezing for Soil Stabilization Case Study

Figure 4.30 Tunnel Construction as Part of the "Big Dig" in Boston. Six lanes of interstate highway traffic needed to be routed under existing railroad tracks, requiring innovative construction methods. The tunnel was built in concrete "box" sections, and the sections were hydraulically rammed (or "jacked") through the soil. In order to stabilize the overlying soil it was frozen by circulating cold brine through a piping system (note the white PVC pipes protruding from the soil surface). The exposed soil surface of the cuts was covered with sprayed-on insulation (seen above and adjacent to the tunnel sections) to keep the soil frozen and prevent collapse into the excavation.

Source: Courtesy of the Massachusetts Department of Transportation.

prevent the accumulation of explosive or asphyxiating gases. The type of boring method will depend on the construction timelines, the cost that the owner is willing to bear, and the type of material through which the boring occurs.

Rail (ASCE 2009 Grade: C–)

While the mileage of freight rail operated in the United States has decreased substantially to approximately one-half the railway miles present in 1960, rail remains the largest mode of domestic goods transport (Figure 4.31) when measured in ton-miles (tons of freight multiplied by the miles transported). Rail has also experienced the greatest percent change in ton-miles since 1990 of all transportation modes (74 percent increase, compared to 53 percent for trucking).

There are many components of rail systems. One component is the rails themselves, designed to guide the flanged wheels of rail cars. These rails are attached to cross ties ("railroad ties") that transfer the load from the train to the substructure. Ties are most often made of timber in North America, but concrete is also used. The rail system substructure consists of ballast material, composed of inch-sized stone. Its purpose is to hold the rails in place, provide drainage, and transfer the loads from the cross ties to the underlying native soil.

Rail systems include **rail yards** (see Figure 4.32), where locomotives or rail cars can be separated for maintenance or repair; incoming cars can be separated (or **classified**) and regrouped for final destination; or locomotives can be refueled with diesel. **Switches** are devices that allow track changes and are found in rail yards as well as at other points in the system where routing is required. Multiple simple switches are seen in Figure 4.32.

High-speed rail is quite popular for passenger travel in Europe and Asia, and several new projects are being pursued in the United States. The only system currently operating in the United States is the Acela Express from Washington, DC to New York. In 2008, California voters approved a $10 billion bond to begin work on a high-speed rail system from San Diego to San Francisco. The final, expanded project is estimated to cost nearly $50 billion. Capable of travelling at over 100 mph, these systems can dramatically reduce

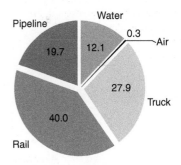

Figure 4.31 Modes of Domestic Freight in the United States as a Percentage of Total Ton-Miles. A ton-mile is equal to one ton of freight transported one mile.

Data source: U.S. Department of Transportation, 2009.

Figure 4.32 A Railway Classification Yard.

Source: M. Penn.

travel times. In 2010, the Obama administration announced $8 billion in new funding to support the development of high-speed rail.

Rails-to-Trails

Thousands of miles of railway are abandoned each year. Since the 1970s, over 10,000 miles of rail corridor have been converted to hiking and biking trails due to the efforts of state agencies, the Rails-to-Trails Conservancy, and other organizations. Because of the low-grade (i.e., flat) nature of railways, the trails are ideal for recreation for a wide variety of users, including the young, elderly, and disabled. There are also many rail corridors that have provided trails in the right-of-way adjacent to operating rail lines, termed rails-with-trails.

The Elroy-Sparta Trail in Wisconsin, one the nation's first rails-to-trails projects, includes three unique tunnels

constructed in the late 1800s that are over one-fourth mile in length each and are unlit (Figure 4.33). The trail has been an "economic engine" for the small, rural towns along its 30-mile route. Over $1 million per year in trail-related tourism is brought to the area.

Rails-To-Trails Conservancy

Figure 4.33 A Tunnel on the Elroy-Sparta Rail Trail in Wisconsin. The original rail tunnel was completed in 1873 and is 1,700 feet in length. If you look closely, you can see "the light at the end of the tunnel" above the middle biker's head.

Source: Courtesy of C. Kennedy.

Outro

Transportation is perhaps the most visible aspect of our infrastructure, and it "connects" various infrastructure components. After reading this chapter, it should be clear that the transportation infrastructure consists of far more than highways. As populations increase and become more dense, transportation needs and challenges become more complex. Transportation also creates significant environmental impacts, which will be discussed in the following chapter.

chapter Four   Homework Problems

Calculation

4.1   Using an online aerial photography service such as maps.google.com or maps.bing.com, view the Port of Los Angeles, which was mentioned in this chapter. Print out a copy of this aerial photograph and compute the land area devoted to the port. Your answers will vary depending on how you choose to delineate the port's boundaries; therefore, it is very important that you show your work.

Short Answer

4.2   What types of conflict points exist for the roundabout in Figure 4.8 (i.e., merging, diverging, or crossing)? How would the type of accidents that occur for a roundabout differ for those that would occur in the intersection of Figure 4.7?

4.3   Between which stations in Figure 4.2 will material be removed (cut) in order to build the new road profile? Between which stations will fill be required?

Discussion/Open Ended

4.4   Imagine a community for which its first roundabout is being built. As the Director of Public Works for the community, how might you educate the members of the community? Which members of the community might be most critical to target with your educational program? What are the most important points about roundabouts to "teach" community members?

4.5   Would you consider being a "slug" if you lived in a large metropolitan area? Why or why not?

4.6   There are many arguments for and against the federal government subsidizing high-speed rail. One such argument against such funding states that by subsidizing high-speed rail with general tax dollars, some people are forced to help pay for a system (through their taxes) that they may never use, perhaps because they live too far away from the rail system. Do you think this is a valid argument? Why or why not?

4.7   How does your university encourage bicycle use? What additional measures could be taken to further encourage bicycle use? What are the benefits to a university encouraging bicycle use? What are some disadvantages to increasing bicycle use?

Research

Note: For Homework Problems 4.8 through 4.13, the *ASCE Report Card for America's Infrastructure* is available at www.wiley.com/college/penn.

4.8   Read the *ASCE Report Card for America's Infrastructure* chapter on roads. Based on this chapter, state the most challenging problems facing the roads sector.

4.9   Read the *ASCE Report Card for America's Infrastructure* chapter on inland waterways. Based on this chapter, explain why inland waterways are important. Characterize their present condition.

4.10   According to the *ASCE Report Card for America's Infrastructure* chapter on aviation, how resilient is the U.S. aviation system?

4.11   List the five most important points (in your opinion) made by the *ASCE Report Card for America's Infrastructure* chapter on aviation.

4.12   Read the *ASCE Report Card for America's Infrastructure* chapter on transit, and list the components that make up a transit system.

4.13   Read the *ASCE Report Card for America's Infrastructure* chapter on transit. Use information from this chapter to write a letter to the editor of the local newspaper, arguing either that the community should invest more (or less) in mass transit.

4.14   Research the National Transportation Statistics website (a link to which is available at www.wiley.com/college/penn). Select data from this site that you find most interesting and create three different graphs using a spreadsheet program. For each graph, write one paragraph that explains the graph and its significance.

4.15   Evaluate the parking on your campus. Is it sufficient? Why or why not? If not, what are some possible solutions?

Key Terms

- alleys
- automated tolls
- classified
- conflict points
- cross-section
- crowned
- cut
- design speed
- draft
- express lanes
- ferries
- fill
- flexible pavements
- frontage roads
- grades
- high-speed rail
- horizontal alignment
- horizontal curves
- interchanges
- intermodal
- intersections
- level of service
- locks
- offsets
- parking structures
- plan view
- profile view
- rail yards
- right of way
- rigid pavements
- roundabouts
- service roads
- sight distance
- slugging
- station
- subgrade
- switches
- tangents
- vertical alignment
- vertical curves

References

Kutz, M. *Handbook of Transportation Engineering*. New York: McGraw-Hill Professional, 2003.

U.S. Department of Transportation. 2009. *National Transportation Statistics*. http://www.bts.gov/publications/national _transportation_statistics/, accessed August 8, 2011.

U.S. Federal Highway Administration. 2004. *Signalized Intersections: Informational Guide*. FHWA-HRT-04-091.

Wang, X., Z. Q. Guo, and J. Meng. "Design of hazard prevention system for Shanghai Yangtze River Tunnel." *The Shanghai Yangtze River Tunnel—Theory, Design and Construction*. London: Taylor & Francis Group, 2008.

chapter Five Environmental and Energy Infrastructure

Learning Objectives

After reading this chapter, you should be able to:

1. List the main subsectors and components of the environmental and energy infrastructure.
2. Explain the function of each infrastructure sector.
3. Identify components related to environmental and energy infrastructure.
4. Explain the function of components related to environmental and energy infrastructure.
5. List some of the important design parameters for components related to environmental and energy infrastructure.
6. Describe some of the fundamental relationships between environmental and energy infrastructure and other infrastructure sectors, subsectors, and components.

Introduction

The environmental infrastructure exists to protect the environment from the impacts of other infrastructure components and to provide clean drinking water and recreational areas. It is linked to the energy infrastructure in many ways; for example, the extraction and use of energy creates many environmental impacts. Moreover, the accumulation of greenhouse gases is inextricably linked to the energy use.

Energy (ASCE 2009 Grade: D+)

ENERGY SOURCES

From a civil infrastructure perspective, "energy" is typically either electricity, fuel for heating and cooling buildings, or vehicle fuel. This section will focus on electrical energy and the "power grid." Electricity is generated from both **renewable** and **non-renewable** sources. Major renewable sources include hydropower, wind, solar, geothermal, and biomass and account for approximately 10 percent of electrical generation in the United States. Nonrenewable (finite) sources include fossil fuels (oil, coal, natural gas) and nuclear material, which account for approximately 90 percent of electrical generation in the United States (Figure 5.1). Energy sources and their environmental impacts are discussed in greater detail in Chapter 12, Energy Considerations.

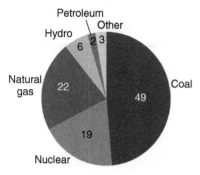

Figure 5.1 Energy Sources for Electrical Generation in the United States, 2008.

Source: U.S. Energy Information Administration.

ENERGY TRANSPORT

The transport of solid (e.g., coal, nuclear material) energy sources is often by rail. Natural gas and petroleum are transported long distances via pipelines (Figure 5.2). Note in Figure 5.3 the dramatic increase in natural gas pipeline length as new gas fields have been tapped.

ENERGY GENERATION

In most cases, energy generation results from an energy source being utilized to turn a turbine that rotates wires within a magnetic field to produce electricity. Conventional "power plants" (e.g., coal or nuclear power generating facilities) use heat from combustion or nuclear reactions to produce steam to turn the turbines. A hydroelectric plant utilizes the potential energy (elevation) of an accumulated water body to turn turbines (Figure 5.4).

Electrical generating facilities require integration with not only the power grid, but also the transportation "grid." A coal-fired power plant requires the delivery and storage of tremendous amounts of fuel (Figure 5.5), as much as 10,000 tons (100 rail cars) per day. Additionally, large amounts of water are

Figure 5.2 Natural Gas Pipelines in the United States. A similar, but less extensive, network exists for petroleum.

Source: U.S. Energy Information Administration, Gas Transportation Information System.

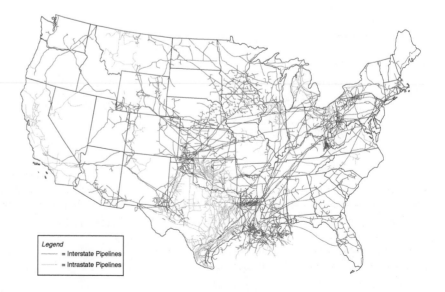

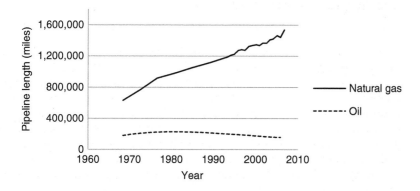

Figure 5.3 Pipeline Length in United States for Natural Gas and Oil.

Data source: U.S. Department of Transportation, 2010.

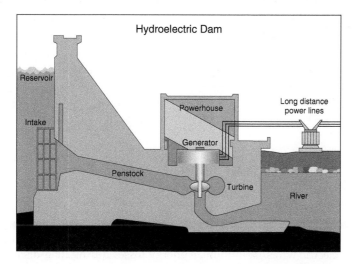

Figure 5.4 A Generalized Cross-Section of a Hydroelectric Dam.

Figure 5.5 A "Small" Coal-Fired Power Plant. The power plant (80 megawatts) serves approximately 50,000 homes. Notice the large coal stockpile to the right of the plant. This several decades-old facility was once surrounded by industries, but today is in the midst of urban redevelopment (offices, hotels, restaurants) as evidenced by the landscaped parking lots in the foreground.

Source: M. Penn.

Figure 5.6 An Electrical Substation. High voltage lines are coming into the substation from the right and lower voltage lines are leaving the substation to the left.

Source: M. Penn.

Figure 5.7 **Distribution Transformers (Circled in Foreground) to Lower Voltage to 240 Volts for Residences.**

Source: M. Penn.

Figure 5.8 **Transmission Lines Crossing the South Dakota/Minnesota Border.** This picture is taken on Interstate 90 at the state line (note the "Welcome to Minnesota" sign in the shape of the outline of the state).

Source: M. Penn.

needed for cooling the steam; thus, these facilities are typically located beside large rivers or lakes.

DISTRIBUTION

The electricity that is generated must be distributed to the end users via electric wires. To minimize energy losses during long-distance transmission, high voltages, on the order of several hundred thousand volts, are used. This voltage is "stepped down" at regional substations (Figure 5.6), typically to 7,200 volts for local distribution in the power lines that are commonly seen above streets. An additional reduction to 240 volts occurs at distribution transformers for most end users (e.g., businesses and residences; Figure 5.7).

The national power grid allows for redundancy, whereby power is supplied from many sources into regional networks. If there is a reduction in production at any one generating facility, electricity is available from other facilities. Most states receive power from neighboring states in addition to their own generation (Figure 5.8).

Drinking Water (ASCE 2009 Grade: D−)

Public water supplies must meet the demands of residential, commercial, institutional, and industrial users. Water must be delivered to homes such that residents always have safe drinking water at an appropriate pressure. The same system that treats and distributes the drinking water to homes must also meet the pressure and quality needs of the non-residential users. Perhaps most importantly, this system must be able to supply large quantities of water for firefighting purposes.

Typically, the entire water supply distribution system carries water that has been treated to drinking water standards, even though less than 1 percent of the water is used for drinking. As drinking water standards become more stringent, the cost for treatment escalates. Some communities have implemented, or are considering the implementation of, *separate* water systems, whereby some users (e.g., industry and landscape irrigation) are supplied with water that is *not* treated to drinking water standards. The cost of savings from lack of (or less thorough) treatment must be weighed against the capital, operation, and maintenance costs of the additional infrastructure (i.e., a "second" water distribution system).

SOURCES AND DEMAND

The residential **per capita water demand** for the United States averages about 60 gal/capita/day. When taking into account all water needed to meet residential, industrial, commercial, and municipal irrigation demands, the average per capita demand doubles to 120 gal/capita/day. Given that landscape irrigation, commercial use, and industrial use varies widely from region to region, so too does the total per capita water demand of any system. Water sources include groundwater (from either shallow or deep wells), lakes, rivers (either directly, or from reservoirs created by constructed dams) and oceans. Another source, albeit one that is highly contentious and unsavory in the public's opinion, is treated domestic wastewater.

Groundwater is used for individual rural residential use as well as for public water systems and industrial users. As discussed in Chapter 2, The Natural Environment, there is a direct relationship between groundwater and surface water. Therefore, large withdrawals of groundwater must consider environmental impacts such as lowering river flows or lake levels. Many communities are facing challenges as their withdrawal of groundwater far exceeds the rate of aquifer recharge; consequently, water levels in their wells are decreasing, often at rates of more than several feet per year. In coastal communities, the withdrawal of groundwater can draw ocean water inward to the previously freshwater wells (termed **saltwater intrusion**), which creates the need for costly additional treatment to remove salts (desalinization).

STORAGE

The need for water storage can be understood by considering that the water demand of a community fluctuates over a 24-hour period, termed **diurnal variation** (Figure 5.9). The average residential demand of 60 gal/person/day is approximately equivalent to 3 gal/person/hour; however, at 3 a.m. when most people are sleeping, the *average* use for a community may be much less than 1 gal/person/hour. On the other hand, at 7 a.m., when most people are showering and using toilets, the demand may be as high as 30 gal/person/hour. Rather than inefficiently trying to supply a water system to meet varying demands over the course of a day, the system pumps can deliver water at the *average* rate, and excess water is stored during periods of low demand.

Storage may be aboveground (Figure 5.10), below ground, or at ground level. Elevated storage tanks are filled during times of lower demand and emptied at times of peak demand. Elevated storage tanks also provide necessary pressure in the system (typically between 50 to 70 pounds per square inch, or psi) and allow for less variation in pressure throughout the day. Also, storage tanks provide a volume of water that can be used for firefighting or for emergency use (e.g., in the event of the failure of a well pump).

Water demand can also fluctuate seasonally, for example in tourist areas and in arid regions where landscape irrigation needs are great. Systems must be designed to account for these seasonal changes. Additionally, long-term variations in climate can affect water availability. Communities depending on surface water reservoirs (dammed rivers) will face potentially severe water

The total water use (demand) of a community (or several communities served by one system) divided by the population is the **per capita water demand**.

Water Reuse

In 2009, Orange County (including Los Angeles), California started reusing wastewater, indirectly, as a drinking water source. After treating the wastewater *beyond* drinking water standards, it is injected underground to recharge aquifers supplying drinking water to over 2 million people. An added benefit to the injection of the treated wastewater underground is that it prevents saltwater intrusion.

Many communities *indirectly* reuse treated wastewater as a drinking water supply. A downstream community that uses a river as a water supply will be utilizing the treated and diluted wastewater of upstream communities.

Figure 5.9 Diurnal Variation in Water Demand for Various Customer Types. These types include: (a) industry, (b) residential, and (c) commercial. The overall community water demand (d) is a combination (the sum) of all individual user demands. The quantity and timing of industrial water demand in communities is highly variable depending upon the types of industries present.

Source: From Mihelcic and Zimmerman, 2010. Reprinted with permission of John Wiley & Sons, Inc.

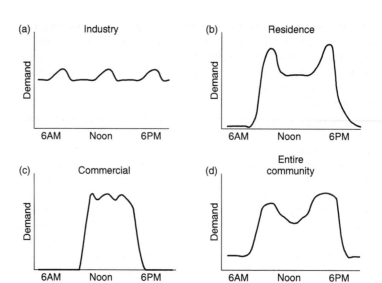

Figure 5.10 An Elevated Water Tank with Two Nearby "Users." The "users" are the car wash in the foreground and fire station behind the car wash. This single 110-foot-tall tank has a capacity of 250,000 gallons and serves a community of 1,000 residents. Larger communities may have many tanks with capacities up to several million gallons.

Source: M. Penn.

shortages if the reservoir was not designed or managed to have sufficient storage during drought periods.

CONVEYANCE

Groundwater wells are often located in close proximity to water users. Surface water reservoirs may, in extreme cases, be 100 or more miles away from end users. New York City has a network of concrete lined tunnels (up to 24 feet in diameter) to bring water to the city for treatment. Southern California uses large concrete lined channels that carry water hundreds of miles from water supplies in northern California and from the Colorado River.

TREATMENT

The treatment of drinking water has two goals: to produce safe (**potable**, pronounced pō-tə-bəl) drinking water and to provide aesthetically pleasing water. The safety of drinking water is carefully regulated, ensuring that toxic chemicals and **pathogens** (disease-causing microorganisms) are removed to appropriate levels. Providing aesthetically pleasing water is typically not regulated nor required, but is driven by public opinion. Examples of aesthetics as they relate to drinking water include tastes and odors, iron/manganese (which stain bathtubs and laundry), and hardness (lime scale in and on plumbing components).

Treatment facilities vary widely depending on the composition of the source water, the volume of water treated, the types of treatment processes utilized, and the extent to which aesthetics are to be addressed. Many groundwater supplies have low contaminant concentrations and require minimal treatment. Surface waters, especially major rivers, contain not only high amounts of particulates (suspended sediment) but are also often contaminated with microbial and chemical pollutants. There are many different feasible design solutions that may be used for a given scenario (just as there are many different types of bridges that could be designed to cross a certain span.) Processes to treat water include contaminant removal by physical methods (e.g., settling of particles, screening/filtering of particles, UV radiation to kill pathogens) and chemical methods.

DISTRIBUTION

Drinking water is distributed in pressurized pipes, or water **mains** (often PVC or ductile iron), to users. Older systems used cast iron and wooden mains (Figure 5.11). The minimum diameter for new pipes is 6 inches, but older systems include 4-inch pipes and smaller.

Water distribution networks are designed to include redundancy, so that water can usually be supplied to a user from multiple routes in case of water main breaks, maintenance work, or reconstruction. Valves are located throughout the system to isolate certain sections of the system (e.g., valves on each side of a main break can stop the loss of water and allow maintenance workers to repair the main). Fire hydrants are an integral part of the distribution system, and are used for fighting fires and the flushing and testing of mains (Figure 5.12).

Figure 5.11 A Portion of a Wooden Water Main Wrapped with Wire.

Source: Courtesy of S. Anderson.

Figure 5.12 Students Test a Hydrant to Determine Flow and Pressure Characteristics of the Water System.

Source: Courtesy of D. Rambo.

Hydrants are cast with a date on them corresponding to the year of manufacture, and thus they may be used to estimate the age of the water main to which they are connected. Installation dates typically closely correspond to the casting date, and hydrants are typically replaced when the water mains to which they are connected are replaced.

Security is a major concern in public water supply and will be discussed in detail in Chapter 18, Security Considerations.

Water Main Breaks

A break in a water main can have wide-ranging impacts. For example, in May 2010, a break in a 10-foot-diameter 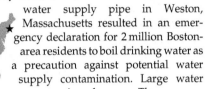 water supply pipe in Weston, Massachusetts resulted in an emergency declaration for 2 million Boston-area residents to boil drinking water as a precaution against potential water supply contamination. Large water main breaks can cause extensive damage. The pressurized water erodes the adjacent soil and can result in road damage.

A typical repair requires isolating the ruptured water main by closing adjacent valves; removing the overlying roadway; excavating to the water main; replacing a section of the main; filling and compacting the excavation; and repairing the roadway. Such repairs are disruptive to traffic and businesses, inconvenient to customers, and costly. For a small system, two or three breaks per year are common, whereas at the other extreme, the city of Los Angeles repairs approximately 200 water main breaks per year.

An Internet search for "water main break" will provide you with many images.

You Never Know What You'll Find Inside a Water Main

While extreme precautions are taken to ensure that water mains are properly installed, surprises do occur (e.g., Figure 5.13). In 2005, pieces of flesh were found in the water supply system of Carroll, Iowa, presumably from an animal that was hiding in a pipe during installation. At the extreme, even treated wastewater may be found in water systems if pipes are improperly installed. In 2007, cross-connections between the water supply and the treated wastewater irrigation supply were found at two residences in Cary, North Carolina.

Figure 5.13 A 4 × 4-Inch Timber Found in a 6-Inch Water Main, Many Years After Installation.

Source: Courtesy of J. Malkowski.

Wastewater (ASCE 2009 Grade: D−)

Wastewater (sewage) is generated by the use of water for residential (shower/bath, toilets, laundry), commercial (car washes, laundromats, restaurants, sinks/toilets from any workplace), institutional (universities, schools, government offices, prisons) and industrial purposes. Prior to discharge to the environment after use (typically to rivers, lakes, estuaries or oceans, and sometimes to groundwater), it must be treated. Treatment removes debris, pathogens, toxic chemicals, organic material (that when degraded can rob a receiving stream of oxygen needed to sustain aquatic life), and unsightly materials such as fats, oils, and grease. Inadequate or non-existing wastewater treatment is a major public health threat in developing countries.

SOURCES

Given the diversity of sources, wastewater may include "clean" water (e.g., a leaky faucet in a home), human waste, cleaning chemicals, illegally disposed material (e.g., used motor oil), and industrial solvents and other byproducts. Wastewater is conveyed by underground pipes, termed **sanitary sewers**, to wastewater treatment facilities. However, hundreds of older cities in the United States have **combined sewers** for wastewater *and* stormwater (storm runoff from streets, roofs, etc.).

Sewer systems leak, especially pipes and manholes that are several decades old (Figure 5.14 and Figure 5.15), and sometimes other pipes (storm drains, basement sumps, etc.) connect into the system. These *unplanned* sources are called **inflow** and **infiltration** (I and I, or I/I), and collectively they may contribute more flow to sewers than the sum of planned sources (residential, commercial, and industrial). Inflow and infiltration (I/I) decreases the capacity of sewers for conveying sewage and adds to treatment costs. As such, considerable efforts are made to identify and minimize these sources.

CONVEYANCE

Sanitary sewers are typically made of PVC or reinforced concrete pipe, and must be sized to handle current and expected future flows. Sewer systems are networks in which smaller pipes (**laterals** from homes) feed into larger

In **combined sewers**, wastewater and stormwater travel together in pipes rather than in separate systems.

Infiltration is water that enters the sewer system below ground, typically through cracks or bad joints in pipes.

Inflow is from direct connections to the pipes and may come from holes in manhole covers or from improperly connected house downspouts or cross-connected storm sewers.

Figure 5.14 **Infiltrating Water Seen from the Televising of a Sewer.**

Source: Courtesy of Carylon Corp.

Figure 5.15 **A Coating Being Applied to Seal a Leaky Brick Manhole.**

Source: Courtesy of Carylon Corp.

Figure 5.16 A Gravity Flow Sewer with Gravity (Lateral) Connections from Homes.

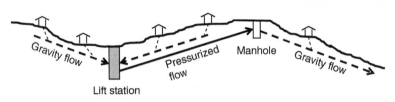

Gravity flow

Figure 5.17 A Combined Gravity and Pressurized Sewer System with a Lift Station to Minimize Excavation.

Gravity flow   Pressurized flow   Manhole   Gravity flow

Lift station

pipes (**collectors**) and ultimately into even larger pipes (**interceptors**). This is analogous to street systems discussed in Chapter 4. Manholes are used to allow access to the sewers for cleaning, inspecting, and other maintenance activities. Sewers are designed to flow with a velocity greater than 2 ft/s in order to ensure that solids do not accumulate in the pipes. If topography (land slope) and geology (soil depth) allow, gravity sewers are preferred because there is no need for pumping. A typical section of gravity sewer is illustrated in Figure 5.16.

Sewers must slope downward in order to convey the sewage by gravity. If the topography is relatively flat, the sewer becomes progressively deeper over long distances. If the depths of excavation become too great, or if shallow bedrock is encountered, excavation may become cost-prohibitive. Deep excavation also occurs when sewage must be conveyed across a ridge. In such situations, the sewage must typically be pumped (Figure 5.17). Pumping the sewage results in pressurized flow, and the pressurized sewage is carried in pipes termed **force mains**. Pumps add energy to overcome energy lost to friction and to overcome elevation changes. Pumps are located in structures known as **lift stations**.

In cities with combined sewers, the conveyance system is often overwhelmed by stormwater during large rain events. These systems have outfalls to release the excess water (Figure 5.18). The release of the untreated sewage and runoff is termed a **combined sewer overflow**. To reduce the public health

Milwaukee's Deep Tunnel System Website

Figure 5.18 A Combined Sewer Outfall. The hinged cover will open when the gravity sewers fill and become pressurized, releasing untreated combined sewage into the river (foreground).

Source: M. Penn.

threat from pathogens, combined sewer systems are being replaced with separate sewers, or large underground storage reservoirs are constructed to store the excess water for later treatment. Sanitary sewer systems with substantial sources of I/I can also be undersized during extreme precipitation events, potentially leading to sewage back-ups in basements and **sanitary sewer overflows**.

A **sanitary sewer overflow** occurs when the system becomes pressurized and manhole covers are "blown" off their bases, thereby releasing untreated sewage into streets and ultimately to nearby surface waters via storm sewers.

TREATMENT

Wastewater treatment facilities (WWTFs; also referred to as wastewater treatment plants [WWTPs] or sewage treatment plants [STPs]) may be publicly owned by a community, regionally owned by a utility, or privately owned by a large wastewater company. Many industries (e.g., a brewery or cannery) have wastewater treatment facilities that pre-treat their waste before discharging to the sanitary sewers. Treatment systems consist of several processes, each with a specific objective, that rely on chemical, physical, or biological means of contaminant removal. Each process is a component within the system; as such, it must be *compatible* with subsequent processes to ensure that the system operates efficiently and effectively.

- *Preliminary treatment* processes, such as screens and settling tanks, are often included at the beginning (**headworks**) of the facility to remove items (e.g., sand and other grit, rags, condoms) that will potentially damage or shorten the operating life of treatment processes or pumping equipment.

- **Primary (physical) treatment** (Figure 5.19) removes floating fats, oils, greases, and settleable inorganic and organic material.

- **Secondary (biological) treatment** utilizes microorganisms, either on attached surfaces such as corrugated plastic or in a suspended slurry termed **activated sludge** (Figure 5.19). The purpose is to transform organic material into carbon dioxide in a controlled setting, rather than in the natural environment where it will rob the receiving water of oxygen.

- **Tertiary (chemical or biological) treatment** is used for nutrient (phosphorus or nitrogen) removal.

- Throughout the treatment system, **clarifiers** (settling tanks) are used to settle solids.

Figure 5.19 Students Tour a WWTF that Treats 40 Million Gallons of Sewage Per Day. This amount of sewage is produced by approximately 400,000 people. Primary clarifiers are in foreground. Activated sludge is in the background. The large pipes running parallel to the activated sludge tanks are carrying pressurized air to the tanks for aeration.

Source: M. Penn.

Figure 5.20 An Anaerobic (Absence of Oxygen) Digester. The rollers and steel columns protruding from the top are support for the "floating" cover under which methane produced during digestion accumulates.

Source: M. Penn.

Figure 5.21 Effluent from a WWTF (Background) Discharged into a Stream. Newly placed gravel (left of the outfall) was placed after scour of the bank during a flooding event.

Source: M. Penn.

Anything You Can Imagine Can be Found In a Sewer

Wastewater system operations and maintenance personnel are accustomed to finding just about anything in sanitary sewers and at the headworks of treatment facilities. Inadvertent cross-connections between storm sewers and sanitary sewers are common. Large rocks, basketballs, and even human corpses have been found.

- **Disinfection** is used to reduce pathogens to safe levels for discharge to the environment.

- **Sludge** (accumulated solids, also termed biosolids) from clarifiers is collected and treated for disposal, often using **digesters** (Figure 5.20) to reduce pathogens and decompose the organic material. Prior to disposal or land application, in order to reduce volume for reduced hauling expenses, water is removed from the sludge via dewatering or drying.

DISCHARGE

Effluent (treated water exiting the wastewater treatment facility) is typically discharged to rivers, streams (Figure 5.21), lakes, estuaries, or oceans. The level of treatment required is based on the desired water quality of the receiving water body with which the effluent mixes. In some cases, effluent is discharged to groundwater through large seepage cells on the land surface.

Stormwater

Stormwater is the runoff that occurs during rain events, or from melting snow and ice. Runoff occurs primarily from impervious surfaces such as paved roads, driveways, parking lots, sidewalks, and roofs. Runoff, to a lesser degree, is also generated by **pervious** surfaces such as lawns, parks, and agricultural land. It should be noted that when properly engineered, the stormwater runoff from developed sites (e.g., a residential subdivision) may be *less* than the pre-development runoff (e.g., an overgrazed pasture with compacted soils and minimal vegetative cover).

COLLECTION AND CONVEYANCE

Stormwater travels across lawns and roadways until it is channelized in gutters, ditches, channels, or streams. Roadway curb and gutter is an integral

Figure 5.22 A Stormwater Inlet Inset into the Curb and Gutter of a Street with Debris Accumulated from a Storm. If inlets are not periodically cleaned of debris they will clog, preventing stormwater removal from the street surface. This picture also graphically illustrates the term "crumbling infrastructure"; note the broken concrete over the inlet and the cracked pavement.

Source: P. Parker.

Figure 5.23 Spread of Stormwater into the Driving Lane During an Intense Rainstorm. Note that the car in the foreground is forced to cross into the oncoming traffic lane.

Source: M. Penn.

part of the stormwater collection system. Due to the crowning of the roadway, rainwater that lands on the road surface travels to the edge of the road where it is conveyed in gutters. The stormwater then travels in the gutters until it is intercepted by an **inlet** (Figure 5.22). Inlets are spaced such that stormwater does not accumulate and spread too far into the travel lanes and thereby create a safety hazard (Figure 5.23). They are also placed at intersections, before bridges, and in other locations to prevent crossflow of stormwater across pavement. Inlets come in several shapes and sizes depending on the flowrate and velocity of water to be intercepted.

 Photo of Eroded Outfall

Once intercepted by an inlet, stormwater is conveyed in conduits called storm drains or **storm sewers**. Most often, stormwater systems flow by gravity, although in some cases pumping stations are necessary. Unlike a sanitary sewer system that conveys all wastewater to a single WWTF, stormwater conveyance systems include many points of discharge (**outfalls**). Outfalls may discharge directly into surface waters (e.g., rivers and lakes) or onto land for conveyance by overland flow (in which case the outfall should be armored with a rip-rap apron to minimize erosion, Figure 5.24).

Although storm sewers are effective in the sense that they convey water away quickly from the point of generation, this benefit is also a drawback. Such rapid conveyance increases the peak flowrates in streams that receive the stormwater that may result in flooding. In addition, the channelization of stormwater into concrete, plastic, or steel pipes does not allow the stormwater to infiltrate into the groundwater as it may have done before development. Moreover, traveling in natural grassed waterways also removes some contaminants from the stormwater, which does not occur in pipes.

Figure 5.24 A Storm Sewer Outfall During Dry Conditions with an Inadequate Erosion Control Apron. Note the erosion (exposed soil) in foreground beyond the rip-rap.

Source: M. Penn.

CULVERTS

Culverts are used to transport channelized runoff under roads or structures (Figure 5.25). They may be the equivalent of short sections of storm sewer, or may be concrete structures called box culverts (Figure 5.26).

Figure 5.25 **Two Corrugated Metal Culverts.** These twin culverts (encased in concrete at entrance and exit ends) allow stormwater to pass under a street. Note the elliptical shape that allows more flow than a circular pipe of the same height.

Source: M. Penn.

Figure 5.26 **A Dual Box Culvert (Left) Adjacent to a Bridge That Spans a River.** The box culverts are located in the floodplain, and provide additional passage of water under the road during flood events.

Source: M. Penn.

STORAGE

A **hydrograph** is a graph of **flowrate** (the volume of flow passing a certain point in a given amount of time, expressed in gal/min or ft^3/s) as a function of time.

The excess runoff that often accompanies land development needs to be temporarily stored. The necessity for storage is illustrated by Figure 5.27, which shows three **hydrographs**. Each hydrograph demonstrates that in response to a storm, the flowrate increases and then tapers off after the end of the rain event. The "pre-development" curve represents the flow of stormwater before development (e.g., before converting agricultural land into a commercial area). As a result of development, the peak flowrate has increased, due to the increased runoff from the newly added impervious area ("post-development" curve). Note also that the time of occurrence of the post-development peak flowrate is sooner than the pre-development peak flowrate because the runoff travels faster across impervious surfaces as compared to pervious surfaces. As shown by the third hydrograph, if some storage, in the form of a detention pond, is used to capture the stormwater before releasing it to the stream, the peak flowrate can be reduced to match the peak flowrate that occurred before development.

Figure 5.27 Hydrographs Demonstrating the Effect of Stormwater Detention on Runoff.

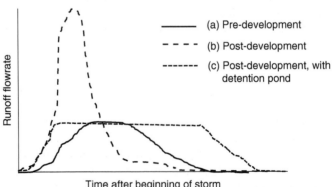

Runoff flowrate

——— (a) Pre-development

– – – (b) Post-development

------- (c) Post-development, with detention pond

Time after beginning of storm

Figure 5.28 **A Stormwater Detention Pond.** Such ponds store and slowly release stormwater runoff. The concrete structure (circled in photo center) is the outlet structure. Openings in the concrete are sized to release the pre-development flowrate.

Source: M. Penn.

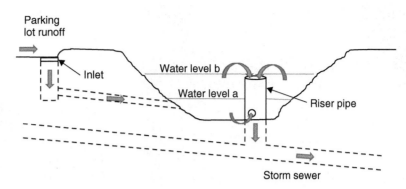

Figure 5.29 **Profile of a Detention Pond and Typical Outlet Structure (Riser Pipe) for a Parking Lot.** Stormwater is captured by an inlet and flows into the detention pond as shown. For small storms, the level of the water in the pond corresponds to "water level a," and the stormwater exits the pond to the storm sewer through the small circular opening at the bottom of the riser pipe. For larger storms, the level of stormwater in the pond corresponds to "water level b," and stormwater exits the pond through the top of the riser pipe as well as the small circular opening at the bottom. Note that without any detention pond, stormwater from the parking lot would go directly to the storm sewer at higher flowrates.

The area under a hydrograph represents the volume (e.g., cubic feet, gallons, cubic meters) of stormwater runoff. Examining Figure 5.27 reveals that, in addition to increasing peak flowrates, land development also increases the *volume* of runoff. Also note that the area under the "post-development" curve is the same as the area under the "post-development, with detention pond" curve. This implies that the detention pond does not lessen the *volume* of runoff leaving the site (other than minor losses through evaporation and perhaps some infiltration) as compared to not having a detention pond. This is consistent with the concept of a hydrologic budget discussed in Chapter 2, The Natural Environment. The detention pond is designed to decrease the *peak flowrate* to that of pre-development conditions. The area under the "pre-development" curve (i.e., the pre-development runoff volume) is *less* than that of *both* "post-development" curves because of significant pre-development infiltration through pervious surfaces.

Reducing the peak flow is beneficial because high peak flowrates may cause flooding, increase stream bank erosion, and lead to loss of habitat. Constructed detention ponds (Figure 5.28, Figure 5.29) are the most common method of stormwater storage. However, these require significant land area. In highly urbanized areas, underground tunnels or tanks (Figure 5.30) or "cube storage" (Figure 5.31) may be used to temporarily store stormwater.

TREATMENT

Stormwater runoff is a potential source of surface water pollution as it carries litter, debris, soil, nutrients, oil, grease, antifreeze, metals, etc. (Table 5.1). State and local codes are becoming increasingly stringent with regard to treating stormwater before it can be released to lakes, rivers, estuaries, or oceans. Still today, in many communities, *little or no treatment* of stormwater occurs; thus,

Figure 5.30 Underground Stormwater Storage.

Source: Courtesy of Contech Construction Products, Inc.

Figure 5.31 Plastic Cubes Providing Structural Support and Pore Space for Underground Storage of Stormwater.

Source: Courtesy of Invisible Structures, Inc.

Table 5.1

Urban Stormwater Pollutants, Sources, and Impacts

Pollutant	Potential Sources	Receiving water impacts
Nutrients (nitrogen, phosphorus)	Animal waste, lawn fertilizer, soil erosion, failing septic systems, sanitary sewer cross-connections	Nuisance plant and algae growth, reduced dissolved oxygen levels
Sediment	Construction sites, lawn runoff, runoff from impervious surfaces, stream bank erosion	Decreased water clarity, sedimentation (alteration of stream bottom habitat), impacts of other pollutants which are associated with sediments (e.g., nutrients)
Pathogens	Animal waste, failing septic systems, sanitary sewer cross-connections	Drinking water supply contamination, contaminated swimming beaches, contaminated shellfish
Organics	Animal waste, lawn clippings, leaves, failing septic systems, sanitary sewer cross-connections	Reduced dissolved oxygen levels, decreased water clarity
Hydrocarbons (oil and grease)	Industrial and commercial processes, vehicles, improper used oil disposal	Toxicity to aquatic organisms, decreased aesthetics
Metals (e.g., zinc, copper)	Industrial processes, vehicle brakes and tires, metal roofs	Toxicity to aquatic organisms
Thermal impacts	Runoff with elevated temperatures from contact with impervious surfaces (e.g., asphalt)	Adverse impacts to cold- or cool-water aquatic organisms (e.g., trout)

Source: Adapted from Connecticut Department of Environmental Protection, 2004.

a cigarette butt or hamburger wrapper tossed by a litterer out a car window or the used motor oil illegally dumped down a storm inlet will be carried by runoff *directly* to the receiving body of water.

There are many methods to treat stormwater and only a few are presented here. Wet detention ponds are constructed reservoirs that have a permanent

Figure 5.32 **A Wet Detention Pond to Store and Treat Stormwater Runoff From a Large Retail Center.** Note the fencing in the foreground, which is required for safety reasons.

Source: M. Penn.

Figure 5.33 **A Newly Constructed Bioretention Area.** There are three depressions (dark areas with plants) connected with small culverts. The light-colored area surround the depressions has been seeded and covered with biodegradable erosion matting.

Source: M. Penn.

pool of water at the base (Figure 5.32). They are designed so that stormwater will remain in the reservoir for sufficient time (e.g., 12 hours) to settle out particulate matter (to which many contaminants are attached). Oil/water separators are sometimes required at gas stations and maintenance yards to remove floating contaminants (gas, oil, etc.). Rain gardens (also termed bioretention areas) are landscaped depressions that store, treat, and infiltrate stormwater runoff (Figure 5.33).

Parks and Recreation (ASCE 2009 Grade: C–)

Parks and recreation are important aspects to the quality of life of many people in urban areas, and are weighted criteria in nearly every ranking of "most livable cities." Portland, Oregon, and the Twin Cities (Minneapolis and St. Paul) in Minnesota are prominent examples of U.S. cities incorporating large "green spaces" in the midst of residential, commercial, and industrial areas.

Ottawa, Ontario has a **greenbelt** (contiguous agricultural and forested land protected from development) that surrounds the city in a "belt" that is approximately 2 miles wide on average (Figure 5.34). In addition to aesthetic and recreational benefits, green spaces can potentially serve as stormwater storage and treatment areas.

Parks that are poorly planned, located, designed, or maintained often experience little public use, whereas some parks serve as major tourist attractions; for example, an estimated 25 million visit Central Park in Manhattan, New York each year.

Property values near parks are typically higher than those of comparable properties located further away. This fact illustrates a major challenge concerning parks—the balance between the value of the park land for potential development (and subsequent tax revenue) versus the public benefit of the land as a park.

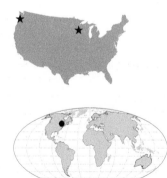

Figure 5.34 Ottawa's Greenbelt Surrounds the Urban Core of Development. Newer suburban development is occuring outside the greenbelt.

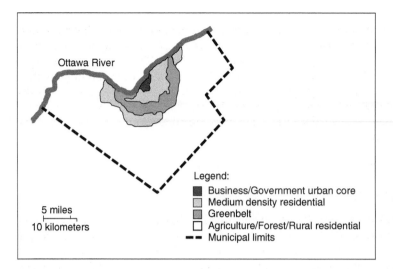

Ottawa River

5 miles
10 kilometers

Legend:
■ Business/Government urban core
▢ Medium density residential
■ Greenbelt
▢ Agriculture/Forest/Rural residential
- - - Municipal limits

Central Park

Central Park (Figure 5.35), now considered a model for successful park management, was not always so revered. In the mid-1970s, the Central Park Conservancy President stated, "Years of poor management and inadequate maintenance had turned a masterpiece of landscape architecture into a virtual dustbowl by day and a danger zone by night." The fact that parks have a positive effect on adjacent property values is dramatically illustrated in Figure 5.36; the real estate value of the land occupied by Central Park is estimated to be over $500,000,000,000.

Figure 5.35 Central Park in Manhattan, New York covers 843 acres.

Source: Copyright © Terraxplorer/iStockphoto.

Figure 5.36 Prime Real Estate Bordering Central Park.

Source: Copyright © travelif/iStockphoto.

Most communities have development ordinances to ensure that additional park space is added as new development occurs. However, if the requirement is based solely on land area (e.g., 3 acres of park per 100 acres of development) and does not include considerations such as location, access, and amenities, the park may go largely unused. Many communities have had great success by linking parks and green spaces with walking/biking trails. At a larger scale, many biking trails exist to link neighboring communities through regional trail systems.

Solid Waste (ASCE 2009 Grade: C+)

Municipal solid waste (MSW) is generated by homes, businesses, institutions and industries. MSW is a "waste" because it no longer has value to the generator. According to the U.S. Environmental Protection Agency (EPA), in 2008, Americans generated about 250 million tons of trash annually, equivalent to an average of 4.5 pounds per person per day. The composition of MSW in the United States is shown in Figure 5.37.

COLLECTION

MSW is typically collected weekly at curbside from residences. Larger generators (e.g., factories, home construction sites) use dumpsters with pick-up scheduled as needed.

TRANSFER

A few decades ago, it was not uncommon for nearly every community to have its own landfill. Today, there are far fewer landfills, and they are typically very large, serving a region rather than a single community. Since curbside collection trucks cannot cost-effectively transport waste long distances,

---

Construction and Demolition Waste

The EPA estimates that 160 million tons of construction and demolition (C&D) waste are generated annually in the United States. This waste often ends up in C&D landfills. In the last decade, significant efforts to reduce, reuse, and recycle C&D waste have been made.

U.S. EPA Recover Your Resources Brochure on C&D Recycling

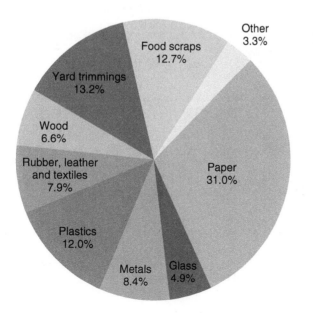

Figure 5.37 Composition of Municipal Solid Waste in the United States, as Mass Generated (Including Recycled Items).

Source: U.S. EPA, 2008.

Figure 5.38 Transfer of MSW from
Collection Trucks to Hauling Trucks.

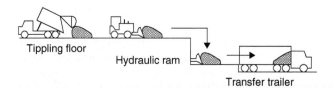

transfer stations (Figure 5.38) are used. Transfer stations accumulate waste, compact it to a higher density, and transfer it to larger trucks for long-distance hauling. Separation of recyclables often also occurs at transfer stations.

RECYCLING AND REUSE

The "three Rs" of waste reduction are reduce, reuse, and recycle. **Recycling** is the utilization of the raw material of a waste to produce a new product (e.g., a fleece jacket from plastic soda bottles). Recycling rates in the United States have increased dramatically due to federal, state, and local regulations, as well as due to increased environmental awareness by the public (Figure 5.39). However, the rate seems to have reached a plateau in recent years at 33 percent, and is much lower than the rates achieved in some other countries (e.g., Austria, Norway, Japan), which are approximately 60 percent.

One challenge with recycling is sorting the recyclables from the waste. Some communities have all waste go to material recovery facilities where recyclables and non-recyclables are separated by hand-sorting, magnets, and other mechanical means. Other communities separate as they pick up the waste (**curbside sorting**) into trucks with bins for each type of recyclable. While most of us think that recycling is a "good thing" and that we should recycle more, the realities of complex and variable markets for recycled materials make it challenging to achieve very high rates of recycling without significant government regulations and subsidies.

Reuse is largely an individual effort. Someone may choose to reuse a baby food jar to store nails on a garage shelf. Donating used furniture and clothing for resale at a local "thrift shop" is another example. "Trash picking" or "dumpster diving" is yet another example.

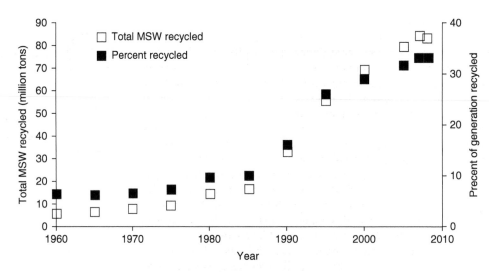

Figure 5.39 **Amount of MSW Recycled and Percent of Total Generation.**

Source: EPA, 2008.

Figure 5.40 **A Composting Site for Yard Waste, Wooden Construction Debris, and Food Scraps.** Finished compost is seen in the large pile at left.

Source: M. Penn.

To **reduce** waste production is the preferred alternative to managing any type of waste (MSW, industrial hazardous waste, etc.). If the waste is not generated, then there is no need to haul, recycle, reuse, or dispose of it.

COMPOSTING

Some people have added a fourth "R" to the three Rs (i.e., "rot"), which represents **composting**. Given that much of MSW is organic, the majority of MSW is potentially compostable (food scraps, wood, yard trimmings, paper; Figure 5.37). However, some items that are potentially compostable (e.g., office paper) have a high recycling value and thus may be separated. Also, plastics, although organic, are not compostable (excluding new biodegradable plastics). Most states have banned yard waste (grass clippings, leaves, garden waste) from landfills, which has dramatically increased the number of composting sites (Figure 5.40). If properly managed to ensure complete breakdown and screened to remove litter and debris, compost can be sold at a profit. Many communities allow residents to haul away finished compost for use in home flower and vegetable gardens.

Composting is the aerobic (using oxygen) microbial breakdown of readily decomposable organic wastes to a stable end product (**compost**) resembling soil.

LANDFILLS

Today's landfills have little in common with the landfills of old (more correctly called "dumps"), which often were abandoned gravel pits or rock quarries with no means of protecting groundwater from the wastes. Modern landfills are highly engineered systems. Engineers must go through a lengthy and challenging siting process to ensure that the potential for negative environmental impact is minimized and to ensure that the landfill will be a "good neighbor" to nearby property owners. Landfills are lined with plastic geomembranes (Figure 5.41) and clay to prevent contaminant-laden **leachate** (liquid generated from decomposition of waste and from rainfall) from entering the groundwater. Leachate is collected and pumped from landfills and treated at either on-site treatment facilities, or is trucked or sewered to municipal WWTFs. Landfills are divided into cells, typically with a fill life of a few years. Once filled, the cells are covered (again with geomembranes and clay) to prevent future precipitation from entering and creating more leachate.

Methane is generated during the decomposition of the waste. This methane can be collected and separated from other landfill gases for use as an energy source (Figure 5.42). In some cases, the methane is simply burned (flared) without energy recovery.

 Virtual Tour of a Landfill

 Freshkills Park (Closed Landfill) Website

Figure 5.41 A Geomembrane Placed Above a Clay Layer at the Base of a Landfill. The kneeling worker is cleaning one sheet of geomembrane (HDPE plastic) in the foreground, preparing it for "welding" to seal the overlapped seam (background).

Source: Courtesy of J. Puls.

Sorting and Disposing of the World Trade Center Debris

Nearly 2 million tons of debris from the 9/11 terrorist attack of the World Trade Centers was sorted in 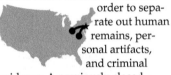 order to separate out human remains, personal artifacts, and criminal evidence. A previously closed landfill was re-opened for staging, sorting, and ultimate disposal. The final closure of the landfill after these operations marked the end of solid waste disposal in New York City. More than 3 million tons of residential solid waste is exported from the city annually to landfills as far away as Virginia and Ohio.

 Sorting Process Example

Figure 5.42 Landfill Gas Extraction Well.
Source: EPA Landfill Methane Outreach Program.

INCINERATORS

Incinerators were once quite common at sites of waste generation (e.g., factories, dormitories, and prisons). However, regulations that require treatment of air emissions from incinerators have virtually put an end to the practice of on-site incineration (except for some very large generators such as hospitals). In densely populated areas with high land values (e.g., the Northeast), large-scale incineration of MSW with energy recovery (waste-to-energy) and air pollution control is common. Incineration is also very common in Japan and Europe; in fact, landfilling is now banned in Norway. Incineration reduces the need for large landfills and is more efficient at energy recovery as compared to recovering methane from landfills. Incineration does not *eliminate* the need for landfills, however; the resulting ash, approximately 20 percent of the original waste mass, is typically landfilled.

Concerns over air pollution remain, and public resistance to new waste-to-energy facilities is often strong. You should realize that engineers *can* design air pollution control devices to virtually eliminate all pollutants; however, as the removal efficiency increases, so does the cost. There is a saying in environmental engineering that "it only takes 20 percent of the cost to remove 80 percent of pollution, and 80 percent of the cost to remove the last 20 percent of pollution."

Hazardous Waste (ASCE 2009 Grade: D)

Federal and state regulations have strict definitions as to what is classified as hazardous waste, nuclear waste, or biohazard waste (all being different than MSW). For purposes of this textbook, consider hazardous waste to be any waste containing high levels of toxic, explosive, or flammable chemicals. Prior to the 1970s, these wastes were largely unregulated and were haphazardly dumped and discharged into the environment, in many cases creating significant environmental and public health risks. The generation of hazardous waste has been dramatically reduced due to the high costs required for management and disposal, as well as increased environmental awareness. If not reused or recycled, hazardous wastes must be disposed of either by incineration or in specially designed hazardous waste landfills.

SUPERFUND

In the previous section discussing solid waste, we described the unlined landfills (dumps) of the past. These abandoned gravel pits and rock quarries had no liners and accepted nearly all waste (including waste that today would be classified as hazardous). As a result, extensive groundwater contamination has occurred. It was also common practice prior to the 1970s for industries to (at the time, legally) dump industrial liquid waste into unlined lagoons and ditches. In 1980, the Superfund was created to clean up the most dangerous waste sites. Approximately 1,000 sites have been cleaned up to date, but the list of sites awaiting cleanup includes many more. The EPA estimates that 1 in 4 Americans lives within 4 miles of a Superfund site. The cost per site for cleanup often is millions of dollars. Additional costs are incurred from lawsuits as companies attempt to free themselves from, or lessen their financial responsibility for site cleanup.

Figure 5.43 Excavation of Contaminated Soil at a Leaky Gas Station Site.

Source: M. Penn.

The Superfund selection process targets the "worst of the worst." There are hundreds of thousands of contaminated sites across the United States that are not contaminated "enough" to be prioritized for Superfund. A large percentage of these sites are gas stations, with leaking gasoline and diesel fuel tanks, and dry cleaners, with improperly handled perchloroethylene. Most states have programs to fund and administer the cleanup of these sites. In some cases, the cleanup may be as simple as the removal of the leaky tanks and the excavation of the soil around them (Figure 5.43). In other cases, when the groundwater has been contaminated, lengthy and costly groundwater treatment activities are also required.

BROWNFIELDS

Brownfields are sites that are vacant, abandoned, or underutilized because of real or perceived contamination. In many cases, sites are literally abandoned by the owner and left to the city, or lie dormant because potential buyers are reluctant to purchase the property due to the extensive cost of cleanup if the site is contaminated. These properties do not generate significant, if any, tax revenues for the city, yet they are often located in central areas that, if it were not for the real or perceived contamination, would be "prime" real estate. Thus, an incentive exists for cities to have these areas developed. An additional incentive, from the perspective of infrastructure, is that brownfields, by virtue of their centralized location, are served by water, wastewater, and other utilities; thus, developing these sites is preferable as compared to developing at the fringes of the cities, in that the latter will require the expansion of utilities.

Video of Brownfield Success Story (Gretna, Louisiana)

Outro

After reading the last three chapters discussing infrastructure sectors and components, we hope that you are excited by the challenges and opportunities that careers in civil and environmental engineering offer. Now that the major sectors and components of infrastructure have been introduced, the remainder of this book will focus on the technical and non-technical aspects of infrastructure components and systems (of which glimpses were provided in this chapter), preparing you for future courses.

chapter Five   Homework Problems

Short Answer

5.1   Why is electricity conveyed at high voltages in transmission lines?

5.2   Does saltwater intrusion decrease the volume of water in an aquifer? Explain.

5.3   How might the level of water in an elevated water storage tank vary over a typical day? Explain.

5.4   How are storm sewer outfalls "armored" to prevent soil erosion?

5.5   Under what situations does wastewater need to be pumped in a wastewater collection system?

5.6   Consider Figure 5.27. Explain why the peaks of the "pre-development" and "post-development, with detention pond" curves are equal to each other.

Discussion/Open Ended

5.7   What do you think are some advantages and disadvantages of curbside sorting of MSW?

5.8   Some have said that "dilution is the solution to pollution." What is the logic behind this statement? How does it apply to combined sewers? Do you believe that dilution is sound practice for "treating" wastewater?

Research

Note: For Homework Problems 5.9, 5.11, 5.13, 5.14, and 5.15, the *ASCE Report Card for America's Infrastructure* is available at www.wiley.com/college/penn.

5.9   According to the energy chapter of the latest *ASCE Report Card for America's Infrastructure*, how resilient is the energy infrastructure in the United States?

5.10   How much energy does the United States export? Use the *CIA World Factbook* (available at www.wiley.com/college/penn) to compare the energy that the United States exports to five other countries of your choice. Explain your rationale for choosing those five additional countries.

5.11   Read the drinking water chapter of the latest *ASCE Report Card for America's Infrastructure*. Based on this chapter, do you think that the near-failing grade is deserved? Explain.

5.12   Tabulate the location and dates on 10 hydrants in your city. Summarize your findings with a map.

5.13   According to the hazardous waste chapter of the latest *ASCE Report Card for America's Infrastructure*, what is the link between brownfield redevelopment and tax revenues?

5.14   Read the wastewater chapter of the latest *ASCE Report Card for America's Infrastructure*. What are the largest challenges facing the wastewater sector?

5.15   According to the *ASCE Report Card for America's Infrastructure*, solid waste received one of the highest grades. Read the chapter on this sector and provide some reasons why such a relatively high grade was earned.

5.16   Using U.S. EPA data (available at www.wiley.com/college/penn), graph the number of existing MSW landfills over the past two decades. What are some reasons for the decline in the number of landfills?

Key Terms

- activated sludge
- clarifiers
- collectors
- combined sewer overflow
- combined sewers
- composting
- curbside sorting
- digesters
- disinfection
- diurnal variation
- effluent
- flowrate
- force mains
- greenbelt
- headworks
- hydrographs
- infiltration
- inflow
- inlet
- interceptors
- laterals
- leachate
- lift stations
- mains
- non-renewable
- outfalls
- pathogens
- per capita water demand
- pervious
- potable
- primary treatment
- recycling
- reduce
- renewable
- reuse
- saltwater intrusion
- sanitary sewer overflows
- sanitary sewers
- secondary treatment
- sludge
- storm sewers
- tertiary treatment
- transfer stations

References

Connecticut Department of Environmental Protection. 2004. *Connecticut Stormwater Manual*. http://www.ct.gov/dep/cwp/view.asp?a=2721&q=325704, accessed August 8, 2011.

Environmental Protection Agency (EPA). 2008. *Municipal Solid Waste Generation, Recycling, and Disposal in the United States: Facts and Figures for 2008*. EPA-530-F-009-021.

Environmental Protection Agency (EPA). *Landfill Methane Outreach Program*. http://www.epa.gov/lmop/basic-info/lfg.html#02, accessed August 8, 2011.

Michelcic, J. and J. Zimmerman. 2010. *Environmental Engineering: Fundamentals, Sustainability, Design*. Hoboken, NJ: John Wiley & Sons, Inc.

US Department of Transportation. 2010. *National Transportation Statistics*. http://www.bts.gov/publications/national_transportation_statistics, accessed August 8, 2011.

US Energy Information Administration. *Energy Explained*. http://tonto.eia.doe.gov/energyexplained/index.cfm?page=electricity_home#tab2, accessed August 8, 2011.

US Energy Information Administration, Gas Transportation Information System. *U.S. Natural Gas Pipeline Network, 2009*. http://www.eia.doe.gov/pub/oil_gas/natural_gas/analysis_publications/ngpipeline/ngpipelines_map.html, accessed August 8, 2011.

chapter Six Construction Sites

Learning Objectives

After reading this chapter, you should be able to:

1. Describe the security and staging needs of a construction site.
2. Identify and describe the functions of various types of construction equipment.
3. Describe various erosion control techniques.

Introduction

Many engineers not only design infrastructure components and systems, but also supervise construction of their designs. This chapter will describe some aspects of the construction of the infrastructure components presented in Chapters 3, 4, and 5. Successful and efficient construction requires significant planning efforts, which will be discussed in later chapters. Construction engineers are responsible for construction site layouts, equipment specification, scheduling, and erosion control.

Security

Securing a construction site is important for public safety and to protect contractor equipment from theft and vandalism. Fencing with locked gates is typically needed at a minimum. At remote sites without fencing, valuable tools are often hoisted and stored suspended from cranes in tool boxes. Surveillance cameras are also used. Checkpoints for vehicles entering the site are common.

Staging Areas

In general, the size of the **staging area** is proportional to the size of the project (Figure 6.1). In high-density urban areas, where available space is limited, off-site staging may be required. Most projects require submittal and approval of construction staging plans prior to starting work. These plans typically include details not only of the staging area, but also of access routes for deliveries and worker parking, security fencing, dumpster locations, traffic control, covered pedestrian walkways (Figure 6.2), etc.

Staging areas are designated portions of construction sites for fabrication and for receiving and storing materials and equipment needed for project construction.

Figure 6.1 A Staging Area at a Building Construction Site.

Source: M. Penn.

Figure 6.2 A Covered Walkway for Safety of Pedestrians During Construction Activities.

Source: M. Penn.

Equipment

A wide variety of equipment is used during construction depending on the site conditions and the nature of the project. Most types of equipment are available in a wide range of sizes. Construction engineers determine which size is appropriate based on capacity, access, and other factors. As a basic introduction, only a few common types of equipment are included in this section. Many undergraduate civil engineering programs have a course to provide more detailed information on construction equipment.

TRAILERS AND STORAGE CONTAINERS

Trailers are used at construction sites for equipment and material storage as well as for on-site offices (Figure 6.1 and Figure 6.3). Portable metal containers are also used (seen in foreground in Figure 6.1).

Figure 6.3 Trailers for Storage (Background) and Offices (Foreground, with Windows).

Source: M. Penn.

Figure 6.4 Cranes for High-Rise Building Construction (Background) in Oslo, Norway. The landmark Oslo Opera House is in the foreground.

Source: M. Penn.

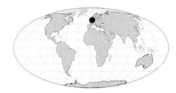

Photos of Construction Equipment and Activities

CRANES

It has been said of city skylines that "where there are cranes, there is money." Indeed large cranes are used for the construction of high-rise buildings (Figure 6.4). Many types of cranes are used; the type used depends on the height of the lift, weight of the objects, and access needs of a project (Figure 6.5, Figure 6.6, and Figure 6.7).

EXCAVATORS

Excavators are used not only for general excavation, but also for trenching of utilities and placement of materials (Figure 6.8). Many attachments, such as grapples (Figure 6.9) and roller compactors, can replace the excavating buckets, making excavators one of the most very versatile types of heavy equipment. For smaller excavations, **backhoes** are commonly used (Figure 6.10).

Figure 6.5 A Lattice Boom Crane Placing a Water Tower Tank on Its Pedestal.

Source: Courtesy of Laramie Equipment Co.

Figure 6.6 A Telescopic Hydraulic Crane Lifting a Truss into Position.

Source: M. Penn.

Figure 6.7 Tower Cranes Used for Skyscraper Construction.

Source: Copyright © Christopher Arndt/iStockphoto.

Figure 6.8 An Excavator Placing Crushed Stone at the Base of an Excavation.

Source: M. Penn.

Figure 6.9 Excavators with Grapple Attachments Clear Debris at Ground Zero After the 9/11 Terrorist Attacks.

Source: Federal Emergency Management Agency/M. Rieger.

Figure 6.10 A Backhoe Utilized for a Small Excavation.

Source: M. Penn.

BULLDOZERS

Bulldozers (or dozers) are commonly used for rough grading of sites, including the placement of fill material (Figure 6.11).

 Remember that full-sized versions of textbook photos can be viewed in color at www.wiley.com/college/penn.

LOADERS

Loaders (Figure 6.12 and Figure 6.13) are used for a variety of purposes such as loading excavated material into trucks, placing of material, lifting equipment, etc.

Figure 6.11 Two Bulldozers Leveling Soil at a Fill Site.

Source: M. Penn.

Figure 6.12 A Loader Placing Crushed Stone.

Source: M. Penn.

Figure 6.13 Front-End Loaders Placing Debris for Sorting at the World Trade Center Site After the 9/11 Terrorist Attacks.

Source: Federal Emergency Management Agency/L. Lerner.

Figure 6.14 Two Skidsteers Performing Different Functions—Street Sweeping (Foreground) and Placing Topsoil (Background).

Source: M. Penn.

SKIDSTEERS

Skidsteers, also termed skidloaders (Figure 6.14), are so named because of the wheels or tracks on one side are fixed during a turning maneuver. Turning is accomplished by braking one side of the vehicle while driving the other. Often generically referred to as "Bobcats" (a popular brand and one of the first four-wheeled skidsteers produced), skidsteers are built by many construction equipment manufacturers. A variety of implements have been designed to attach to the skidsteer (sweepers, jack hammers, etc.).

DUMP TRUCKS

Dump trucks are routinely used to remove excavated material or bring fill material to a site, as well as to haul away construction debris. There is a wide variety of truck designs to match site conditions and project needs (Figure 6.15 and Figure 6.16).

Figure 6.15 An Articulated Dump Truck for Use in Rough Terrain, with a Capacity of 40 Tons or 30 Cubic Yards.

Source: M. Penn.

Figure 6.16 A Standard Dump Truck Placing Fill.

Source: M. Penn.

EARTHMOVERS

Scrapers (Figure 6.17) are used for rough grading and leveling sites. Unlike dozers that push soil, scrapers *remove* surface material into a hopper. Scrapers are analogous to planers used for woodworking. Graders (Figure 6.18) are used for final (smooth) grading.

Figure 6.17 A Tow Scraper Pulled by a Tractor (Left), and a Self-Propelled Motor Scraper (Right).

Source: Courtesy of R. Schmitt.

Figure 6.18 A Grader Repairing a Road Damaged by Hurricane Rita.

Source: Federal Emergency Management Agency/M. Nauman.

LIFTS

Whereas cranes are used for high-elevation lifting, lifts (Figure 6.19) can be used for lower elevation lifting of material or as short duration working platforms. Scaffolding is used for elevated working platforms for longer duration (Figure 6.20).

Figure 6.19 A Telescopic Lift Supporting Workers.

Source: M. Penn.

Figure 6.20 Three-Tier Scaffolding for Masonry Work.

Source: M. Penn.

BLASTING EQUIPMENT

Blasting with explosives is used for demolition as well as rock excavation (e.g., a road cut, whereby holes are drilled into the rock and filled with explosives; Figure 6.21, Figure 6.22, and Figure 6.23).

Figure 6.21 Two Drill Rigs Boring into Bedrock (Shallow Topsoil has Been Removed) for Blasting.   The bent over worker (far left) is placing explosives into a bored hole.

Source: M. Penn.

Figure 6.22 The Site from Figure 6.21 After Blasting.
Note the rock has been fragmented and has "heaved." The rock will be excavated to lower the existing grade for a new roadway.

Source: M. Penn.

PILEDRIVERS

Piledrivers (Figure 6.24) are used to drive pilings for foundations either by raising and lowering a weight, hydraulic ram, or vibration.

Figure 6.23 Final Road Grade Preparation After Completed Blasting and Excavation at the Site in Figure 6.21 and Figure 6.22.

Source: M. Penn.

Figure 6.24 The Piledriver (A), Attached to a Crane, Drives Steel Piles into the Subsurface. The piles (B) are lifted vertically into position and driven to the design depth, in this case approximately 50 feet. Once in the ground, the piles (C) are then cut at the surface (D) for integration into the building foundation.

Source: M. Penn.

DEWATERING

Dewatering is not only required for underwater foundation construction, it is also necessary at sites where excavation depths exceed the depth to the local water table. In these situations, a series of shallow vertical groundwater wells are placed around the excavation, and pumped to temporarily lower the water table (Figure 6.25 and Figure 6.26). Failure to anticipate the need for dewatering can cause project delays and added construction expense. After large rain events, dewatering of excavations via portable pumps may be required. Proper grading of a site to minimize runoff into an excavation can minimize this need.

Figure 6.25 Preparing to Dewater an Excavation. Note that the excavation depth is below the water table. Hoses connected to the header pipe are connected to groundwater wells that will be pumped to lower the water table during construction.

Source: Courtesy of Peterson Geotechnical Construction, LLC.

Figure 6.26 Cross-Sections of an Excavation, (a) Before Dewatering and (b) During Dewatering. After construction work is complete, pumping will be halted and the water table will return to the original elevation.

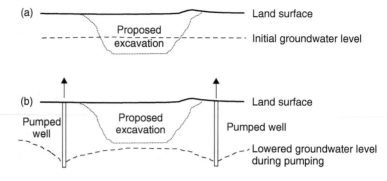

(a) Land surface
Proposed excavation
Initial groundwater level

(b) Land surface
Pumped well
Proposed excavation
Pumped well
Lowered groundwater level during pumping

READY-MIX CONCRETE TRUCKS

Perhaps the easiest way to be labeled as a "greenhorn" at a construction site is to refer to a ready-mix concrete truck (Figure 6.27) as a "cement mixer." **Cement** is the bonding agent in **concrete**; concrete also includes aggregate (sand and gravel) and water. Ready-mix trucks deliver and place well-mixed concrete at construction sites.

Figure 6.27 A Ready-Mix Concrete Truck Placing Concrete with a Chute for a Sidewalk.

Source: M. Penn.

CONCRETE PUMPING

Bridge Deck Pour Video

When the concrete placement location cannot be reached by chute of a ready-mix truck because of access restrictions or elevation, a concrete-pumping truck may be used (Figure 6.28 and Figure 6.29).

CONCRETE FORMS

Forms are moulds that are placed so that concrete can be poured into place for curing (Figure 6.30). After curing of the concrete is complete, temporary forms are removed. Formwork may be as simple as supported sheets of plywood, or may involve prefabricated modular units.

Figure 6.28 A Concrete Pumping Truck Places Concrete at the End of a Moveable Boom (the "M" Shape in the Photo). A ready-mix truck (right) supplies concrete to the pump.

Source: P. Parker.

Figure 6.29 Concrete Being Placed Via Pumping. Note the reinforcing steel matrix through which the concrete is poured.

Source: Courtesy of Caltrans, District 4 Photography/J. Huseby.

Figure 6.30 Removal of Modular Forms from a Cast-in-Place Reinforced Concrete Wall.

Source: M. Penn.

FALSEWORK

Falsework is a temporary structure to support a structure during construction, until the structure is self-supporting. Concrete forms and supports are, in essence, simple types of falsework. For bridges and elevated roadways, more elaborate falsework may be required (see steel towers supporting elevated concrete roadway in the background of Figure 6.29), often supporting formwork for the concrete (Figure 6.31 and Figure 6.32).

Figure 6.31 Falsework for an Elevated Roadway. The steel beams in the foreground are used to support the formwork for the concrete roadway deck. After the concrete cures, it will be self-supporting and the falsework will be removed. Note the permanent concrete structural support columns seen protruding through the falsework. These columns will be integrated into the deck structure.

Source: Courtesy of Caltrans, District 4 Photography/J. Huseby.

Figure 6.32 Falsework Seen from Below. The finished, self-supporting section of the elevated roadway is seen at right. Two permanent concrete supporting columns are seen within the temporary steel columns of the falsework.

Source: Courtesy of Caltrans, District 4 Photography/J. Huseby.

Utilities

Active construction often requires connecting to, or safely working around, existing utilities (e.g., natural gas, electric, fiber optic). "Call before you dig" is the mantra of utility providers. Location of utilities via spot painting and flagging; Figure 6.33) is critical before excavation commences. In fact, the location of utilities may affect design considerations (e.g., site layout to avoid conflicts with utilities). Do not rely on completed plans of previous projects or "hearsay" as to where utilities are located, as both of these sources are often incorrect. Even when utilities are "located," surprises do occur; we have had several phone outages on our campus due to lines severed during construction of new buildings. The extension of utilities into new developments is an

Figure 6.33 Flags and Paint Identifying Location of Utilities.

Source: M. Penn.

important phase of construction scheduling. Routing is typically parallel to the street in the right-of-way (Figure 6.34). Directional drilling is often used for utility placement under streets at intersections (Figure 6.35).

Figure 6.34 **Natural Gas Line Placement in a Trench.**

Source: M. Penn.

Figure 6.35 **Directional Drilling of Utilities Under a Street.** The worker (bent over) is using a sensor to locate the drilling head to relay information to the driller (foreground) to control the line of drilling. The drill will stop at the excavation on the opposite side of the street (see circled soil pile in background), where trenching will continue.

Source: M. Penn.

Erosion Control

Stormwater must be managed during active construction (*post*-construction stormwater management was discussed in Chapter 5, Environmental Infrastructure), and is often termed **erosion control**. Effective erosion control is imperative during construction because the existing vegetation and topsoil is generally removed at sites, leaving exposed subsoil that is prone to erosion during rainstorms. Without control measures, soil loss from construction sites can be *orders of magnitude* (e.g., 10x or 100x) greater than post-construction losses at the same site after new vegetation is established.

The most common erosion control method is the use of **silt fences**. When properly installed, it is a very effective means of preventing soil loss from a site. It does not prevent soil erosion, but merely keeps eroded soil from migrating off site during light to moderate storms. Very intense (low probability) storms will create high runoff rates, in which case the silt fence is often "uprooted" or "overtopped" and therefore ineffective (Figure 6.36). Unfortunately, without proper site inspection and enforcement to ensure correct installation and maintenance, it is not uncommon to see marginally effective or completely ineffective silt fences at construction sites (Figure 6.37). In some cases, contractors may be aware of damaged or inappropriate silt fences but choose to ignore them.

A second erosion control method is a **sedimentation basin (or sediment trap;** Figure 6.38). Once filled with stormwater, the particles will settle to the bottom and accumulate in the basin. Basins are cleaned out with backhoes when they fill with sediment.

Silt fence acts as a filter, allowing water to pass through it while capturing soil particles.

Sedimentation basins are depressions or dammed areas for the accumulation of storm runoff and the eroded soil carried with it.

Figure 6.36 Soil-Laden Runoff Flowing Over Silt Fencing that was Overtopped During a Storm.

Source: M. Penn.

Figure 6.37 Improper Silt Fence Installation. The bottom of the silt fence should be trenched into the soil and backfilled. Note that stormwater can simply pass underneath this silt fence (which overlaps the sidewalk, see arrow) and no particulate removal will occur.

Source: M. Penn.

After rains, construction sites can become a muddy mess. Trucks and other equipment leaving the site carry mud with them onto the roads (called tracking; Figure 6.39). **Tracking pads** are areas of placed rock, typically at the

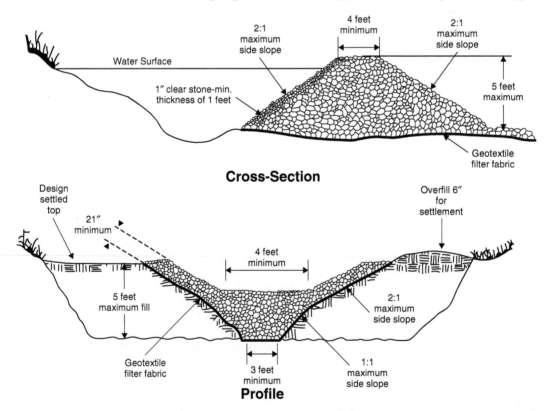

Cross-Section

Profile

Figure 6.38 Cross-Section and Profile of a Sediment Trap. Geotextile filter fabric is placed under the stone dam to prevent erosion of the underlying soil as water seeps through the stone. Note that for cross-section view, water is moving from left to right, while for the profile view, water is moving "out" of the page.

Source: Dane County Land and Water Resources Department, 2007.

Figure 6.39 Excessive Tracking of Material from a Construction Site onto Adjacent Road.

Source: M. Penn.

Figure 6.40 A Tracking Pad at the Exit of a Construction Site. Note the clumps of mud that lie on the pad and were removed from the tires of exiting vehicles. Also note minimal tracking onto the street.

Source: M. Penn.

site exit, to remove mud from vehicle tires (Figure 6.40). The rock needs to be replaced periodically to remain effective.

Another erosion control method is the use of mulch or compost. **Mulch** is material placed on soil to prevent erosion. Without mulch, the raindrops easily dislodge soil particles from the soil surface making them free to erode (Figure 6.41). Mulching is typically applied after construction is complete and final topsoil has been placed and seeded (Figure 6.42). However, mulching can be used during construction as well.

Mulches include hay/straw and woodchips that dissipate the energy of falling raindrops.

Figure 6.41 Erosion Channels Causing Soil Loss on an Unmulched Slope. Also note the unmaintained silt fencing at the base of the stairs.

Source: M. Penn.

Figure 6.42 Straw Mulch Being Placed by a Bale Chopper/Mulch Spreader at a Road Construction Site Nearing Completion.

Source: M. Penn.

Site phasing is a method of construction whereby only areas of active construction are disturbed on an as-needed basis, rather than removing all topsoil from the entire site initially.

There are many other approaches to minimize construction site erosion, including non-physical practices (those that do not require construction or installation as needed with the above examples). For example, at a large site, **site phasing** may be used.

Outro

The success of the designs of engineers depends upon proper construction. Some engineers oversee the construction of their designs, whereas others rely on fellow engineers and construction managers. The cost and timeline of construction activities depends upon proper planning and construction equipment selection. Proper erosion control techniques can minimize the environmental impacts of active construction. Most employers of civil and environmental engineers look favorably upon construction experience because "having fought in the trenches" results in engineers that consider construction aspects of their designs.

chapter Six Homework Problems

Calculation

6.1 Based on the cross-section drawing in Figure 6.38, calculate the minimum base width of a 4-foot-tall stone dam for a sediment trap.

6.2 Consider the cross-section drawing in Figure 6.38. Using a 3:1 (horizontal-to-vertical) side slope and a 6-foot top width, calculate the base width of a 4-foot-tall stone dam for a sediment trap.

6.3 When soil is excavated and placed in a dump truck, the volume increases because the soil expands from its compacted natural state, typically by a factor of 1.25. How many dump truck loads (20-cubic-yard capacity) are needed to haul the soil from an excavation that is 30 feet wide by 50 feet long by 10 feet deep? Note: round your number of loads up to the next whole number.

6.4 Estimate (roughly) the capacity, or volume, of the bucket on the loader in Figure 6.12. Express your answers in cubic yards. Explain your assumptions.

Short Answer

6.5 Why is it important to "call before you dig?"

6.6 Give two examples of projects that would require a covered pedestrian walkway during construction. Give two examples of projects that would not require such a walkway.

6.7 Why might security check-ins be more likely at large construction sites than at small construction sites?

6.8 What is the primary advantage of directional drilling for utility placement?

6.9 Summarize methods to minimize soil loss from a construction site.

Discussion/Open Ended

6.10 Why do many cities have requirements for contractors to provide a detailed construction staging plan?

6.11 Summarize the requirements of the City of Sarasota, Florida, regarding construction staging plans (available at www.wiley.com/college/penn).

6.12 Why is scaffolding used to support construction workers, rather than the lift, in Figure 6.20?

6.13 As a summer intern for a city, one of your job responsibilities might be inspecting erosion control at construction sites. Describe your possible activities for this task.

6.14 A large dump truck has a greater hauling capacity and thus requires less hauling loads for a given excavation volume. What site conditions might cause a construction engineer to choose a smaller dump truck (requiring more hauls)?

Research

6.15 Use a program such as Microsoft PowerPoint to create a photo gallery that includes photos of each of the equipment types listed in this chapter. Make sure to provide references for photos obtained online.

6.16 Investigate and summarize the various types of attachments commonly used on excavators.

6.17 Investigate and summarize the various types of attachments commonly used on skidsteers.

Key Terms

- backhoes
- blasting equipment
- cement
- concrete
- concrete forms
- cranes
- dewatering
- dozers

- dump trucks
- earthmovers
- erosion control
- excavators
- falsework
- lifts
- loaders
- mulch

- piledrivers
- sedimentation basin
- silt fences
- site phasing
- skidsteers
- staging area
- tracking pads
- utilities

References

Dane County (WI) Office of Lakes and Watersheds. 2007. *Dane County Erosion Control and Stormwater Management Manual*, 2nd Edition.

http://danewaters.com/business/stormwater.aspx, accessed August 8, 2011.

chapter Seven Infrastructure Systems

Learning Objectives

After reading this chapter, you should be able to:

1. Explain how infrastructure sectors and components act as systems.
2. Describe infrastructure system interactions.
3. Explain the need for a systems-wide view of engineering.

Introduction

In Chapters 3 through 5, we presented several "sectors" into which the various components of civil infrastructure could be categorized. These sectors were: transportation, energy, water supply, wastewater, parks and recreation, stormwater, structures, solid waste, and hazardous waste. Components included landfills, bridges, and roads to name a few.

In this chapter, we will explore how these components and sectors do not "stand alone," but rather are integrated together into systems. Being mindful of how the components of infrastructure interact with each other is essential to designing a safe and effective infrastructure. Engineers have often been accused of not looking at the "big picture" or not having a holistic viewpoint. Our hope is that this book will help you view your future work as a civil or environmental engineer in the context of a much larger system (the infrastructure), a system that directly affects the quality of people's lives.

Introductory Case Study: New Orleans Hurricane Protection System

New Orleans, like many coastal cities, is protected from hurricanes by a Hurricane Protection System (HPS). When Hurricane Katrina devastated New Orleans, it represented a massive failure of the HPS. As a result, over 1,000 people were killed, a regional economy was temporarily devastated, thousands of homes and businesses were destroyed, and hundreds of thousands of people were relocated. Had the HPS not failed, casualties would still have occurred and the regional economy would still have suffered; however, the magnitude of these impacts would have been much less.

The HPS for New Orleans consists of levees, floodwalls, and pumping systems. A series of levees and floodwalls are built around much of New Orleans, necessary because most of the city lies several feet below sea level. New Orleans has been described as a "bowl" surrounded by water (Lake Pontchartrain to the North, Mississippi River to the South, and the Gulf of Mexico to the East; Figure 7.1).

Some of the levees are earthen, and the height of some of these are raised with concrete walls. Given the bowl-shaped topography of the system, any stormwater runoff generated within the city and any floodwaters that enter the city must be pumped out. This "interior drainage" consists of some very large pumping stations and their associated piping and canals. Some of these pumping stations are nearly 100 years old. For this reason the system must protect not only against hurricanes, but also large rainstorms and is titled the Hurricane and Storm Damage Risk Reduction System (HSDRRS).

The levee- and floodwall-lined canals convey water from north to south, as depicted in Figure 7.2. The levees and floodwalls that line the canals are connected to east-west levees along the shore of Lake Pontchartrain. The water level of Lake Pontchartrain increases due to hurricane **storm surge**, and consequently the water elevation in the canals also rises, which limits the ability to pump water out of the city (Figure 7.3).

Post-Katrina improvements to the system include the construction of floodgates and pumping stations at the canal outlets (Figure 7.4). These floodgates will be closed during storm surges and the pumping systems will bypass canal water around the gates out to the lake.

 U.S. Army Corps of Engineers Video Overview of the New Orleans HSDRRS

 Summary Map of New Orleans HSDRRS

A **storm surge** refers to the increase in water levels as high winds from a hurricane literally "pile up" water against the shore.

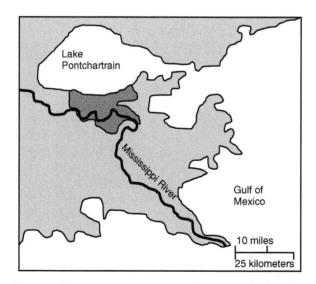

Figure 7.1 **Metropolitan New Orleans Area (Shaded) and Vicinity.**

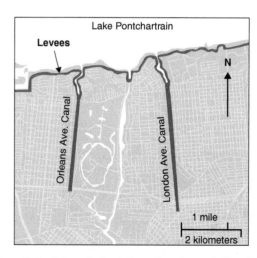

Figure 7.2 **Two of the Many Levee- and Floodwall-Lined Drainage Canals and the Lakefront Levee in New Orleans, Louisiana.**

Source: Redrawn from the U.S. Army Corps of Engineers.

Figure 7.3 A Diagram of a Pre-Katrina Canal Outfall. Note that the water level in the canal is equal to that of the lake.

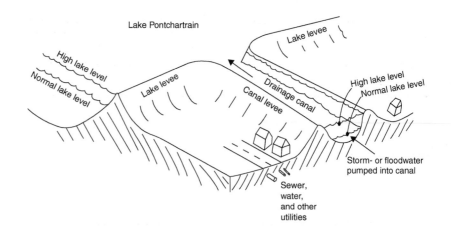

Lake Pontchartrain

High lake level

Normal lake level

Lake levee

Lake levee

Drainage canal

Canal levee

High lake level

Normal lake level

Storm- or floodwater pumped into canal

Sewer, water, and other utilities

U.S. Army Corps of Engineers Video Gallery of SDRRS Improvements

Figure 7.4   Floodgate and Pumping Station Construction at the 17th Street Canal Outfall. This aerial view is looking south (from Lake Pontchartrain toward the city). The pumping capacity of this station is approximately 9,000 cubic feet of water per second; at this rate, an Olympic-sized swimming pool would be filled in less than 10 seconds! The pumps are located adjacent to the floodgate on the inland side. The large pipes bypass the floodgate and transport water to the Lake Pontchartrain side of the floodgate when it is closed. The floodgate prevents a storm surge from entering the canal, and provides additional canal capacity for stormwater pumped from inland pumping stations. This is one of several *temporary* floodgate systems installed after Hurricane Katrina, which are to be replaced with permanent systems in 2014.

Source: U.S. Army Corps of Engineers/Digital Visual Library.

Engineers use **factors of safety** to take into account uncertainty in the design. A factor of safety of 2, for example, will result in a structure that is, depending on the application, twice as large, twice as strong, twice as rigid, etc., as compared to the size, strength, or rigidity obtained from initial design calculations.

A **datum** is a benchmark to which critical points in a construction project are referenced; points include horizontal and vertical coordinates.

Subsidence is the gradual sinking or settlement of the ground surface due to subsurface physical or biological processes.

There are many reasons that the HPS failed in the face of 12 inches of rain in 24 hours, peak winds of 140 miles per hour, and an increase of 12 feet in the water level of Lake Pontchartrain due to the storm surge. **Factors of safety** for the levees were too low in some instances and soil strength of the levees was overestimated. Some walls were built too low based on the use of an incorrect **datum**. Critical equipment at some pumping stations was inundated, and widespread loss of power occurred which rendered virtually all of the pumping stations inoperable. The soils of New Orleans are very susceptible to **subsidence** (land surface sinking as much as 1 inch per year in some locations), and differential settlement of portions of the levees resulted in elevations lower than design elevations.

However, the root cause for much of the failure of the HPS was the fact that *the HPS was neither designed nor operated as a system*. In other words, the HPS was a "system" in name only. A more thorough analysis of the failure of the system can be obtained from the report "The New Orleans Hurricane Protection System: What Went Wrong and Why."

The New Orleans Hurricane Protection System: What Went Wrong and Why

Systems Overview

In this chapter, we will demonstrate that the civil infrastructure behaves as a **system**, and does so on many different levels. For example, an individual infrastructure component (e.g., a roadway) may be viewed as a system. At the next higher level, components within an infrastructure sector are part of a more complex system, what we will call an **intra-sector system**. For example, a roadway is part of the transportation infrastructure sector and relates to other components (e.g., parking, mass transit) in the transportation sector. Infrastructure sectors can also interact with one another (**inter-sector systems**). For example, a roadway interacts with stormwater conveyance systems, solid waste management systems, and portions of the electric grid. At the most complex level, these infrastructure sectors interact with society, the environment, and many other non-engineering "systems."

This integrated nature of infrastructure sectors and associated components is illustrated in Figure 7.5. The arrows represent the interrelationships between sectors, and between the infrastructure and external "systems." The arrows are double-ended to emphasize interdependency. For example, not only does the civil infrastructure affect economic well-being, it also affects the state of the economy, and in turn, affects how much investment in the infrastructure occurs. In this drawing, every possible arrow has not been included, because doing so would unnecessarily clutter the drawing. Also note that in this drawing, the natural environment is unique in the sense that it is shown in Figure 7.5 as a system that is affected by and affects the infrastructure, yet as we have

A **system** is a set of interacting parts.

Intra-sector system is a system composed of multiple components within the same sector.

Inter-sector systems are systems composed of multiple components from different sectors.

Case Studies on Transportation Systems from IBM

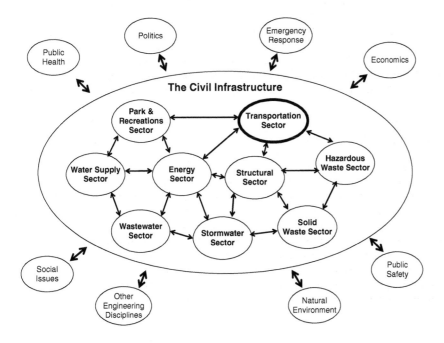

Figure 7.5 The Civil Infrastructure System, Its Sectors, External Considerations, and Interconnections. Note that for ease of reading, all interconnections are NOT shown with arrows. For example, the Transportation Sector is linked to ALL infrastructure subsystems, not only those adjacent (Hazardous Waste, Energy, Structural, Parks & Recreation) as shown. Note also that each sector is linked to each external consideration (not shown for ease of reading). Also, it is important to note that the external considerations are interconnected (not shown for ease of reading); for example, social issues and economics influence politics and public safety. The Transportation Sector (bold circle) is shown in more detail in Figure 7.6.

explained in Chapter 2, Natural Environment, the environment is also an integral "component" of many infrastructure sectors. Many of the external interactions shown in the figure will be discussed in more detail in Chapters 11 through 19 of this book.

The "civil infrastructure" system of Figure 7.5 contains many sectors. The transportation sector is expanded and illustrated in Figure 7.6 with its components and external considerations. Additionally, each component can be further subdivided into its subcomponents, as shown for street systems in Figure 7.7.

Any of the street subsystems (e.g., lighting) could be further subdivided into components, each of which has potential interconnections with other street subsystems, other transportation systems, other infrastructure sectors

Figure 7.6 **The Transportation Sector of the Civil Infrastructure and its Components.** Note that for ease of reading, all interconnections are NOT shown with arrows. For example, Mass Transit Systems are linked to ALL transportation components, not only those adjacent (Waterway, Rail, Ports) as shown. Note also that each component is linked to each external consideration (not shown for ease of reading). Streets (bold circle) are presented with more detail in Figure 7.7.

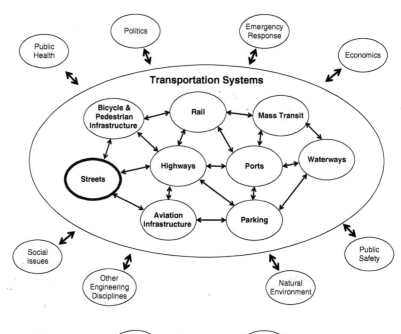

Figure 7.7 **Street Subsystems of the Transportation Infrastructure.** Note that for ease of reading, all interconnections are NOT shown with arrows. For example, Lighting Systems are linked to ALL subsystems, not only those adjacent (Traffic Control, Local Street, Collector Street) as shown. Note also that each subsystem is linked to each external consideration (not shown for ease of reading).

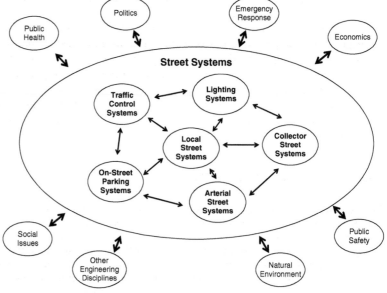

and external "systems." For example, a seemingly simple street lighting system has many interconnections.

- Unless consisting of individually solar-powered light-emitting diode (LED) lights, the lighting system constitutes a significant portion of the energy demand of a municipality.

- The source of fuel used to generate electricity may, through pollution, have an impact on human health and the environment.

- The lights and the electrical grid are designed by other engineers.

- The brightness and spacing of lights can have an impact on public safety (for pedestrians and motorists) and may contribute to "light pollution."

- The choice of decorative lighting fixtures may be costly, but adds aesthetic value and potential economic benefit to attract business and tourism; however, necessary funding may not be available or the project may not be politically "sellable" in a community.

Infrastructure systems are large and complex. Ensuring that they are designed and operated as a system is "much easier said than done." However, a systems viewpoint is essential if the public health and welfare are to be protected. Indeed, American Society of Civil Engineers (ASCE) has recommended that, to protect the nation's infrastructure, engineers must "employ an integrated systems approach." The report, "Guiding Principles for the Nation's Critical Infrastructure," published by ASCE elaborates on this guiding principle:

*Critical infrastructure must be planned, funded, designed, constructed, and operated as a system that is appropriately integrated with all other interdependent systems. Critical infrastructure systems must also be resilient and sustainable throughout the system's **life cycle**.*

The systems must be properly maintained, operated, and modified, as necessary, to perform effectively under changing conditions. A life cycle systems management approach—as developed and endorsed by the project stakeholders—will help ensure that appropriate political will, organizational structures, and funding mechanisms are established and implemented throughout the entire life of the project.

When planning, designing, and managing systems, it is imperative that you take into account the compatibility between components and subsystems. For human kidney transplants, considerable testing is performed to ensure that the donor kidney will be compatible with the host. Likewise, compatibility evaluations are necessary for infrastructure systems. For example, a new bridge over a river must have proper clearance for ships expected to pass beneath it; a proposed inter-city bus system must evaluate beneficial and adverse impacts on automobile traffic.

The Gulf Oil Spill of 2010

The explosion of the Deepwater Horizon off-shore oil platform in the Gulf of Mexico resulted in 11 fatalities, many injuries, and billions of dollars in environmental damage to the Gulf of Mexico ecosystem from the leaking crude oil. In an investigation of the cause of the explosion, the National Academy of Engineering stated: "Of particular concern is the lack of a *systems approach* to integrate the multiple factors impacting well safety, to monitor the overall margins of safety, and to assess various decisions from a well integrity and safety perspective." [emphasis added]

Guiding Principles for the Nation's Critical Infrastructure

The **life cycle** of a portion of the infrastructure refers to its entire "life," including its construction, use, and reconstruction or other end-of-life fate.

Teamwork and Systems Engineering

The ability to function on teams is crucial as engineers become more systems-oriented and projects become more complex. Indeed, the engineer of the future cannot succeed alone. This claim is supported by a report published by the American Society of Civil Engineers entitled "The Vision for Civil Engineering in 2025." The report stresses that civil (and environmental) engineers need to "collaborate on intra-disciplinary, cross-disciplinary, and multi-disciplinary traditional and **virtual teams**." Virtual teams are teams that work without meeting face to face; rather, meetings are conducted through electronic means such as the Internet. Virtual teams will be required for potential **outsourcing** and international projects. Outsourcing

Teamwork and Systems Engineering (Continued)

is the subcontracting of work to a third-party company. **Off-shoring** is the contracting of technical work to workers in other countries. Although offshoring and outsourcing have not affected civil and environmental engineering as much as it has other engineering disciplines (e.g., software engineering), the potential exists for increased offshoring in the future.

The Vision for Civil Engineering in 2025

Compatibility

There are "two sides" to the compatibility issue. A new component or subsystem must (1) not be adversely *affected by* other system components or subsystems, and (2) it must not adversely *affect* other system components or subsystems. In other words, the addition or modification to a system must work with the system, and the system must continue to work after the addition or modification.

Case in Point: The Benefits of Highway Systems

In developing countries, the need for improved transportation systems is considered vital to fostering economic growth. Furthermore, this need exists not only in individual countries, but in developing *regions*. That is, the highway system serving multiple countries needs to be treated as a *system*.

Sub-Saharan Africa serves as an excellent example. With 48 countries and approximately 800 million people, this region represents fragmented markets with inadequate infrastructure. Road conditions are generally poor, but even good roads are designed to different standards in neighboring countries, thus having different weight limits for trucks. Consequently, a truckload of goods traversing several countries is only partially loaded in order to meet the lowest weight limit.

The impact of this reality is summarized by the following statement: "Would America have flourished if trade and movement of people, goods, and services were subjected to different rules, regulations, standards, border inspections, customs procedures, transit charges, different road conditions and the many and myriad problems facing Africa's infrastructure systems all interpreted and administered differently from State to State, across the 50 States? America would come to a standstill. Is it little wonder therefore, that Africa is lagging behind the rest of the world in development" (Simuyemba, 2000).

Infrastructure Components as Systems

Individual infrastructure components can (and should be) viewed as systems. For example, a roadway is a system of sub-base, base course, pavement, curb and gutter, signage, and traffic signals. A wastewater treatment facility is a system that includes several treatment steps.

The components of the Hurricane Protection System for New Orleans may also be viewed individually as systems. For example, the levees themselves comprise a system that potentially integrates geotechnical engineering, environmental engineering, transportation engineering, and structural engineering, as well as non-technical considerations. Components of this system include the earthen embankments, native soil, groundwater, armoring, vegetation covering the earthen embankment, and in some case an "I-wall" or "T-wall" concrete structure integrated within the embankment (Figure 7.8).

Intra-Sector Systems

The infrastructure sectors (e.g., water, transportation) may also be viewed as systems in which the various subsystems and components are connected. For

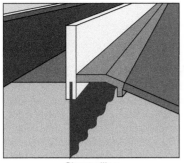 

| Sheet piling | Diagonal support pilings |
| Pre-Katrina: I-Wall | Post-Katrina: T-Wall |

Figure 7.8 Pre- and Post-Katrina Floodwall Designs. The height of the concrete wall varies, but is typically about 12 feet. The depth of sheet piling (interlocking steel panels) also varies, but may be as deep as 65 feet. The diagonal support pilings of the post-Katrina walls are driven between 70 and 135 feet below sea level.

Source: Redrawn from U.S. Army Corps of Engineers, 2006

example, a port must be integrated with other subsystems of the transportation sector such as rail or roadways. Wastewater treatment facilities must be integrated with the wastewater collection system and the sludge disposal process. The HPS in New Orleans was composed of components (levees, floodwalls, pumps, and canals) that unfortunately were not fully integrated together as a system.

Case in Point: The Proposed Twin Cities–Chicago High-Speed Rail

The proposed nine-state Midwest Regional Rail System includes a high-speed rail system between Chicago, Illinois, and Minneapolis/St. Paul (Twin Cities), Minnesota. Several alternative routes between the cities are being evaluated for this proposed line. Currently, Amtrak passenger rail has a line between Chicago and the Twin Cities. Using this existing rail line will minimize costs as compared to building a new rail line, but it does not travel near the major population and business centers of Eau Claire, Wisconsin or Rochester, Minnesota (Figure 7.9).

Planners also must consider the tradeoffs of adding stops; as more stops are added, ridership may increase, but the speed of travel will decrease.

Additionally, the route chosen will have significant impacts on the transportation (and other infrastructure) systems and the economies of cities along the path. Cities and regions must consider these potential impacts when planning and designing other transportation systems.

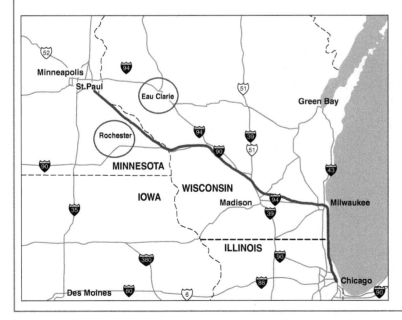

Figure 7.9 Existing Amtrak Rail Line (Shown as Bold Line) From Chicago to the Twin Cities. Major population and business centers of Eau Claire, Wisconsin, and Rochester, Minnesota, (circled) are not directly served by the existing rail line.

Case in Point: The Interconnectedness of Global Shipping

Changes to one component of a complex system can impact other components in the system. For example, consider the effect of an ongoing $5 billion expansion of the Panama Canal on global shipping. The expansion will increase the number and size of ships that can pass through the waterway. The economic benefits from increased toll income are expected to dramatically increase the wealth of the country of Panama after the construction costs have been paid off. The expansion of the canal could potentially double the amount of cargo at U.S. seaports.

Largely as a result of this expansion, the State of Louisiana has proposed the construction of a new port, the Louisiana Deep Water Gulf Transfer Terminal, at the Mississippi River's primary outlet to the Gulf of Mexico. Construction started in 2010. The port is designed to handle trans-oceanic "megaships" that are too large for ports along the Mississippi River. Unlike most ports that are designed to transfer cargo from ships to rail or truck, this port will be designed to transfer cargo to other, smaller ships. It will be financed through private funds via public–private partnerships (discussed in more detail in Chapter 14, Economic Considerations).

When originally proposed, concern surfaced over the State of Louisiana funding a new port that would compete with existing Louisiana ports. In response, State Senator Crowe, chairman of the Louisiana International Deep Water Gulf Transfer Terminal Authority, stated, "We are going after the big, giant ships that go to *other* ports. We are not going to do one thing that will jeopardize one nickel that flows to the ports (now operating). We are not going to use public money. End of story. We (state entities) are not going to be involved in its building" (Anderson, 2009).

This sidebar illustrates not only the interconnectedness of global transportation systems, but also the economic and political considerations. As should be expected, there are also environmental concerns and challenges associated with the siting and construction of a new port.

Case in Point: The Tres Amigas Project

The United States has three major power grids as noted in Figure 7.10. These three grids are not managed as a system due to limited connections between them. The proposed Tres Amigas Project is a $1 billion superstation that will allow buying and selling of power between the three grids. "Basic economics tells us increasing trade opportunities between markets of any kind tends to increase overall economic efficiency. In electric power systems, the benefits from transmission links and increased trade can include more efficient use of generation capability, increased system stability, and the ability to economize on stand-by reserves. More efficient use of generation results when the joined systems can reduce reliance on high cost peaking units by importing cheaper power from their neighbor instead. The joined system will have more generators available to exert stabilizing influences in response to short term disturbances, aiding reliability" (Giberson, 2009).

Figure 7.10 The Proposed Location for the Tres Amigas Project and the Three U.S. Power Grids.

Tres Amigas LLC Website

Inter-Sector Systems

Sectors of infrastructure also interact with each other to a large extent. For example, the wastewater treatment and conveyance sector interacts with the transportation sector, as roads must be used to supply chemicals, haul away

Figure 7.11 A Portion of the Collapsed I-35 W Bridge, Blocking a Rail Line and a Local Road.

Source: Minnesota Department of Transportation.

sludge, and get workers to and from the WWTF. The wastewater sector also interacts with the natural environment, as the effluent (treated wastewater) is discharged to the environment, most often a river or ocean. In many older cities, combined sewers (conveying stormwater and wastewater) represent further integration between infrastructure sectors.

The collapse of a portion of the I-35 W bridge in Minneapolis in 2007 demonstrates inter-sector interactions. The collapse obviously shut down a major interstate highway, but also directly affected parking lots, a nearby lock on the Mississippi River, railway lines, and local roads as demonstrated in Figure 7.11.

The failure to manage and design the New Orleans HPS as a system is one of the primary causes for its failure. At its most fundamental level, the HPS was not an interconnected system, but rather, according to the ASCE Expert Review Panel, "a disjointed agglomeration of many individual projects that were conceived and constructed in a piecemeal fashion. Parts were then joined together in 'make-do' arrangements."

The HPS interfaced with many infrastructure sectors, but there was a failure to integrate all components together as a system. Penetrations through the levees to accommodate roads, railways, and pipelines had gates, but these gates were either poorly designed or inoperable, and in some cases the closure systems were non-existent. Electric power failed; consequently, pumping systems failed. Moreover, the pumping system was primarily designed for "routine" storms, not for flood protection. As another example, there was no integration with land use planning, so people built right next to the earthen levees (see Chapter 3, Figure 3.21). This limited the means by which the levees could be raised; in order to raise the levee height, the levee base must also be widened, but this widening could not occur given that homes had been built so close to the base. The following quote from the ASCE External Review Panel summarizes the causes of damage: "A storm of Hurricane Katrina's strength and intensity is expected to cause major flooding and damage. A large portion of the destruction from Hurricane Katrina was caused not only by the storm itself, however, but also by the storm's exposure of engineering and engineering-related policy failures. The levees and floodwalls breached because of a combination of unfortunate choices and decisions, made over many years, at almost all levels of responsibility."

Prior to Katrina, the relationship between the natural environment and the New Orleans HPS had been largely overlooked. For navigational purposes,

The Cost of Gulf Coast Resiliency
A report recently released by the Entergy Corporation estimates that the Gulf Coast from Alabama to Texas could experience more than $350 billion in cumulative hurricane damages to residences, commerce, industries and infrastructure in the next 20 years. Current *annual* average damage values are $14 billion, with expected increases to $18 billion without climate change and $23 billion when extreme climate change impacts are incorporated into estimates. Several measures to improve resiliency are outlined in the report; in particular, measures costing $50 billion are identified that will result in $135 billion in averted losses.

the Mississippi River has been "extended" into the Gulf of Mexico. As a result, the river sediment that has historically fed coastal wetlands, thereby counteracting natural erosion, is now transported far into the Gulf. The extent of coastal wetlands and the barrier islands that protect them has dramatically decreased, with an estimated loss of nearly 2,000 square miles since the 1930s. These wetlands could have acted as "sponges" to absorb much of the storm surge, and their absence exacerbated the flooding during Hurricane Katrina. The U.S. Army Corps of Engineers, the U.S. Fish & Wildlife Service, the State of Louisiana, and other agencies and organizations are currently implementing projects as part of a multi-billion dollar, multi-decade plan to restore the coastal wetlands, not only for flood protection, but for habitat, fisheries, recreation, and other purposes.

 Louisiana Coastal Area homepage

Should New Orleans Have Been Rebuilt?

Following Hurricane Katrina, there was some debate over whether New Orleans should be rebuilt. There are many valid arguments against rebuilding: New Orleans is below sea level and prone to flooding; the land is unstable, subsiding, and ill-suited for building upon; channelization of the Mississippi River has exacerbated the flooding potential; New Orleans will likely someday experience a hurricane of even greater magnitude than Katrina. These and many other arguments suggest that the cost of rebuilding is simply not worth it. Of course,

New Orleans is being rebuilt (still to this day) after Katrina, due to a host of economic, social, and political reasons.

Some have proposed a new future for New Orleans, recreating it as a floating city!

 Brief Videos on Mega Engineering: "Floating New Orleans" and "Building a Floating City"

Case in Point: The Trinity River Corridor Project

The Trinity River runs through Dallas, Texas, (Figure 7.12), and recently the city has begun an ambitious redevelopment project. The project will enhance environmental management, provide recreation opportunities, improve transportation efficiency, and provide future flood protection.

The project will include construction of new levees, bridges, a tollway, wetland and wildlife restoration, trails, and sports facilities.

These components are being *simultaneously* designed as a system, taking into consideration existing facilities, future growth and land use/urban planning. These types of projects, involving many disciplines beyond civil and environmental engineering, are increasingly common. A similar project is underway in neighboring Fort Worth, Texas.

 Trinity River Corridor Project Website

 Trinity River Corridor Video

Figure 7.12 **The Trinity River (Arrow), Adjacent Undeveloped Greenway (Floodplain), and Downtown Dallas, Texas (Background).**

Source: Courtesy of trinityrivercorridor.org.

Stand-Alone Systems

Basic infrastructure needs are similar for a small town, a large city, an isolated military base, an off-shore oil rig, or a large industrial complex. Differences arise from specific needs of each system.

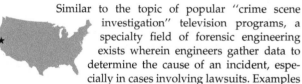
The John Deere facility in Dubuque, Iowa, manufactures large construction and forestry equipment. At its peak employment, more than 4,000 people worked at this facility. The facility is described as "its own city," including a coal-fired power plant, a landfill, a wastewater treatment facility, water supply wells, water storage, fuel storage, fire protection, and security. The decision to provide these services rather than utilizing city services is based on several factors, but the two primary drivers are cost and control. By providing these services, John Deere can control them without needing approval from the city. Due to the several mile separation distance, the cost of extending city services (e.g., water mains, sanitary sewers) would be excessive. If the facility location was adjacent to the city, a scenario is possible where the cost to extend services would be economically favorable; however, the "cost" of not having control of services would be factored into the decision. It is possible that when designing these systems, engineers would have to meet different regulatory requirements for an industrial classification rather than a municipal classification. Another consideration is the additional costs to address regulatory compliance of their own services.

Forensic Engineering

Similar to the topic of popular "crime scene investigation" television programs, a specialty field of forensic engineering exists wherein engineers gather data to determine the cause of an incident, especially in cases involving lawsuits. Examples include structural collapses (famously in the recent I-35 W bridge collapse) as well as train, aviation, and automobile accidents.

Consider the 1995 case of an affluent neighborhood in San Francisco, California, in which a combined sewer collapsed and resulted in damage to several properties, including the destruction of an entire home. Recall from Chapter 5 that a combined sewer is one that conveys stormwater *and* wastewater. The sewer was a 100-year-old brick pipeline. During a routine rainstorm, the combined sewer became pressurized. Water exfiltrated (exited) the sewer through cracks in the aging pipe. At high pressures and velocities, this "jet" of water washed away the overlying sand, which in turn decreased the overlying soil pressure that was acting to keep the brick pipe intact. Eventually the pipe imploded and created a washout several hundred feet wide and approximately 40 feet deep. A large multi-million dollar home collapsed into the washout.

A forensic engineer was contracted to determine the cause of the incident. The cause was not immediately evident, given that the pipe had effectively carried similar storms in recent years. It was concluded that the sewer would likely not have failed had the water flow not been constricted (thus increasing pressure) by a partially closed metal gate. The gate had been recently installed to temporarily close the sewer and hold back wastewater (during non-storm events) to allow construction work to proceed on downstream portions of the sewer. The gate could be raised and lowered, but apparently was unintentionally left in a partially closed position, thus restricting flow. This forensic engineering study relied on computer models to simulate flows in the sewer, eyewitness accounts, on-site observations, and videotapes taken of the inside of the pipe during routine maintenance prior to the incident.

This incident also serves as an example of potentially disastrous unintentional effects of modifying a component in an infrastructure system.

Newspaper Account of the Combined Sewer Collapse

Summary Report of Storm Sewer Collapse Investigation, with Photographs

The "Big Picture" System

The infrastructure is part of a much larger system, which includes the many items shown in Figure 7.5 such as the economy, human health and welfare, politics, or emergency response. For example, consider that the recreational use of a beach may depend on the ability of a wastewater treatment facility to effectively treat its wastewater, or for contaminants to be removed from stormwater. Also, the ability of firefighters to extinguish fires depends in large part on the ability of the water distribution system to supply the needed fire flow; the roadway system to provide lane and turning clearances for fire trucks; and perhaps the emergency response systems to clear roads of debris (caused by hurricanes, tornadoes, etc.) and snow/ice accumulation. This book will investigate the relationship between the infrastructure and many of these other systems in more detail. The purpose of this chapter is to raise your awareness of the complexity and interconnectedness of the infrastructure, and to foster an appreciation that a system-wide view is necessary.

Even the simplest components of the infrastructure must be viewed as parts of systems. For example, consider the design of a cul-de-sac. Cul-de-sacs are very popular in many modern subdivisions, as they carry no through traffic, and thus afford the residents living on the cul-de-sac with a quiet and relatively private home site. When considering how to design a cul-de-sac, you might first consider that the radius of the cul-de-sac is large enough to allow cars to easily traverse the cul-de-sac. Indeed, automobile drivers will be the most regular users of the cul-de-sac, but the design must consider how the cul-de-sac fits into other systems. For example, the cul-de-sac will be utilized by fire and safety personnel and maintenance workers, and thus must be readily accessible to ambulances, fire trucks, garbage collection trucks, and snow plows in cold climates. Additionally, cul-de-sac design must consider pedestrians and bicyclists.

In the past, engineers, planners, and other policymakers failed to foresee the impact of cul-de-sacs on the larger transportation system of an entire region. Consider that a new subdivision will generate increased traffic rates in the surrounding area. When subdivisions rely on many cul-de-sacs, the additional traffic generated by that subdivision will often *not* use the streets of the subdivision but will use (and increase traffic loads) on the original surrounding streets. For example, consider Figure 7.13, which shows a neighborhood with many cul-de-sacs. In order for a person living at point A to travel to points B or C, that person will most likely use one of the pre-existing roads (e.g., N. High Point Road, Old Sauk Road, or N. Gammon Road) to get there, rather than the curved roads of the neighborhood.

These examples consider various transportation uses of the cul-de-sac, yet you must also consider the cul-de-sac as part of a larger system than just the transportation system. For example, the environment is part of the system affected by the cul-de-sac design. Specifically, a large amount of stormwater can be generated by cul-de-sacs; when completely paved, cul-de-sacs are a

Figure 7.13 A Neighborhood with Many Cul-De-Sacs.

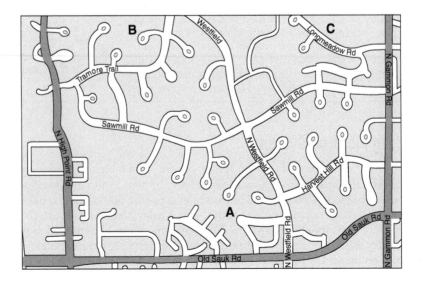

Figure 7.14 A Vegetated Depression in a Cul-De-Sac. There is no curb and gutter on this cul-de-sac, which allows runoff to flow into the center depression.

Source: Courtesy of the University of Connecticut, Center for Land Use Education and Research.

significant source of stormwater runoff. The increased stormwater runoff (as compared to the runoff before development) can increase the volume and flowrate of runoff as well as the concentration of pollutants in lakes and streams that ultimately receive the runoff. This runoff can be mitigated by creating pervious areas in the center of the cul-de-sac and sloping the cul-de-sac such that the runoff enters the pervious central area. Such pervious areas may be termed rain gardens, bioretention areas, or infiltration basins, and can potentially improve the aesthetics of the cul-de-sac (Figure 7.14).

Also of great importance is the fact that the cul-de-sac can influence social systems as well. Some sociologists suggest that the conventional design of subdivisions, with its heavy reliance on cul-de-sacs, can affect social interactions between people (e.g., decreasing interactions with neighbors passing by during walks).

The HPS in New Orleans failed to consider the "big picture" system, and as a result, its failure was wide-ranging and affected: emergency management activities; society (e.g., mass migration; human health, both mental and physical; crime; personal safety; cultural aspects); and personal property value (Figure 7.15).

Figure 7.15 A Breach (Circled) in the 17th Street Canal and Resultant Flooding Damage, Resulting in Significant Damage to Private Property. Note the lack of flooding in right side of photo.

Source: U.S. Army Corps of Engineers/Digital Visual Library.

Remember that full-sized versions of textbook photos can be viewed in color at www.wiley.com/college/penn.

Moreover, the HPS was strongly linked to a multitude of other systems such as financial and political systems. Unfortunately, the HPS was funded in a piecemeal fashion over several decades due to the fact that portions of the HPS had to be built as funding became available. Additionally, no single entity was in charge; the U.S. Congress, the U.S. Army Corps of Engineers, local levee districts, and local sewer and water utilities were all in charge of certain portions, but without a single leader, their actions were not fully coordinated.

The fact that some sections of the levees were as much as 2 feet lower than intended elevations also highlights the fact that the HPS was not constructed and operated as a system. The levees were too low either because of subsidence and/or because of the use of incorrect datums. Surely, engineers, surveyors, geologists, and public works officials knew that New Orleans was subsiding, but this information was not properly integrated into the levee designs since there was no systems-level oversight.

Outro

The devastation that occurred in New Orleans from Hurricane Katrina was first met with sadness and shock by the nation, but this quickly turned into anger and frustration toward President George W. Bush, the Director of FEMA (Michael Brown), and other public officials. In large part, engineers escaped much of the public's wrath. However, for not designing and operating the HPS as a true system, engineers, perhaps more than the President or the Director of FEMA, must shoulder some of the blame also. As summarized in the *What Went Wrong and Why* Report, "No one person or decision is to blame. The engineering failures were complex, and involved numerous decisions by many people within many organizations over a long period of time." Although a proper systems-view of the HPS would not have eliminated flooding, the extent of the impacts would have been greatly reduced. For example, systems modeling suggests that two-thirds of the deaths would not have occurred if the levees, flood walls, and pumping stations had not failed. Moreover, a properly functioning HPS would have resulted in less than half of the value of property losses. Congress has authorized over $7 billion to restore and improve the New Orleans HPS. Although this represents a significant investment and should preclude a disaster from an event of a similar magnitude to Hurricane Katrina, this authorization is yet another unfortunate case of being *reactive* rather than being *proactive*. Henry Rollins, a pioneer of punk music, succinctly stated, "Katrina was more than just Mother Nature reminding us of her supremacy; it was *human* nature reminding us of our fallibility."

As you will learn in Chapter 17, Ethical Considerations, the first "canon" of the ASCE Code of Ethics is this: "Engineers shall hold paramount the safety, health, and welfare of the public . . . in the performance of their professional duties." As a result of not having a systems-wide viewpoint, the safety, health, and welfare of the public was not protected as it should have been.

Short Answer

7.1 In what ways was the HPS not designed and operated as a system?

7.2 What is compatibility, with respect to infrastructure? List some examples.

7.3 How do cul-de-sacs interact with the transportation sector as a whole?

7.4 How do cul-de-sacs interact with the environment?

7.5 What are some of the components of the New Orleans HPS?

Discussion/Open-Ended

7.6 Why is forensic engineering inherently systems-based?

7.7 What are the compatibility issues that planners/engineers must consider when adding a new light rail mass transit system into a downtown area?

7.8 How will the Trinity River Corridor Project benefit from systems thinking?

7.9 Which of these infrastructure components can be considered a system? Briefly explain your reasoning for each one.

a. Interstate bridge
b. Bicycle lane
c. Port
d. Commuter rail station
e. Fire hydrant
f. School bus stop
g. Detention pond

7.10 For each of the following (or as designated by your instructor), give at least two examples of how the infrastructure sectors/systems interact.

a. Water treatment and wastewater treatment
b. Water distribution and wastewater conveyance
c. Water supply and hazardous waste
d. Structural systems and transportation systems
e. Energy systems and transportation systems
f. Stormwater systems and transportation systems

Research

7.11 Perform research on the Deepwater Horizon explosion. Prepare a one-page report on how the lack of a systems approach may have contributed to the failure.

7.12 Research the "infrastructure" needed to support workers on an offshore oil rig. Summarize your findings in a one-page report.

7.13 Research the I-35 W bridge collapse. Summarize the impacts of the collapse on the infrastructure systems of the Minneapolis/St. Paul region in a one-page report.

7.14 Perform research on the experience of New Orleans residents that were sheltered in the Superdome during Hurricane Katrina. As a system, how was the Superdome inadequate? Summarize your findings in a one-page report.

Key Terms

- datum
- factors of safety
- inter-sector systems
- intra-sector system
- life cycle
- storm surge
- subsidence
- system

References

Anderson, E. 2009. "Work on Southwest Pass Terminal Could Be Under Way in Early 2010, Chairman Says." *The Times-Picayune*, September 23, 2009.

ASCE Hurricane Katrina External Review Panel. 2007. *The New Orleans Hurricane Protection System: What Went Wrong and Why.* Reston, VA: American Society of Civil Engineers.

Entergy Corporation. 2010. *Building a Resilient Energy Gulf Coast: Executive Report.* http://entergy.com/content/our_community/environment/GulfCoastAdaptation/Building_a_Resilient_Gulf_Coast.pdf, accessed August 8, 2011.

Giberson, M. 2009. "Tres Amigas Proposes Three-way Transmission Link." *Alternative Energy Stocks*. Published online November 11, 2009. http://www.altenergystocks.com/archives/2009/11/tres_amigas_proposes_threeway_transmission_link.html, accessed August 8, 2011.

Simuyemba, S. 2000. *Linking Africa Through Regional Infrastructure.* http://www.afdb.org/fileadmin/uploads/afdb/Documents/Publications/00157662EN-ERP-64.PDF, accessed August 8, 2011.

U.S. Army Corps of Engineers. 2006. *Task Force Hope Status Report*, April 19, 2006. http://www.mvn.usace.army.mil/hps/Status%20Report%20Newsletters/status_report_page_april_19_2006%5B2%5D.pdf, accessed on August 8, 2011.

U.S. Army Corps of Engineers. 2007. *Permanent Protection System for Outfall Canals—Report to Congress.* http://www.mvn.usace.army.mil/pdf/hps_reporttocongress.pdf, accessed on August 8, 2011.

chapter Eight History, Heritage, and Future

Learning Objectives

After reading this chapter, you should be able to:

1. List the landmark events in the historical evolution of roads, canals, bridges, water distribution, water treatment, wastewater conveyance, and wastewater treatment.
2. Explain how technology has been integral in the historic development of infrastructure.
3. State examples of lessons that modern engineers can learn by studying the history of infrastructure.

Introduction

In prehistoric times, there was no need for an infrastructure, and no ability to build one. There was no need to transport goods because there were no goods to transport. Before agriculture, there was no need to build irrigation systems. Without permanent residences, there was no need to try to control floods by building levees and dams.

Reinventing the Wheel

The history of the wheel is somewhat obscure, but it most likely was "reinvented." The wheel was used in Europe thousands of years before it was used in Mesopotamia, with no apparent link between the two. A similar story can be told of many of the early infrastructure components, including roads, water conveyance, dams, etc. Significant technological advances were accomplished by a few talented people in one part of the world, but due to the lack of communication and travel, the same advances would have to be invented completely separately in other parts of the worlds, perhaps centuries later.

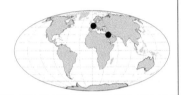

This all changed about 10,000 years ago with the Neolithic (agricultural) Revolution. People transitioned from hunting and gathering to raising crops and forming settlements. As these ancient people started living together in settlements, the need for infrastructure became apparent. For example, there was a need for defense, such as a wall or fence surrounding the settlement. As more crops needed to be grown, irrigation projects were undertaken. As settlements continued to grow to become towns and then cities, people needed a way to deliver increasing quantities of drinking water to the urban center; the nearby springs or streams that had been used for drinking water and cleaning purposes no longer supplied enough water to support the larger and more dense population. Also, removing refuse became necessary in order to reduce odors and the number of insects. Some ancient cities became so advanced that rather than dumping human fecal material in a pit or on the roadside to be washed away by the next rain storm, infrastructure was developed to convey this waste away. Roads were built to move troops, trade goods, and collect taxes.

This chapter is intended to be a very brief introduction to the history and heritage of infrastructure development throughout the world. In no way is it meant to be a comprehensive review. Books could be (and have been) written on the history of any one of the infrastructure sectors or components, such as bridges or water conveyance. Even the history of various components associated with water conveyance (e.g., pumps, hydrants, pipes, aqueducts) could be told in volumes rather than in paragraphs or pages. We hope that the few examples we have selected and our brief overview will illustrate the need for infrastructure and help you to better understand why infrastructure is so important to society.

Suggested Reading

The following publications are recommended for additional reading on the history of civil and environmental engineering:

E. Trantowski et al. *A History of Public Works in Metropolitan Chicago*. 2008. Kansas City, MO: American Public Works Association, 2008.

Civil Engineering Magazine (Special issue: Celebrating the Greatest Profession). 2002, Volume 72, Number 11. American Society of Civil Engineers.

L. H. Berlow. *The Reference Guide to Famous Engineering Landmarks of the World*. Westport, CT: Greenwood, 1997.

D. L. Schodek. *Landmarks in American Civil Engineering*. Boston, MA: MIT Press, 1987.

H. Petroski. *Remaking the World: Adventures in Engineering*. New York: Vintage, 1998.

The Builders. Washington, DC: National Geographic Society, 1998.

Introductory Case Study: The National Road

Also known as the Cumberland Road, the National Road demonstrates the tremendous foresight of early American elected officials. One reason for the road (and a reason that President Washington was a great proponent of another means of transport, namely canals) was to unify the country; that is, to connect the various states and help them work together as a nation.

It was intended that the National Road travel 1,000 miles from Cumberland, Maryland (which was already connected by road to Baltimore) to St. Louis, Missouri. Without such a road, the alternative was to use rivers (which often were impassable

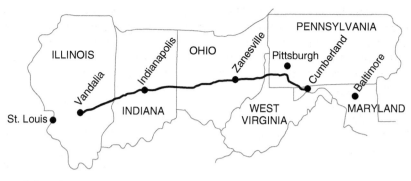
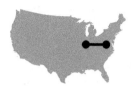

Figure 8.1 The National Road.

during the winter and during dry seasons) and rutted and winding trails. Construction began in 1811 and finished in 1839, although the road never did extend to St. Louis as originally envisioned. The road's path as of 1839 is shown in Figure 8.1. Much of current U.S. Route 40 uses the original alignment, and Interstate 70 in turn follows U.S. Route 40 closely.

Specifications for the road included a 66-foot right-of-way with a 20-foot wide travel lane. Drainage ditches were designed for both sides with a maximum road grade of 8.75 percent. The route was set based on ease of construction (parts of it followed a 12-foot wide path that British General William Braddock had cleared in 1755) as well as politics; the route for the first portion was chosen to travel through three states (Virginia, Maryland, and a corner of Pennsylvania, the latter meant to obtain support from that state for the road's construction). Advances were made in construction techniques, and several notable bridges were built (e.g., see Figure 8.2). The road to Ohio was completed in 1818 at a cost of $13,000 per mile (approximately $200,000 per mile when adjusted for inflation).

Figure 8.2 Blaine Hill Bridge (Foreground) on the National Road, was Built in 1828 and Rehabilitated in 2005. An arch bridge carrying Interstate 70 is in the background.

Source: Courtesy of B. Powell.

The completion of the National Road had a tremendous influence on the early American nation. Despite some terrible road surface conditions and varying quality, the cost of transporting goods from Baltimore to Ohio was halved, as was the travel time; 6 weeks before completion, 3 weeks after completion (compared to 6 hours travel time on today's interstates). Manufactured goods from the east could travel to the west, whereas flour, whiskey, bacon and other goods from the west could travel east. However, with rails and canals being increasingly built, the need for roadways was not seen as of great importance, as the former were faster, smoother, and more cost effective.

This investment by the U.S. federal government in transportation infrastructure is one of a very few such investments in the 1800s. Strict constructionists of the Constitution felt that due to the concept of states' sovereignty, the federal government should have no role in such activities, and therefore assistance was not provided in building the Erie Canal or further expanding the National Road.

Themes

In the remainder of this chapter, we will trace the history and evolution of a few selected infrastructure sectors. As you read about the history of the sectors, you will note that various themes emerge:

The human dimension: We realize that many students find reading about history to be boring. We feel that this is so because the focus is often on memorizing facts. Yet history should be viewed as a set of stories about *real people*, in our case, the people that made infrastructure history happen, but also the people that were affected by the evolving infrastructure.

Technological innovations: The improvement to infrastructure has been an evolutionary process. Many of the improvements took place gradually as new technology was invented. In many instances, that new technology was not created with infrastructure improvement in mind, but was adapted by innovative and thoughtful engineers. In some instances the technology developed "in the field" as the projects were being constructed.

Government responsibility: The monetary investment needed to build and maintain the civil infrastructure is tremendously large, so large that infrastructure is typically funded by governments (the **public sector**) rather than by individuals or corporations (the **private sector**). In contrast, the rail, power, and communication infrastructure is largely privately owned and operated. Moreover, when a government is ruling effectively, it can fund and, equally importantly, maintain the infrastructure in order to benefit its citizens. When governments fail or cannot otherwise rule effectively, the infrastructure (and society) often suffers, as is experienced in many developing countries.

Longevity is possible: Infrastructure components exist today that were built hundreds of years, or in extreme cases, millennia ago. They remain because of their functionality, the heritage they represent, and in many instances, their *beauty*. In many cases, they remain because they were excessively *over*-designed to ensure safety; when constructed, engineering knowledge did not exist to design and build the components to exacting specifications.

Transportation

ROADS

The earliest transportation mode by humans was on foot, using worn-down paths often originating as wildlife paths. Consequently, the only things transported were the people themselves and whatever they could carry or drag. However, as goods needed to be distributed, as soldiers and their equipment needed to be moved, and as taxes needed to be collected, roads were needed. In many cases, roads were not "engineered" or built to meet this need, but rather paths were gradually widened as animals (e.g., donkeys, oxen) replaced humans for burden bearing. Such roads have been around for more than several thousand years, and indeed were found throughout the United States into the early 19th century and are still found throughout lesser-developed areas of the world.

However, roadways are often not truly useful for moving large amounts of materials until they are properly engineered. Simply packing down the native soil is not effective for many reasons. First, native soil throughout the world is rarely strong enough to support the repeated and frequent loadings associated with a wagon's wheels, a horse's hooves, etc. Moreover, water will not drain properly from a non-engineered road or path, and with frequent use and rainfall, will become rutted, muddy, inefficient, unsafe, and ultimately, unusable. Consequently, goods cannot be moved as rapidly and their price increases, or armies cannot pillage the surrounding countryside as effectively.

The Romans are history's most vaunted road builders, and their roadways were *engineered*. Many of these roads are still in existence and in use (e.g., Figure 8.3). Nearly 2,000 years ago, the Roman empire constructed 50,000 miles of paved roads (for comparison, the U.S. Interstate Highway System is 42,500 miles) and their entire road network was approximately 250,000 miles long. The most important roads, termed viae ("viae" is the plural of "via"), ranged between 14 and 20 feet wide, which permitted the passage of two carriages. The Roman road systems were built to enhance commerce and to unify their vast empire, but its primary purpose was for moving troops. Armies could move nearly 20 to 30 miles per day on paved roads as compared to 8 miles per day on unpaved roads (in dry weather), and virtually zero miles on unpaved roads (in wet weather).

The first Roman roads were built with the idea of strength. These were extremely time-consuming, expensive, and labor-intensive to build. They had similarities to modern roadways, in that a series of layers were incorporated into the roadway structure (Figure 8.4). From the bottom to the top, the layers

Figure 8.3 **Remains of a Portion of the Appian Way.**

Source: Copyright © Paolo Cipriani/iStockphoto.

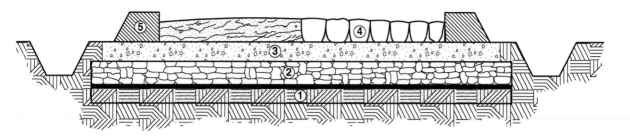

Figure 8.4 **Cross-Section of a Roman Via.**   A Roman via consists of: (1) compacted earth, (2) gravel aggregate, (3) aggregate of lime, gravel, broken tile or ironwork slag, (4) polygonal stone slabs (shown right) or gravel (shown left), and (5) curbstones.

Source: Redrawn from Sandström, 1970.

consisted of well-compacted dry earth, crushed rock on mortar, concrete, and aggregate. This top layer was often crowned (as seen in Figure 8.4) to shed rainwater.

The Appian Way (or Via Appia) is seen in Figure 8.3 and was one of the most important Roman roads, the main portion of which was completed in 312 B.C.

It was very successful in meeting its prime objective, which was to move troops. Thanks to disrepair over the millennia, much of it is no longer usable while some portions are no longer in existence. This is a theme we wish to stress in this book, namely the need for maintenance of infrastructure. We hope that by now you realize that engineers don't just design and build things, but they are also involved in managing infrastructure through appropriate maintenance plans.

You may find it surprising that not much happened for the next two millennia as far as advances in road building. But without a stable and unified government, construction of and innovation in major infrastructure projects and innovations are not possible. Perhaps the next important advance in road building occurred in the early 1800s, which was the creation of "Macadam" or macadamized roads, named after their designer, John McAdam. Macadamized roads were much less expensive than other types of road building and were also much longer lasting. Like earlier roads, they were constructed of several layers, but McAdam realized that if the soil underneath the road was *well-drained*, it would be stronger, and thus the pavement structure itself would not need to provide as much strength. His roads included ditches on both sides to aid in drainage, and, as compared to his contemporary road builders, the bottom layer was composed of cobbles rather than of manually placed stones. Construction is depicted in the painting in Figure 8.5. The top layer was composed of small, angular stones that, as they were compacted by wheels and hooves, packed together and became interlocked, thereby creating a surface that was largely impervious to water. This method was soon widely used throughout the world and was the method chosen for the National Road.

Macadam roads and gravel roads suited the horses that were the primary mode of providing motive power. The surfaces provided some "give" that were easier on hooves. However, the advent of self-propelled vehicles riding on rubber wheels at relatively high speeds had two effects on these early roads: rubber wheels did not get adequate traction and at high speeds vehicles "sucked" dust out of the road, leading to degradation of the pavement. In response, tar was used to grout the macadam, leading to the term tarmacadam (or "tarmac"). Ultimately road construction evolved to today's "sealed" asphalt and concrete pavement structures.

The Interstate Highway System

Officially titled the Dwight D. Eisenhower National System of Interstate and Defense Highways, the Interstate Highway system is considered one of the greatest public works projects in history. Originally designated by the Federal-Aid Highway Act of 1944, it was not destined to become a reality until the Federal-Aid Highway Act of 1956 (which included the declaration that such a system was "essential to the national interest" of private, commercial, and military ground transport). President Eisenhower strongly promoted the system that he correctly envisioned would "change the face of America." Expansion of the originally planned 40,000 miles of expressways continues to this day.

Figure 8.5 **Laying of the First Macadam Road.** Workers in the foreground are seen breaking rocks for the base. Rocks needed to be small enough "so as not to exceed 6 ounces in weight or to pass a two-inch ring."

Source: Painting by Carl Rakeman. This image is provided courtesy of the Federal Highway Administration.

CANALS

Although canals are currently not a prime means of moving goods in most of the developed world, for much of history, moving goods by water was the most economical means to do so. Canals came into prominence in the late 1700s and early 1800s, spurred on by the Industrial Revolution's need for resources and ability to move products to market. However, they have been used for millennia, with China's Grand Canal being the oldest (large portions constructed in the 5th century B.C.) and longest canal (over 1,000 miles in length).

Constructing canals was a challenging venture, as they needed to cross rivers and mountains, all while maintaining a relatively consistent slope (1 to 2 inches in elevation change per mile). A significant amount of earth needed to be moved, and in many cases, locks were needed. You might wonder how these early engineers learned these skills. In many cases, they had no formal education; indeed, there was no such thing as civil engineering discipline. Many of the early canal engineers were surveyors that learned on the job, in many instances from their own mistakes.

Perhaps no infrastructure project had a greater impact on America than the Erie Canal, which opened in 1825. By connecting the Hudson River to Lake Erie (Figure 8.6), it provided a route for commerce between the midwestern United States and the Atlantic Ocean, and therefore the entire world. Indeed, some have said that the Erie Canal started "globalization."

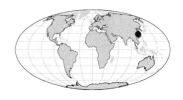

Images of the Erie Canal

Figure 8.6 **Map of the Erie Canal.**

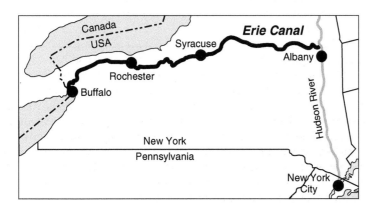

Figure 8.7 Aqueduct on the Erie Canal. Canal boats would travel across the river, which can be seen below the aqueduct, while being pulled by mules that walked on the adjacent towpath (now a recreational trail) in the foreground.

Source: Courtesy of G. Frysinger.

The canal width was 40 feet and boats could carry 30 tons of freight. Although this is relatively small compared to ships that traverse the Panama Canal today (cargo capacity greater than 50,000 tons), it was substantially more than roads at the time could bear. Boats were pulled along by mules or horses that walked on towpaths (see Figure 8.7).

Like any canal, the Erie Canal required a large capital investment, and its construction in the United States was initially hindered as the federal government was reluctant to finance the canal and private investors could not raise enough capital. Eventually, the canal was constructed primarily by public money from New York state. To accomplish this funding, strong political leadership was required, which was supplied by New York Governor DeWitt Clinton. Like many visionaries, Governor Clinton was scoffed at, and his detractors famously referred to the proposed canal as "Clinton's Ditch."

The Erie Canal was very successful by nearly every measure and affected the nation in lasting ways. The costs of transportation (of products and migrating pioneers from east to west and natural resources from west to east), which before relied on a combination of boats and horse-drawn wagons, were cut 95 percent. The canal was also responsible for making New York City the primary port in the United States. Moreover, it changed the face of the nation as it opened up the western frontier in ways that the National Road never did.

Although canals had a tremendous positive impact for a few decades, railroads made them obsolete by the mid-1800s. Currently, many inland canals such as the Erie Canal are used primarily for recreation and tourism. China's Grand Canal is still used, and of course, the Panama Canal and Suez Canal remain integral components of the modern economy.

Case in Point: Suez and Panama Canals

The Suez Canal (completed in 1869) and the Panama Canal (completed in 1914) are historic canals that have had tremendous impacts on the global economy. About 18,000 ships a year move through the Suez Canal and more than 14,000 through the Panama. Their construction saved thousands of miles for shipping and made the trips (e.g., from New York City to India, Italy, or the west coast of the United States) much safer or cost-effective than alternative routes, which were either around the southern tips of South America or Africa, or across land. As a result, the cost of transporting goods was greatly reduced.

Construction of these canals was a tremendous undertaking. Although the Suez Canal is about twice as long as the Panama Canal, the latter was much more difficult to construct. Unlike the Panama Canal, the Suez Canal does not have any locks—it is simply a single-level waterway connecting the Red Sea to the Mediterranean Sea. Also, construction workers of the Panama Canal had many tropical diseases to contend with such as malaria and yellow fever. Without the discovery that mosquitoes were the carriers of these diseases and the resulting protective measures, it is unlikely that the canal would have been completed. Indeed, construction of the Panama Canal had been abandoned by the French in the 1890s due to more than 20,000 worker deaths.

Construction of both canals had unsavory aspects. For example, the Suez Canal was built with many slaves, with much of the digging by hand (some of the digging was literally "by hand," as the slaves were not equipped with shovels). Panama was originally governed by Columbia. When Columbia refused to allow the United States to build across Panama, the United States supported a Panamanian revolution, and when the new country declared independence, U.S. troops blocked Columbian ships from Panama. Perhaps not surprisingly, the new country of Panama accepted the treaty with the U.S. allowing the U.S. to build the canal.

BRIDGES

The story of the history of bridges is in one sense the story of the introduction of new technology, especially of new materials. Through the ages, bridges have been built primarily of two materials: timber or stone. These materials have been used to create many beautiful and functional bridges, some of which have stood for thousands of years. Indeed, few materials are as beautiful or as able to withstand the elements of nature as stone. However, newer materials are much less costly and easier to handle and to build with, and allow for taller, longer, stronger bridges that can cross gaps that previously were not economically, and sometimes, not technically feasible.

Iron was used as a bridge material beginning in the late 1700s, first in the picturesque (and aptly named) Iron Bridge (Figure 8.8), located in England. Previously, iron had been too expensive to use for building a structure as large as a bridge, but became cost-effective thanks to technological advances in iron making. The first iron bridge built in the United States was built on the National Road.

The use of iron in bridges continued and was adapted to suspension bridges. Suspension bridges offered the ability to traverse wide spans without needing central piers. This improved constructability and river navigability, and provided new opportunities to build beauty into bridges. The most notable early suspension bridge is Thomas Telford's Menai Straits Bridge, located in North Wales. Approaching 200 years old, the bridge still carries traffic.

With increasing technological progress in the making of iron, continued advances to bridge design occurred, notably the Britannia Bridge (Figure 8.9), which also crossed the Menai Straits. As with so many early engineering accomplishments, engineers did not have textbooks to rely upon, or even much in the way of a theoretical understanding. The Britannia Bridge was notable in that, although it did not rely on theoretical structural engineering, it at least relied on testing of different material shapes prior to being constructed. The unique rectangular shape (see Figure 8.9) was found to be most efficient, and it is through the middle of this iron "tube" that rail traffic moved. Although five times as expensive as the nearby Menai Straights Bridge, the Britannia Bridge was much more rigid.

Engineers soon came to realize that the rectangular box section of the Britannia Bridge wasted a significant amount of iron, and therefore was

Figure 8.8 The Iron Bridge.

Source: Copyright © John Hallett/iStockphoto.

Figure 8.9 Britannia Bridge.

Source: Williams, F.S. (1852).

unnecessarily costly. The truss was developed in the late 1800s and was used widely in bridges for the next 100 years. A truss is a lattice that consists of many triangular shapes. As such, imagine cutting triangular shapes of iron out of the Britannia box section; in essence, the remaining iron is a truss. Most railroad bridges were composed of trusses as they offered exceptional rigidity and strength at a relatively low cost. Combined with the development of steel, a number of innovations were possible, as seen in one of the earliest and most remarkable bridges that utilized trusses, the Firth of Forth bridge (Figure 8.10); it is named for the estuary, or "firth," it crosses where the River Forth flows into the North Sea. In time, the box cross-sections of the Britannia Bridge were replaced with steel truss arches.

The use of materials continued to evolve, and today, one of the most common materials for bridge construction is concrete. Concrete is notable because it is very strong in compression, but relatively weak in tension. This means that when you "push" on concrete, or place it under compression, it can withstand tremendous forces; however, when you "pull" on concrete (place it under tension), it can easily fail. Structural members of bridges are subject to both compression *and* tension. Thus, the use of concrete in large bridges was minimal until the widespread development of **reinforced concrete**, which is reinforced with steel. Steel is very strong in tension, and thus when properly placed within the concrete structure, can carry the tensile forces that concrete cannot.

Figure 8.10 Firth of Forth Railway Bridge.

Source: Copyright © Bridget McGill/iStockphoto.

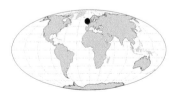

The development of bridges also led to many new construction techniques to address unique challenges. For example, the tubes for the Britannia Bridge weighed 3 million pounds, and needed to be lifted into place, over water. The famous Brooklyn Bridge needed to have 3,600 miles of wire (which was bundled together into cables) strung across it. Caissons and cofferdams had to be built to allow the construction of piers to occur safely in and under water.

Video of Construction of the Brooklyn Bridge

Consequences of Failure

If, in your future engineering career, a project you design or construct fails, the consequences may be grave. Depending on the extent of the failure and the extent (if any) of your negligence, the consequences *to you* may be a reprimand, loss of your job, or loss of your professional engineer (PE) license (plus the personal anguish of the responsibility for any victims of the failure).

But consider the engineers involved in moving Xerxes' troops (Xerxes was the king of Persia in the fifth century B.C.) across the Dardanelles into Greece. When their attempts to do so by anchoring and tying 614 ships together failed, they were beheaded. The picturesque Stari Most, or "Old Bridge,"

(Figure 8.11) in Bosnia/Herzegovina has a similar story. Constructed in the 1560s, legend has it that the bridge collapsed when the **falsework** was removed from the bridge. (Falsework is the temporary structure used to help builders construct the bridge). The engineer was then ordered to build it again, with the threat of beheading if the bridge failed. One story has it that the engineer fled before the falsework was removed the second time, while another story has it that the engineer prepared for his own funeral "just in case." In any event, the bridge stood until 1993 when it was destroyed by mortar fire during the Bosnian War.

Figure 8.11 Stari Most.

Source: Copyright © Tanja Sulič/iStockphoto.

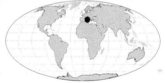

Water Distribution and Treatment

WATER DISTRIBUTION

The earliest form of water conveyance was in containers on the backs and heads of people. As civilizations became more advanced and as water had to be carried greater distances, conveyance systems using human-made channels and later, pipelines, were constructed.

Qanats have been used for millennia throughout Asia and North Africa to convey water, and are still widely used today. Qanats are human-made underground tunnels that intersect the groundwater table and convey the water to the surface some distance away. They function as shown in Figure 8.12. You can imagine how dangerous digging these wells and the connecting tunnels must have been; indeed, some workers wore their funeral clothes while digging in case they were buried alive.

Figure 8.12 Cross-section of a qanat.

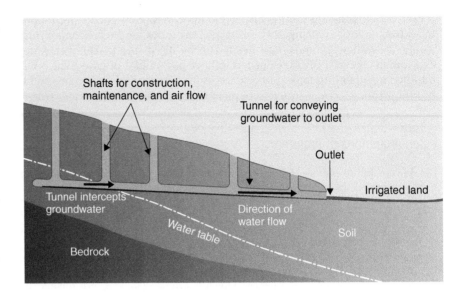

Shafts for construction, maintenance, and air flow

Tunnel for conveying groundwater to outlet

Outlet

Irrigated land

Tunnel intercepts groundwater

Water table

Direction of water flow

Soil

Bedrock

Figure 8.13 Example of a Roman Aqueduct, Built in Segovia, Spain.

Source: Copyright © Alan Crawford/ iStockphoto.

In addition to being outstanding road- and bridge-builders, the Romans were highly capable at water distribution. They built an infrastructure that allowed fresh water to be brought into Rome via aqueducts (Figure 8.13), and built aqueducts throughout their empire.

The water that flowed into ancient Rome did not do so solely in elevated aqueducts. It was also conveyed in tunnels, streams, pipes, and channels constructed at ground level. But, when a gully or gorge needed to be crossed, or where a long, gentle slope was desired that could not otherwise be obtained by building at ground level, elevated aqueducts were used. Pipes could also be used for such situations, but the technology of constructing pipes able to withstand the high pressures was limited at the time. Lead pipes were used throughout the empire, but the stone aqueducts were technologically and economically more feasible to construct for conveying large amounts of water.

The aqueducts supplying Rome conveyed a tremendous amount of water, estimated at about 50 million gallons per day at their peak, which corresponds to approximately 50 gallons per person (which is much more than the water use in most developing counties today). This water was conveyed to some private residences and public buildings, although most private residents obtained water from public fountains. The large amount of water transmitted was not all actually used by the Romans, however, as much of it simply passed through the city.

Another notable event in the history of water conveyance occurred in 1237, when lead pipes were used to convey water from Tyburn Springs to the City of London. The water was transported nearly 3.5 miles to a central location where city dwellers could obtain the water for free. The ability to expand these early systems was limited by technology—the ability to pump water and to deliver it cost-effectively via pipelines. The first water distribution systems in the United States were built in the mid-1700s. The turn of the 20th century brought about a tremendous increase in the number of cities with public water distribution, and by the early 1900s, nearly all cities in the United States had water distribution of some sort.

As cities became larger, water was seen as essential for firefighting purposes. Indeed, that was the sole purpose of some of the earliest municipal systems. Other uses were for domestic use, and, at the start of the 20th century, to flush human waste from "water closets."

Early pipes were made of lead or of bored out logs. If larger pipes were needed that could be made from logs, wood stave pipes (a stave is a slat of wood) were used (see Chapter 5, Figure 5.11). Surprisingly, some wooden pipes are still in service today in U.S. cities. During the 1800s, advances in ironworking led to the popularity of wrought iron and cast iron pipes. The later discovery that lead piping created a significant health hazard because of leaching of lead into drinking water resulted in the ban of its use. However, as with wooden pipes, lead pipes are still in service in many communities.

WATER TREATMENT

Humans have not always treated water prior to drinking, and even today, groundwater and surface water is directly consumed by many people worldwide. Much surface water would be safe to drink if not contaminated by fecal material. As humans began to congregate into larger groups, increased contamination of groundwater and surface water occurred. People polluted the surface water by defecating directly into the surface water or emptying their chamber pots into the surface water. Early wells were quite shallow due to available well-digging methods (i.e., a shovel, a bucket, and a rope), and thus were also easily contaminated; waste that had been dumped into cesspools or cesspits could readily travel underground into these wells.

The result of the mixing of wastewater with drinking water resources resulted in countless deaths due to water-borne diseases. The cause of these much-feared diseases (e.g., cholera and typhoid) are pathogens that originate from human **fecal material**.

A cholera epidemic in London in the mid-1800s caused more than 600 deaths. When this epidemic occurred, an estimated three-fourths of the city's residents fled to the countryside. This epidemic is notable not due to the number of fatalities, but rather that for the first time, a link was made between the cholera cases and drinking water. Prior to this, people didn't know if cholera (and other diseases that we now know to be water-borne) was water-borne or air-borne. Dr. John Snow, by plotting cholera deaths on a map, found that they were clustered around a single well. Removing the well handle (and thus access to the water) effectively ended the epidemic. Later, the water source was found to be contaminated from a nearby cesspit.

The fact that cholera is water-borne is one of the most important public health discoveries in history. But even at the time of its discovery, Dr. Snow and his contemporaries did not fully understand the cause of the cholera disease. Indeed, he writes in one of his reports on the Broad Street case of the "cholera poison" (implying a chemical nature). It was not until the end of the 19th century that the "germ theory" was discovered by Dr. Louis Pasteur; at this time, people first understood that the cause of cholera was microbiological in nature.

Early treatment of water by municipalities consisted of filtration through sand beds. These were common beginning in the mid- and late 1800s, and are still commonly used today. It was found that filtration effectively removed particles from the water, many of which caused objectionable tastes and odors. Later analyses showed that the filters also removed some (but not all) pathogens. However, it wasn't until the advent of chlorination that many water-borne diseases were virtually eliminated. The relationship between filtration and chlorination on typhoid outbreaks is shown in Figure 8.14.

The dramatic effect of water treatment on human health cannot be understated. Now, engineers continue to work at decreasing the possibility of disease transmission by water and there are increasingly strict regulations.

The World's First Engineer

Vitruvius was a Roman writer and architect who lived in the first century B.C. He has been called the world's first engineer. He wrote *The Ten Books on Architecture*, three of which were devoted to technology. He stated that a structure should have three characteristics (*firmitas, utilitas,* and *venustas* or durable, useful, and beautiful), advice that modern engineers should heed. He appears to have designed war machines (e.g., catapults) and wrote much about the aqueducts and the state of the art of surveying.

Fecal material is also known as poop.

Thomas Crapper

The toilet transformed the way homes functioned and smelled, and contributed to sanitary conditions. Legend has it that the toilet was invented by Englishman Thomas Crapper; however, despite the fitting name, the toilet was not invented by Mr. Crapper, and the actual inventor is unknown. Flush toilets gained popularity in the late 1800s to early 1900s, but as recent as 1950, 25 percent of American homes still used outhouses.

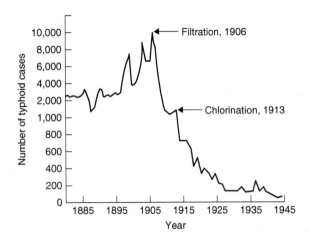

Figure 8.14 **Effect of Water Treatment on Number of Typhoid Cases in the United States.**

Source: Republished with permission of Taylor and Francis Group LLC Books, from *A History of Engineering and Technology: Artful Methods*, E. G. Garrison, 1998; permission conveyed through Copyright Clearance Center, Inc.

Figure 8.15 **Roman Latrine.**

Source: Copyright © Pascal RATEAU/iStockphoto.

Moreover, much work needs to be done worldwide to prevent millions of deaths annually due to water-borne diseases in developing nations.

Wastewater Conveyance and Treatment

WASTEWATER CONVEYANCE

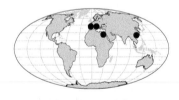

The earliest form of wastewater conveyance, similar to early water conveyance, was by people carrying the waste in containers and dumping them outside their settlement. This activity was not to prevent water-borne diseases that were not understood at the time, but rather to decrease odors and the number of insects. This method continues to be used in many developing countries today.

Wastewater conveyance via pipes and channels can be traced to before 2000 B.C., with systems in ancient Greece, Rome, China, and Egypt. Many of these systems functioned by having a large quantity of water continuously flush through them. In some cities there were connections to elite residences, but otherwise people relieved themselves into chamber pots that they dumped in street openings to the sewer or, in Rome, by using one of the pay latrines (Figure 8.15)

During the 19th century, many cities collected human sanitary waste, but not with pipes. Rather, people would defecate in chamber pots and leave them outside at night. The contents were delicately referred to as **night soil**. The contents would be collected and dumped in cesspits or in water bodies outside of city boundaries.

Once indoor water closets were available with flush toilets, the toilets needed to be connected to conveyance pipes; the most readily available pipes were the pipes in the *storm* sewer system. The resulting sewer system is known as a combined system (combining sanitary and stormwater), and were commonly constructed in the early 20th century. Indeed, many U.S. cities still are served by combined sewers. However, all newly constructed storm sewer and wastewater systems must be separated.

WASTEWATER TREATMENT

Civilizations have conveyed their wastewater for many centuries, but it is only until relatively recently that they have considered treating the wastewater. The philosophy seemed to be "out of sight (or out of nose-range), out of mind."

The first treatment method was to precipitate out the organic content using ferric sulfate. Around 1900, **trickling filters** were first used—bacteria attached to a substrate (e.g., rocks in the early stages of trickling filter construction and corrugated plastic media today) consume the dissolved organics in the wastewater as the wastewater trickles over the substrate. The activated sludge treatment process, which is very common today, was first introduced in 1916 with the first large-scale facility starting operation in Milwaukee, Wisconsin, in 1925.

Today, much of the world lacks proper sanitation, and significant public health risks are common as discussed in greater detail in Chapter 13, Sustainability Considerations.

And Then There Was More . . .

The examples presented thus far in this chapter of the history of transportation, water, and wastewater infrastructure illustrate the lessons to be learned from infrastructure development. Other infrastructure sectors and components could also have been highlighted but for space constraints have not been. This is not to say they are not historically important.

For example, much could be (and has been) written about railroads. Like much of the U.S. infrastructure, their burgeoning success in the mid- and late 1800s (from 10,000 miles of railway in 1850 to over 100,000 miles in 1890) was the result of a variety of previously unrelated technologies working together, namely the development of steam engines to pull the heavy loads up the relatively steep grades of the western U.S. railroads and the economical production of steel that was needed for the thousands of miles of rails. Also, like canals, railroads were a key component in increasing the trade capabilities of the nation and helped unify the country. Additionally, the government role in helping promote the railroad was significant, including issuing of government bonds (bonds will be presented in Chapter 14, Economic Considerations) and by giving 10 square miles of land to the railroads for each mile of track constructed.

Ports also are a very important part of the history of infrastructure. Even when the Romans had their exceptional road system in place, water was still the most cost-effective means of transporting goods. The infrastructure for water transport is relatively simple; no connecting infrastructure such as roads or railroad tracks are needed. But a port at each end of the journey is essential.

A summary timeline of several of the important infrastructure accomplishments and events mentioned in this chapter is presented in Figure 8.16.

The Future

The concept of **diffusion of technology** (Figure 8.17), also termed adoption of technology, describes the rate at which new technology is adopted by society, and is useful when contemplating the future of infrastructure engineering. The shape of the "bell curve" (solid line) is typical of the adoption rate of personal computers, telephones, microwave ovens, etc. The x-axis represents the time that has passed since the technology was originally invented and adopted. Depending on when you "adopt" the new technology, you may be labeled as an "innovator," an "early adopter," or even a "laggard."

Cloacina, Goddess of the Sewers

The Cloaca Maxima was constructed in Rome starting in perhaps 500 b.c. It was the "main drain" for the city and was flushed out by water brought to the city in the aqueducts. The Romans believed that the Goddess Cloacina oversaw the sewer systems. She was honored with a shrine and coins; some have even suggested the following poem might have been offered in prayer to Cloacina!

"O Cloacina, Goddess of this place,

Look on thy suppliants with a smiling face.

Soft, yet cohesive let their offerings flow,

Not rashly swift nor insolently slow."

The Worst Job Ever?

Cesspits had to be cleaned out regularly. This job was relegated to "gong farmers." According to the *Oxford English Dictionary*, the word "gong" refers to a privy (outhouse) or its contents. If only television were available at the time, certainly these laborers would have been highlighted in the show "Dirty Jobs."

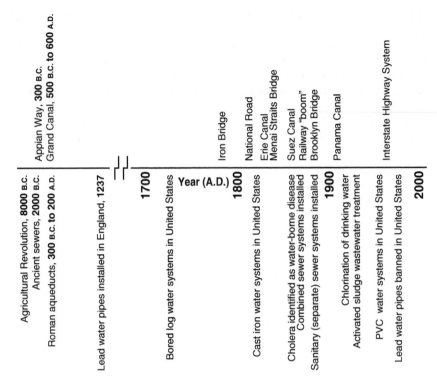

Figure 8.16 A Timeline of Infrastructure History as Presented in This Chapter.

This theory can also be applied to the adoption of new infrastructure systems. For example, the historical (and projected future) growth in system length for various modes of transportation is presented in Figure 8.18; the diffusion of innovation ''bell curve'' is seen for each transportation mode.

As you consider the future of infrastructure in light of Figure 8.18, you should be asking yourself, ''what comes next?'' The peaks in Figure 8.18

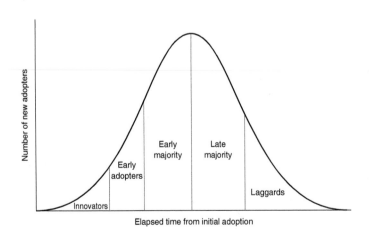

Figure 8.17 The Diffusion of Technology.

Source: The names of the adopter categories are borrowed from Rogers, 2003.

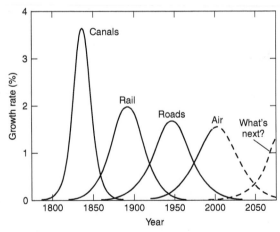

Figure 8.18 The Historical and Projected Future Growth Rates of Various Transportation System Lengths Since 1800.

Source: Redrawn from Grubler, 1990.

arrive at intervals of approximately 60 years. The future of transportation may be maglev, or magnetic levitation, trains, or it may be a technology that is currently in even earlier stages of development. It is easy to scoff at this prediction and to believe that in 50 years we will still be using automobiles as our primary personal transportation mode for example, or to say "we will never use trains regularly in place of automobiles." But recall that this same outlook in the mid-1800s was one of the reasons that the National Road was not completed: many people felt there was no future in roadways, but rather that canals and railways were the future!

Figure 8.18 can also be instructive when combined with the labels of Figure 8.17 and by considering *your* future in infrastructure development. Just as we can think of different people adopting technologies at various rates, we can also consider the people who have invented and marketed the new technologies. Will you be an "innovator" in the new transportation technology that may begin to appear in the very near future? Such a role is a "riskier" role, and by its very nature, will most likely be a role not desired by many people. Or, perhaps you will wait, working in areas that have reached "saturation." In such areas, you will still have opportunity for some innovation, but the work will most likely be involved in maintenance and rehabilitation rather than in new construction.

Figure 8.18 can also be applied to any sector of infrastructure and at many different levels, from the sector level to the subsector level to the component level. For example, a similar graph could be drawn for various types of dams that have been constructed through the ages, different types of wastewater treatment processes, different types of pumps, etc.

The future holds many exciting projects awaiting infrastructure engineers. A very large pumping station, the West Closure Complex Pump Station, is currently being designed for New Orleans; $5 billion is being spent to expand the Panama Canal; high-speed rail is being planned throughout the United States. In your lifetime the tallest building will be completed, as will the longest pipeline, the fastest train, and the most resilient bridge.

The National Academy of Engineering recently identified several "Grand Challenges" for the upcoming century. Those challenges, as well as the great challenges that were met in the past century are listed in Table 8.1. Note how many of these challenges directly relate to the infrastructure, and how many of them are indirectly dependent upon effective infrastructure.

> ### The Importance of Information Technology
>
> Tremendous advances in data collection and data processing have been made, and will continue to be made. There is a new focus on "smart grids," "smart roads," etc. Sensors are being developed to very accurately measure environmental pollution, precipitation events, stresses on structural members, and traffic patterns at remote locations with real-time feedback. This information allows engineers and scientists to not only evaluate the functionality of existing or prototype infrastructure systems, it also contributes to an ever-increasing data set that aids future research and development.

Video on Future Bridges

Grand Challenges

Outro

By reading this brief introduction to the history of infrastructure, we hope that you more fully understand the importance of infrastructure and how it is necessary for the functioning of society. By revisiting our four themes in historical contexts, we hope you better appreciate the human dimension of infrastructure, the need for engineers who are innovative and eager to adopt new technologies, the need for a strong government and visionary leaders. Moreover, we hope you consider longevity and aesthetics in your future engineering projects.

Note that these themes will be revisited in other chapters. The human dimension will be explored in more detail in Chapter 16, Social Considerations. New technologies will be covered in many chapters, including the Design and Analysis chapters (Chapters 9, 10, 20, and 21). Chapter 19, Other Considerations, will discuss the politics involved in infrastructure construction and management, and also aesthetics.

Table 8.1

The Grand Challenges and Achievements of Engineering	Challenges for the Next Century	Achievements of the Past Century
	Make solar energy economical	Electrification
	Provide energy from fusion	Automobile
	Develop carbon sequestration methods	Airplane
	Manage the nitrogen cycle	Water supply and distribution
	Provide access to clean water	Electronics
	Restore and improve urban infrastructure	Radio and television
	Advance health informatics	Agricultural mechanization
	Engineer better medicines	Computers
	Reverse-engineer the brain	Telephone
	Prevent nuclear terror	Air-conditioning and refrigeration
	Secure cyberspace	Highways
	Enhance virtual reality	Spacecraft
	Advance personalized learning	Internet
	Engineer the tools of scientific discovery	Imaging
		Household appliances
		Health technologies
		Petroleum and petrochemical technologies
		Laser and fiber optics
		Nuclear technologies
		High performance materials

Source: National Academy of Engineering.

chapter Eight Homework Problems

Short Answer

8.1 Why was construction of the National Road so important to the young American nation?

8.2 How did John McAdam's roads differ from the Roman roads?

8.3 How did the federal government help finance the construction of the transcontinental railroad?

8.4 What new technologies were important in the history of bridges?

8.5 What is the "germ theory" and how is it important to the history of water treatment?

8.6 Why did the Romans use elevated aqueducts instead of pipes to convey their water?

8.7 What are combined sewers?

Discussion/Open Ended

8.8 One reason that some infrastructure is preserved by society is that society finds it beautiful. Predict the infrastructure on your campus that will be preserved for 100 more years, in part because of its beauty.

8.9 Which of the Grand Challenges for the 21st century do you believe is most important? Why? Which challenge do you believe to be the least important? Why? What challenge do you believe should be added to the list? Why?

8.10 New infrastructure may be needed in the near future to allow the "refueling" of electric vehicles. Should the public sector provide this infrastructure, or should the private sector? Explain your reasoning.

Research

8.11 For one of the following historic infrastructure projects, (1) describe the project and explain why it was needed; (2) If possible, state how much it cost, in today's dollars using an Internet "inflation calculator"; (3) list its beneficial and negative effects; (4) explain the role of new technology; (5) list the ways it affected people; (6) describe the role of government (federal, state, local) in the project.

a. St. Lawrence Seaway
b. China's Grand Canal
c. Paris historic sewer system
d. New York City's historic water treatment and distribution system
e. Brooklyn Bridge
f. Cloaca Maxima
g. Eads Bridge
h. Severn Tunnel
i. Grand Coulee Dam
j. Hoover Dam
k. Machu Picchu irrigation system
l. Rideau Canal
m. Lincoln Tunnel
n. London Bridge
o. Transcontinental Railroad
p. Trans-Siberian Railway
q. Trans-Alaska pipeline
r. "Chunnel"
s. Empire State Building

8.12 List the milestone events in the history of the evolution of one of the following infrastructure components.
a. Airports
b. Ports
c. Locks
d. Tunnels
e. Bicycle transit
f. Skyscrapers
g. Dams
h. Hazardous waste management
i. Landfills
j. Railroads

8.13 For one of the components in Homework Problem 8.12, describe the role of technology in the component's history.

8.14 Select one of the following famous infrastructure engineers. Address (1) main contributions; (2) educational experiences; (3) obstacles faced; (4) risks taken; (5) information they added to the knowledge base.

a. Thomas Telford
b. Benjamin Wright
c. Isambard Kingdom Brunel
d. Karl Terzaghi
e. Gustave Eiffel
f. Ellis Chesbrough
g. Robert Fulton
h. Theodore Judah
i. John Roebling
j. Emily Roebling
k. Squire Whipple
l. Abel Wolman
m. James Eads
n. Ellen Swallow Richards
o. John Smeaton
p. Elsie Eaves

Key Terms

- diffusion of technology
- fecal material
- government responsibility
- grand challenges
- human dimension
- longevity
- private sector
- public sector
- reinforced concrete
- technological innovation

References

American Public Works Association. *Top Ten Public Works Projects of the Century.* http://www.apwa.net/Resources/Reporter/Articles/2000/11/Top-Ten-Public-Works-Projects-of-the-Century, accessed August 8, 2011.

Garrison, E. G. *A History of Engineering and Technology: Artful Methods.* CRC Press, 1998.

Grubler, A. *The Rise and Fall of Infrastructures: Dynamics of Evolution and Technological Change in Transport (Contributions to Economics).* New York: Springer-Verlag, 1990.

National Academy of Engineering. *Grand Challenges for Engineering.* http://www.engineeringchallenges.org/cms/8996/9221.aspx, accessed August 8, 2011.

National Academy of Engineering. Greatest Engineering Achievements of the 20th Century. http://www.great achievements.org/, accessed August 8, 2011.

Rogers, E.M. *Diffusion of Innovations.* New York: Free Press, 2003.

Sandström, G.E. *Man the Builder.* New York: McGraw-Hill, 1970.

Williams, F.S. *Our Iron Roads: Their History, Construction, and Social Influences.* New York: Ingram, Cooke, and Co, 1852.

Wright, K.R. and A.V. Zegarro. *Machu Picchu: A Civil Engineering Marvel.* Reston, VA: American Society of Civil Engineers, 2000.

chapter Nine Analysis Fundamentals

Learning Objectives

After reading this chapter, you should be able to:

1. Describe why engineering analyses are conducted.
2. Explain the steps in the analysis process.
3. Conduct a simplified engineering analysis in the major sub-discipline areas of infrastructure engineering.

Introduction

To many people, "engineering" and "design" are nearly synonyms. Herbert Hoover, a civil engineer and 31st president of the United States, had this to say about engineering:

"It is a great profession. There is the satisfaction of watching a figment of the imagination emerge through the aid of science to a plan on paper. Then it moves to realization in stone or metal or energy. Then it brings jobs and homes to men. Then it elevates the standards of living and adds to the comforts of life. That is the engineer's high privilege."

This eloquently worded description fits many people's image of engineering, namely an occupation centered on design and creation of something that did not previously exist. Indeed, a large portion of an engineer's work is focused on designing new infrastructure or designing a retrofit of an existing portion of the infrastructure. However, a significant portion of the work performed by civil and environmental engineers consists of analysis, which is a distinctly different task than design. Design is covered in more detail in the next chapter, while this chapter introduces the concept of analysis, which often precedes design.

Introductory Case Study: Bridge Inspection

Bridge inspection is an analysis task carried out by technicians and engineers for federal and state highway departments. Engineers and technicians inspecting bridges are shown in Figure 9.1 and Figure 9.2. Such inspections are carried out regularly and have been subjected to increasing scrutiny following the collapse of the I-35 W bridge in Minneapolis in August 2007.

U.S. National Bridge Inventory

PBS Video—"The Price of Decay": Bridges

Figure 9.1   The Eagle Canyon Bridge on I-70 Near Green River, Utah.   For this analysis, engineers used an Under Bridge Inspection Vehicle (see square) to inspect the deck underside, and rope access (see circled worker) to inspect the columns and arches.

Source: Courtesy of Ayres Associates.

Figure 9.2   Engineers (Circled) Use Rappelling Techniques to Inspect a Bridge.

Source: Courtesy of Ayres Associates.

The Nature of Analysis

In its most generic form, analysis is the process of identifying and characterizing a problem. Problems may be identified by elected officials, residents, public servants, or investigative journalists, among others. However, if engineers are to hold paramount the safety, health, and welfare of the public, as they are adjured to do by the ASCE Code of Ethics,[1] they also must be active in identifying problems. We believe that engineers should be **citizen engineers**, actively taking part in problem identification and decision-making at local, state, and national levels. For example, a citizen engineer that notices an unsafe highway intersection would alert the appropriate government entity of the problem.

Once a problem has been identified, the extent of the problem (or equally important, whether or not the problem truly exists) must be assessed by the engineer in coordination with professionals from many other disciplines. Characterizing the extent of a problem most likely requires the collection of data. Engineers *must* base their decisions on logical arguments and evidence and *not* on emotion or hearsay; in other words, the decision-making must be objective rather than subjective. Consequently, the collection of data is essential to ensure that the engineer recommends a solution to the problem that is defensible and is in the public's best interest. Of course, engineers often do not have the final say in a matter and in some cases, the final decision may be made by elected officials with no engineering expertise.

An engineering analysis might be conducted for several reasons:

- determine which component of a system is the weakest link;

- recommend whether an existing component should be upgraded, replaced, or left alone;

[1] The ASCE Code of Ethics will be presented and discussed in more detail in Chapter 17.

- use data that characterize a system to create a model of a process/system/phenomenon that can be used for other similar systems;

- evaluate whether an infrastructure system (or component of the system) is functioning according to its designed capability and meeting the needs of the population it is serving today;

- evaluate whether an infrastructure system (or component of the system) will meet the needs of the population at some point in the future; or

- predict how long an infrastructure system (or component of the system), proposed or existing, will function effectively.

The Analysis Process

The design process is well defined and is often described as a cyclical process (see Chapter 10). The analysis process is more difficult to define, as one analysis project can differ in many ways from another. We have identified eight possible components of an engineering analysis (see analysis process sidebar). An actual analysis might be composed of one or two of these components, all eight components, or possibly others not identified.

1. *Definition of scope:* All analyses must have a well-defined scope. Quite often, the scope is limited by available funding and by time constraints. The purpose of defining the scope is to answer the question, "How far should we go?" Without defining the scope, an engineer will not necessarily know when he or she has completed his or her analysis to a satisfactory degree of specificity.

   In terms of bridge inspection, defining the scope is of utmost importance. The scope will vary depending on whether the inspection is a routine inspection (e.g., an inspection that is to be carried out once every 24 months), an underwater inspection, or a damage inspection (e.g., an inspection following a hurricane or a vehicle crash).

2. *Background research:* Engineers are routinely faced with many tasks for which their education did not fully prepare them. This occurs not necessarily because of a deficiency in their education, but rather because of the incredible breadth and complexity of our modern infrastructure. As a result, engineers must be prepared to continually direct their own learning and conduct research on new technologies, processes, and best practices. Vendors will often provide information, but the engineer must be wise enough to distinguish between facts and a "sales pitch." Similarly, the Internet contains many potential resources, but engineers must be careful to assess each of the sources they use.

   The Federal Highway Administration provides a strict definition for the qualifications required of a bridge inspector. The bridge inspector may need to complete additional background research on a site-specific basis, and also may need to augment knowledge gained from undergraduate and/or graduate courses in structural analysis and design. This background research may come in the form of continuing education courses and seminars.

3. *Data collection:* The purpose of data collection is to characterize the component of the infrastructure that is being analyzed. Data may be collected by physical observation (e.g., counting cars passing through an intersection, or noting brick liner failures that could lead to groundwater infiltration

The Analysis Process

1. Definition of Scope
2. Background Research
3. Data Collection
4. Data Organization
5. Data Analysis
6. Model Application
7. Model Development
8. Recommendation

Evaluation of Web Page Content

Figure 9.3 A Young Engineer Inspecting a Brick-Lined Sewer Enters Data into a Handheld Computer.

Source: Courtesy of M. Davis.

Concrete Problems

Delamination: This is a condition where the bond between the concrete and the reinforcing steel is weakened. It may or may not be visible from the surface, but can be detected by lightly tapping on the concrete with a hammer and listening for a "hollow" sound.

Cracking: Cracks can appear for a number of reasons. An experienced bridge inspector can differentiate between non-critical cracks and cracks that may pose danger to the users of the bridge.

Spalling: Spalling is noted as loose or missing pieces of concrete; it may occur as a result of delamination or cracking.

Microsoft Excel Tutorial

in storm sewers as shown in Figure 9.3); measurements (e.g., measuring the thickness of a structural component or the temperature of a river); non-destructive testing (e.g., locating water leaks) or invasive testing (e.g., drilling sample cores of pavement cross-sections).

For bridge inspection, data collection consists of examining the bridge for any defects or problems that are present. In steel components, defects include cracks, rust, and corrosion. In concrete, the inspector looks for delaminating, spalling, and cracking (see "Concrete Problems" sidebar). Other typical problems include compaction/settlement of the approaches leading up to the bridge, scour around piers in the water, and joint failure. Data are entered into an inspection report, and photographs are taken of any deficiencies.

4. *Data organization:* Engineers can compile a very large amount of data as they analyze infrastructure systems or components. Organizing this data such that it can be retrieved and analyzed in an efficient manner is a key component to ensuring effective analysis. Spreadsheets are perhaps the most common tool, although databases are also widely used.

A GIS (Geographic Information System) database is a type of database that greatly facilitates storage and retrieval of data. GIS links data to spatial location; for example, a city map of streets is linked so that information about each individual street (e.g., type of pavement and date of most recent resurfacing, utilities under the pavement, traffic counts) can be easily retrieved. Information that was previously stored in several different computer files or folders in a file cabinet is now much more accessible. Applications are far ranging, including the management of utilities, parking structures, and cemeteries.

Data from the reports generated by a bridge inspection are typically organized using a database. These inspection reports are available to the public, often via the Internet (e.g., the National Bridge Inventory published by the U.S. Federal Highway Administration).

5. *Data analysis:* Given the large sets of data (perhaps on the order of hundreds or thousands of pieces of data) that engineers often use, the use of a spreadsheet or database is essential. Spreadsheets allow engineers a means of organizing their data; more importantly, a spreadsheet facilitates repetitious calculations on the large sets of data by copying and pasting of mathematical formulas. Spreadsheets also make it easy to create graphics, which should be used whenever possible to present data in an easily understood manner.

Typically, the analysis of the data collected by a bridge inspector does not require advanced statistical analysis. Rather, data can be analyzed using the basic tools found in the "Simple Analysis Tools" sidebar.

6. *Model application:* In its simplest form, a model is a representation of a process or product. That is, a model answers the question, "How does this thing work?" In engineering, many models are expressed mathematically as a formula or an equation. Models may be descriptive (to explain what occurs) or predictive (to predict what may occur given some future change in conditions). In reality, many models used by engineers are quite complex, consisting of many equations and are typically manipulated using software. Models are available for nearly every aspect of the infrastructure. For example, models exist to predict the available water pressure at various points within a drinking water distribution network; the deflection in

a bridge due to loading by motor vehicles; and the efficiency with which traffic moves through an intersection.

Models are not used typically in a bridge inspection, although conceivably a model could be created based on the data collected. For example, a predictive model could be created to help engineers and policy makers predict when a bridge will need a major refurbishment, based on the results of the bridge inspection.

Simple Analysis Tools

Many engineers must use advanced statistics to analyze their data, but some of the simplest statistics are used nearly every day by practicing engineers. These include the **mean, median, mode, maximum, minimum, range,** and **standard deviation.** The mean is the average obtained by dividing the sum of the data entries by the number of entries while the median is the middle number in a set of data, corresponding to the 50th percentile value. The mean is more likely than the median to be affected by outliers. The mode is the value that occurs with the greatest frequency. The maximum represents the largest value in the data set while the minimum represents the smallest value. Range is defined as the mathematical difference between the maximum and the minimum. Standard deviation is a measure of the "spread" of the data relative to the mean value.

7. *Model development:* Although practicing engineers often rely on existing models developed by other engineers and scientists, the goal of an analysis project might be to create a new model. Upon examining a process or technology and characterizing it with appropriate data, an engineer can create a model to predict future performance of the process as it relates to other conditions or other sites. This is described in more detail in a case study later in this chapter.

Model Complexity

Often, engineers have access to more than one model that can potentially meet their needs. One criterion that engineers must consider when selecting a model is the trade-off associated with model complexity: a simple model may not incorporate all of the necessary variables to adequately describe a system whereas a very complex model may have so many variables that determining appropriate input values is not cost effective. This tradeoff is shown schematically in Figure 9.4; for some situations, an overly simplistic model may be as useless as a model that is needlessly complex. Furthermore, complex models can only be used by highly trained and experienced engineers, making this work costly and almost impossible to "hand over" to the client for their subsequent use.

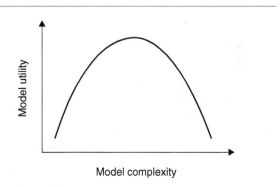

Figure 9.4 Model Utility as a Function of Model Complexity.

In some cases, physical rather than mathematical models are created. For example, scaled down physical models of bays or ports are created to study water movements, such as the 1.5-acre (1:1,000 scale) model of San Francisco Bay built by the Army Corps of Engineers in 1957 to study hydrodynamics as well as the impacts of pollution, dredging, and filling. Scaled-down versions of treatment processes, termed **pilot studies,** are also common. A pilot study can be used to verify results expected from a mathematical model.

Garbage In Equals Garbage Out

With any modeling exercise, the mantra "garbage in equals garbage out" applies; in other words, the model output can be of no higher quality than the model input. Sometimes, models must be based on a limited quantity of data, or even in some cases, data of questionable quality (e.g., data that is very old or that is from an unknown or untrusted source). If, out of necessity, an engineer is thrust into a situation with questionable data, he or she must make appropriate and defensible assumptions, document those assumptions, and ensure that the model results are presented with an appropriate disclaimer. Moreover, many young engineers are quick to realize that because of the large number of input variables and the large range of "acceptable" values for those input variables, models can generate a wide range of results. The ability to pick and choose values that can produce a "favorable" result may pose an ethical challenge to engineers, in that changing the value of a single variable may result in an answer that will either make a solution accepted or rejected by a client or a regulatory agency.

8. *Recommendation:* Often, an engineering analysis concludes with a recommendation. The recommendation should be objective and based on the data collected.

   Results from a bridge inspection are used to make important recommendations. These recommendations may: state when and to what extent a certain bridge needs maintenance; revise a county's or state's ongoing maintenance plan; create a list of bridges, ranked in order of need for refurbishment, in order to determine which bridge gets funding or not.

numerical example 9.1 Data Analysis with Spreadsheets—Water Mains for a Small City

Spreadsheet of Pipes for a Small Water Distribution System

A portion of a spreadsheet that lists all of the water mains for a small city is shown in Figure 9.5. The city is approximately 4 square miles in size and has a population of 10,000. A spreadsheet containing the entire data set (1,324 water mains) is used to answer the following questions, and can be downloaded from the textbook website.

1. What is the total length of pipes in the city?
2. Four-inch diameter pipes are no longer allowed to be installed for municipal water systems, yet older systems often contain 4-inch pipes. How many of these undersized pipes are located in the city's system?
3. What is the most common pipe size for the city, based on the number of pipes?
4. What is the most common pipe size for the city, based on the overall length of the pipe?
5. What is the maximum pipe size?

solution

1. What is the total length of pipes in the city?

   a. Using the SUM command, the total length is found to be 303,143 feet (or about 175 people per mile), which is approximately 57 miles. This may seem to be a surprisingly large number given the relatively small size of the city. However, water mains are needed to serve all residences and businesses, and therefore are located along the length of nearly every street. You can approximate the length of water mains of a much larger city by direct ratio; for example, a city of 100,000 people might have 570 miles

numerical example 9.1 Data Analysis with Spreadsheets—Water Mains for a Small City (Continued)

	A	B	C
1	Label	Length	Diameter
2		(ft)	(in.)
3	P-1411	25	6
4	P-1410	697	6
5	P-1409	92	6
6	P-1408	370	6
7	P-1407	85	6
8	P-1406	674	12
9	P-1405	278	12
10	P-1404	27	6
11	P-1403	12	12
12	P-1402	36	12
13	P-1401	26	6
14	P-1400	19	8
15	P-1399	120	8
16	P-1398	20	6
17	P-1397	393	6

Figure 9.5 A Portion of a Spreadsheet of Water Mains for a Small City.

of water mains. However, this approximation is likely an overestimate because the number of people per mile of water main increases with population, as larger cities have higher population densities (people per square mile).

2. How many 4-inch pipes are located in the city's system?

   a. One way to answer this question is to first sort the data by pipe diameter. Then the number of pipes can be manually counted. Alternatively, you can investigate and use the COUNTIF function if you are using Excel; other spreadsheet tools may have a similar command. In any case, the water system contains 88 4-inch pipes.

   b. Note that if the city's piping system was organized using a GIS, this question could have been answered very efficiently by building a "query" to count all the 4-inch pipes. Additionally, a GIS database that has been built correctly would allow the user to identify on a map, the location of all 4-inch pipes.

   c. You may be wondering why a city would have so many 4-inch pipes if regulations do not allow them in new construction? This is a common scenario, in which today's engineer is faced with upgrading engineering work that was once "state-of-the-practice," but is now obsolete. The reason is most likely a cost issue. Cities have increasingly tight budgets, and tearing up streets for the sole purpose of replacing sections of 4-inch pipe would be a very costly undertaking. It would also create traffic problems, disrupt neighborhoods, and negatively affect businesses. However, given that sewer lines and water mains are typically located under the roadway, a municipality will consider replacing and upgrading the water mains (and sewers, if outdated) when the pavement is replaced.

3. What is the most common pipe size for the city, based on the number of pipes?

   a. This question is asking for the mode of the diameter data set. In the spreadsheet, the MODE function returns an answer of 8 inches.

Lots and Lots of Pipes
San Antonio, Texas, which has a population of 1.4 million, has over 5,100 miles of water mains (275 people per mile of water main). Lined end-to-end, the pipes would reach from San Antonio to Paris, France.

Data Analysis with Spreadsheets—Water Mains for a Small City (Continued)

4. What is the most common pipe size for the city, based on the overall length of the pipe?

    a. This question can be answered by sorting the data and summing up the lengths of pipe for each pipe size. A more elegant technique is to use the SUMIF function in Excel. A summary of the results is provided in Figure 9.6.

5. What is the maximum pipe size?

    a. Using the MAX function in the spreadsheet, the maximum pipe size is found to be 80 inches. However, this answer raises a "red flag" for many reasons. First, 80 inches is not a standard water main size, which an experienced municipal engineer would recognize, and a water main of that magnitude would only be found in very large systems. Second, this pipe size is the only pipe of its size, and is nearly seven times larger than the next largest pipe. Most likely, this pipe size is the result of a data entry error, and the pipe size is actually 8 inches in diameter. Unfortunately, data entry errors are quite common in spreadsheets and databases that describe infrastructure components, and engineers should carefully analyze large data sets for such errors.

    b. This data entry error explains why the sum of the pipe lengths in Figure 9.6 (302, 986 feet) does not equal the answer in part "a" (303,143 feet). The difference between the two sums (157 feet) is the length of the supposed 80-inch pipe.

Diameter (in.)	Length (ft.)
4	23,158
6	74,216
8	101,048
10	42,744
12	61,820

Figure 9.6 **Results Summary.**

The Importance of Significant Figures

The importance of **significant figures** is routinely discussed in introductory chemistry courses. Unfortunately, it is often neglected thereafter. It is very important for engineers to consider significant figures (or "sig figs") in their analyses. You are encouraged to review the rules of significant figures, which you can find in an introductory chemistry or physics textbook, or by using a trusted Internet source.

One of the fundamental rules of significant figures is that when multiplying several numbers, the final result has no more significant figures than that of the number with the least significant figures. Thus, engineers should make sure that they have the best (or most significant) data available. As a brief example, the average pressure (force per unit area, P) acting on a floodwall is determined by the following equation.

$$P = \rho \cdot g \cdot h/2$$

where

$\rho =$ the density of water (ρ, 998 kg/m^3 at 20 deg C)
$g =$ acceleration due to gravity (9.8 m/s^2)
$h =$ the height of the water column at the floodwall (m)

For the units provided, pressure will have units of Pascals (1 Pa = 1N/m^2 and 1N = 1 kgm/s^2). The gravitational constant is typically assumed to be 9.81 m/s^2, but actually varies with latitude and elevation (e.g., 9.802 m/s^2 in New York City and 9.796 m/s^2 in Denver) and is provided to two significant figures. The density of water (which varies with temperature) is given to three significant figures. Thus, the height becomes the controlling dimension with respect to significant figures. The calculated pressure will have only as many significant figures as the given water height. If the water height is 2 m and the values for g and ρ provided are used, a calculator will return an answer of 9,790.038 Pa (9.790038 kPa), but *must* be expressed as 10 kPa. If the water height is given as 1.9 m, a calculator will return a value of 9,300.861 Pa but must be expressed as 9.3 kPa. These answers differ by 7 percent, which is far from trivial.

Undisciplined engineering students (and some practicing engineers) have a tendency to report *too many* significant figures. This can be very misleading—an answer with several significant figures *suggests* a high level of certainty in the calculation. Thus, it is important for the engineer to clarify the conditions or requirements of any given situation.

Case Study: Creation of the Rational Method

One purpose of presenting this case study is to demonstrate that civil and environmental engineers can analyze data with the result of creating a model to be used by other engineers. For this case study, the creation of a method to characterize stormwater runoff rates is discussed.

Civil and environmental engineers often need to predict the runoff created by a storm event for a project. The engineer may be interested in the *volume* of flow (e.g., gallons, cubic feet, or acre-feet) or the peak **flowrate**.

One of the most widely used methods for predicting peak stormwater runoff rates is the rational method. This model was published over 100 years ago, in 1889, by Emil Kuichling, a civil engineer employed by the city of Rochester, New York. Despite the age of this method and its many simplifying assumptions, this simple model is widely used today because of its applicability (see sidebar on Model Complexity). The steps of the analysis process are highlighted below for this example:

1. Identify the Problem
   Kuichling recognized that the models of the day underestimated flows for storm events of short duration; using the existing models could lead to undersizing of storm sewer pipes, which in turn could lead to flooding.

2. Define Scope
   Most likely, Kuichling started his work by defining a scope of more clearly understanding the relationship between runoff and rainfall to develop an improved model.

3. Collect Data
   Although recording rain gauges were much more rare at the end of the 19th century than they are today, Kuichling was able to collect relatively large amounts of precipitation data. Moreover, he collected data on the flowrate of stormwater in sewers for a large number of storms by placing whitewashed wooden rods in the sewers; the whitewash was removed when submerged by stormwater, and thus the high water mark could be readily deduced based on the length of whitewash remaining at the top of the rod. The water depth combined with the geometry of the sewer could then be used to estimate a flowrate.

4. Analyze the Data
   Kuichling carefully analyzed the data and found that the volume of runoff was directly proportional to the amount of impervious area (e.g. paved surfaces) and to the intensity and duration of the rain.

Flowrate is the volume of water that passes a certain point in a given amount of time, commonly expressed in units of cubic feet per second (cfs) or gallons per minute (gpm). Predicting this runoff volume or flowrate for a given watershed area and a given rainfall intensity (e.g., 1 inch/hour) is a very common task for engineers.

5. Model

Based on his analysis, Kuichling created a mathematical model that is now known as the "Rational Method." The Rational Method is defined by Equation 9.1.

$$Q_p = C \cdot i \cdot A \qquad (9.1)$$

where
Q_p = the peak flowrate (ft^3/s, or cfs)
C = runoff coefficient, dimensionless
i = intensity (in./hr), by convention the "i" is not capitalized
A = area of the contributing catchment (acres).

The product of C, i, and A using the units provided is acre · in./hr. By coincidence, 1 acre · in./hr is approximately equal to 1 cfs. The units of cfs are commonly used by engineers, and the resulting error of this approximation is minimal (less than 1 percent).

Runoff coefficient values have been determined for various types of land cover (e.g., parking lots, housing subdivisions, agricultural land) by using the above model with known values of flow, rain intensity, and catchment area. Values for the runoff coefficient are provided in Table 9.1; note that a range of values are presented, and the engineer must use professional judgment in deciding whether to use the minimum, maximum, average, or some other value.

6. Recommendation

Kuichling summed up his article with this recommendation: "Much room for improvement in this direction is still left, and it is sincerely hoped that the efforts of the writer will be amply supplemented by many valuable suggestions and experimental data which other members of the Society [American Society of Civil Engineers] may generously contribute." Today, many more complex mathematical models exist to estimate storm runoff; however, the Rational Method is still commonly used because of its simplicity.

Table 9.1

Rational Method Runoff Coefficients for Various Land Uses and Surface Types

Land Use	C	Surface Type	C
Downtown business	0.70–0.95	Asphalt and concrete	0.70–0.95
Single-family residential	0.30–0.50	Roofs	0.75–0.95
Multi-family residential	0.40–0.75	Lawns, sandy soil, flat (<2%)	0.05–0.10
Light industrial	0.50–0.80	Lawns, sandy soil, moderate slope	0.10–0.15
Heavy industrial	0.60–0.90	Lawns, sandy soil, steep (>7%)	0.15–0.20
Agriculture	0.10–0.50	Lawns, silty/clay soil, flat (<2%)	0.13–0.17
		Lawns, silty/clay soil, moderate slope	0.18–0.22
		Lawns, silty/clay soil, steep (>7%)	0.25–0.35

Source: Data extracted from ASCE and WPCF, 1969.

Analysis Applications

For the remainder of this chapter, we will present sample analysis tasks undertaken in six sub-discipline areas of civil and environmental engineering: construction engineering, environmental/water resources engineering, geotechnical engineering, municipal engineering, structural engineering, and transportation engineering.

CONSTRUCTION ENGINEERING ANALYSIS APPLICATION—GANTT CHARTS

One task widely undertaken by engineers is the management of engineering projects. You may have thought that management would be the responsibility of people trained in business or management; although this sometimes occurs, engineers are often the most qualified to manage infrastructure construction projects. A common career path for a consulting engineer leads from design support and field work to more detailed engineering design work to engineering project management and then to team management (responsibility for overseeing several project managers).

One very important aspect of project management is **scheduling**, which is an analysis of the sequence of various project components. The first step in completing a project schedule is to determine the project tasks. As an "everyday" example, the tasks involved in painting a room include buying the paint, removing the furniture from the room, filling any holes in the walls, applying the paint, cleaning up, and replacing the furniture upon completion.

The next step in creating a project schedule is to determine the precedence for each task. In other words, for a certain task to occur, what other tasks must occur before it? To continue with the painting example, you cannot apply paint until you have purchased the paint, removed the furniture, and filled the holes. But you can remove the furniture and fill holes in the wall without buying the paint first.

Before developing a project schedule, you should also analyze the project tasks to determine how long each task will take. The dependency of one task on another determines potential start dates. This dependency may also have varying levels of importance; in other words, it may be *preferred* that a task is complete before moving to the next, or it may be *absolutely necessary* that the task be completed.

A Gantt chart is a graphical representation (Figure 9.7) of project tasks as a function of time. Each task is represented by a separate bar, noting start and end dates, and if known, the tolerance or variability of these dates. Important events (or **milestones**) are often also included in Gantt charts. Arrows are often shown to represent direct linkages between various tasks.

These charts are routinely used by engineering project managers to track project progress. Often they are updated so that when a task is completed, the bar color will be changed for quick visual reference. Gantt charts allow engineers to see all of the important steps of a project, often on a single page. They are continually referred to and regularly updated, as changes are inevitable in any large project. Critical tasks are often highlighted to note their significance. For example, if the delivery of a key piece of material or machinery is delayed by 2 weeks, the impact of this delay may be readily determined from the chart. Another example would be construction delays due to extreme weather events. Alternative solutions to accommodate these delays can be thoughtfully evaluated in advance, rather than making a "rash" decision in the field that may be potentially problematic.

Example Gantt Chart for a Large Transportation Project.

Project: The Taco Place—Bedford, SC		Month	Jan				Feb				Mar				Apr				May				Jun				
Task	Duration (days)	Week	1	2	3	4	1	2	3	4	1	2	3	4	1	2	3	4	1	2	3	4	1	2	3	4	Responsible Party
Site topographic survey	2		▓																								ABC Surveyors
Geotechnical survey	3			▓																							JN&T Geotech
Site plan	7				▓																						MRP Engineers
Building permit	7					▓																					City
Soil erosion plan	2					▓																					MRP Engineers
Soil erosion plan review and approval	21						▓	▓	▓																		DNR
Structural design	28						▓	▓	▓	▓																	KMT Structures
Foundation design	14									▓	▓																KMT Structures
Parking design	2										▓																MRP Engineers
Design review	3											▓															Client
Earthwork	14												▓	▓													Badger Excavating
Foundation construction	7														▓												BOB Builders
Structural construction	30															▓	▓	▓	▓								BOB Builders
Plumbing, heating, ventilating, lighting	7																			▓							BIL HVAC
Exterior façade	7																				▓						Bricks-R-Us
Parking lot construction	5																					▓					BOB Builders
Finish Interior	14																						▓	▓			Multiple
Final walk-through with client	1																								▓		MRP/Client

Figure 9.7 A Simplified Hypothetical Example of a Project Gantt Chart.

While our discussion has focused on the usefulness of a Gantt chart after it has been made, it should also be noted that the process of making the Gantt chart for a project may be equally helpful. This analysis of a project requires the project manager to focus on the various tasks and their relationships and to "get all of the ducks in a row."

ENVIRONMENTAL/WATER RESOURCES ENGINEERING ANALYSIS APPLICATION—OPEN CHANNELS

The conveyance of stormwater, wastewater, and drinking water is a critical task for infrastructure engineers to undertake. Stormwater is most often collected by inlets in roadway gutters and must be conveyed to a discharge point. Wastewater must be conveyed from homes and businesses to a centralized wastewater treatment facility. Drinking water must be transported from a drinking water treatment facility to sinks, showers, and toilets. Oftentimes, the conveyance is achieved in pressurized systems; such systems are closed systems and pressure is provided by pumps. In other cases, the water is not pressurized, and is conveyed by gravity in **open channels**.

There are many instances of open channel flow in the engineered and natural world. For example, finished drinking water was transported in open channels by the famous Roman aqueducts. Supply water is delivered to Southern California water treatment facilities via the 400-plus mile long California Aqueduct (Figure 9.8). Wastewater and stormwater that flows in underground pipes often flow as unpressurized flow, subject only to gravity. Stormwater and irrigation water are often routinely transported in human-made canals and ditches.

The flow in open channels depends on many variables, including the channel slope; the channel's cross-sectional shape and cross-sectional area; and the type of material with which the channel is constructed. One often-used equation to predict the flow in open channels is the Manning Equation (Equation 9.2).

$$Q = \frac{\alpha \cdot A}{n} R_h^{\frac{2}{3}} S_o^{\frac{1}{2}} \tag{9.2}$$

where

$\alpha = 1.486$ for English units or 1 for SI units

$A = $ cross-sectional area of the channel (ft^2 or m^2)

Figure 9.8 California Aqueduct. This is one of many canals in the system which are typically 6 to 25 feet deep, and 25 to 100 feet wide.

Source: Copyright © Steve Jacobs/iStockphoto.

Remember that full-sized versions of textbook photos can be viewed in color at www.wiley.com/college/penn.

n = Manning's roughness coefficient (see Table 9.2 for values)
R_h = hydraulic radius (ft or m)
S_o = slope of channel bottom (ft/ft or m/m)

The **hydraulic radius** is the ratio of the cross sectional area of flow (A) to the wetted perimeter (P) of flow. The wetted perimeter is the length of the cross-section's perimeter with which the flowing fluid comes in contact.

You must be careful when using Equation 9.2 to use the specific units provided in the definition of terms because Manning's Equation is not dimensionally consistent. This is an example of an **empirical** model, as contrasted to a **rational** (or mechanistic) model. Empirical models are developed based on experimentally derived results, whereas rational models are derived mathematically based on fundamental scientific relationships.

The Manning Equation is widely used by engineers for analysis. (It can also be used for design, which is covered in the next chapter.) For analysis, an engineer can evaluate a given channel or stream cross-section, material type, and slope, and estimate the flow that can be conveyed by that channel. Or, given a certain flow in a channel, the engineer might estimate the corresponding water depth.

Table 9.2

Manning's Roughness Coefficients

Material	Manning's n
Asphalt	0.016
Brickwork	0.015
Concrete	0.013
Clay tile	0.017
Corrugated metal	0.022
Gravel	0.029
Natural stream*	0.02–0.2

*Note that for streams, natural variability in bottom material and in-stream vegetation leads to an order of magnitude range of coefficient values.

numerical example 9.2 Analysis of an Open Channel

A concrete channel is used to convey stormwater. The channel is rectangular in cross-section with a width of 2 feet and a depth of 3 feet. The channel is 1,000 feet long, and drops 5 feet over that length. If 6 inches of **freeboard** is desired, what flowrate (in cfs) can the channel convey? In the case of a channel, freeboard is defined as the vertical distance between the top of the channel edge and the top of the anticipated high water mark.

numerical example 9.2 Analysis of an Open Channel (Continued)

solution

This example will utilize Manning's Equation (Equation 9.2). A schematic of the channel's cross-section is provided in Figure 9.9; note that the freeboard is identified in this figure. From the problem statement and Figure 9.9, we can directly calculate a cross-sectional area of flow in the channel (2 feet · 2.5 feet = 5 square feet) and a slope (5 feet/1,000 feet, or 0.005 foot/foot). Additionally, from Table 9.2, we can look up a roughness coefficient for concrete (0.013).

Figure 9.9 Channel Cross-Section for Example Problem.

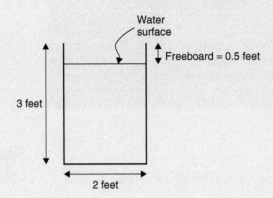

The only other information needed for Equation 9.2 is the hydraulic radius, which depends on the cross-sectional area of flow and the wetted perimeter. The wetted perimeter can be obtained by examining Figure 9.9, and equals the bottom width plus the submerged depth on each side (2 feet + 2.5 feet + 2.5 feet) or 7 feet. Thus, the hydraulic radius equals

$$R_h = A_{\text{flow}}/P = (2\,\text{ft} \cdot 2.5\,\text{ft})/7\,\text{ft} = 0.7\,\text{ft}$$

From Equation 9.2, we can calculate the flowrate.

$$Q = \frac{\alpha \cdot A}{n} R_h^{\frac{2}{3}} S_o^{\frac{1}{2}} = \frac{1.486 \cdot 5\,\text{ft}^2}{0.013} (0.7\,\text{ft})^{\frac{2}{3}} 0.005^{\frac{1}{2}} = 31.862 \frac{\text{ft}^3}{s}$$

Reporting this number to the correct number of significant figures results in an answer of 30 cfs.

GEOTECHNICAL ENGINEERING ANALYSIS APPLICATION—SOIL SIZE ANALYSIS

Geotechnical engineers make design decisions based on many different soil properties. The most fundamental soil property is particle size (diameter). Most soils are mixtures of the three primary size classes of particles: sands, silts, and clays. These particle classifications are defined differently by the American Association of State Highway and Transportation Officials (AASHTO) and the U.S. Department of Agriculture (USDA), as tabulated in Table 9.3.

Table 9.3

Particle Size Classifications	Classification	AASHTO	USDA
	Sand	2–0.075 mm	2–0.05 mm
	Silt	0.075–0.002 mm	0.05–0.002 mm
	Clay	<0.002 mm*	<0.002 mm

*Note that 0.002 mm is equal to $2\,\mu m$ (micrometers, or microns, or 10^{-6} m), which is far smaller than can be seen by the human eye without magnification. This is roughly the size of a bacterial cell. For reference, the diameter of human hair ranges from 20 to $200\,\mu m$.

We include both definitions not to confuse you, but rather to illustrate the fact that it is not uncommon for engineering terms to have variable definitions.

Many engineering properties of soils depend on the relative amount of coarse material (sands) or fine material (silts and clays). A particle size analysis is performed to determine the amount of sand, silt, and clay present in a soil. The analysis is based on dry mass of material of each classification. Sands are further subdivided into coarse, medium, and fine sands, each with differing geotechnical properties. The sand fractions of a soil are determined by a sieve analysis, whereby the mass of particles retained on a sieve with openings of known size is measured.

Silts and clays cannot be sieved because of their very small size, thus their distributions must be determined by other means. The traditional approach is to use a hydrometer, in which the density of a solution of water and soil particles is measured over several hours. Soil particles have a density approximately 2.5 times greater than water; thus the mixture of water and soil will be more dense than pure water. Larger particles have higher settling velocities. Over time, as more particles settle to the bottom of the hydro-meter, the density of the water/soil solution decreases, approaching that of water. From this information of density as a function of time, the relative amounts of silt and clay can be determined. Advances in analytical instrumentation have led to devices that utilize lasers to determine particle size distributions instantaneously. These instruments can provide highly accurate and detailed information about particle size distributions, but may cost hundreds of thousands of dollars. By comparison, a hydrometer costs less than $100.

Particle size data are typically plotted as a distribution curve. The y-axis is the percent of mass of the sample that is "finer" (or smaller) than the diameter (x-axis). Since gravels are removed (by sieves) before a particle size analysis is performed, 100 percent of the sample, by definition, is less than or equal to 2 mm (the upper size limit of sand). A typical distribution curve is shown in Figure 9.10 for two different soils. Note that Soil B in Figure 9.10 is much coarser (0 percent of particles finer than 0.05 mm, thus 100 percent of particles >0.05 mm) than Soil A (50 percent of particles finer than 0.05 mm, thus 50 percent of particles >0.05 mm). Given the distribution of sizes, the soil texture can be determined from a soil texture triangle shown in Figure 9.11.

Using the particle size distribution curve (Figure 9.10) and the USDA classification system, we can conclude that Soil A is 50 percent sand. That is, 100 percent of Soil A is finer than 2 mm and 50 percent of Soil A is finer than 0.05 mm; consequently, 50 percent of Soil A has grain sizes between 0.05 mm and 2 mm. Using the same logic, 35 percent of Soil A is silt (50 percent

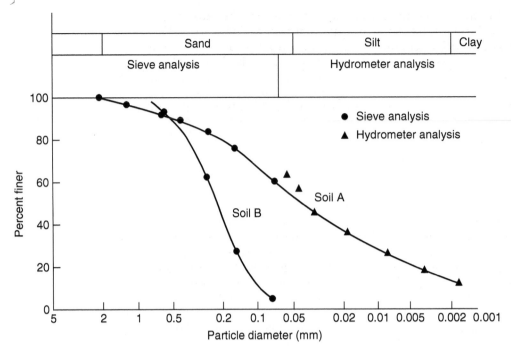

Figure 9.10 Particle Size Distribution Curves for Two Soils Based on USDA Classification.   Note that the *x*-axis is logarithmic scale, while the *y*-axis is arithmetic.

Figure 9.11 Soil Texture Triangle, Showing Soils A and B from Figure 9.10.

Source: Soil Survey Division Staff.

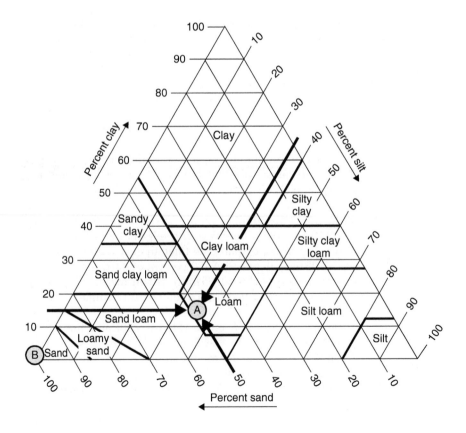

finer than 0.05 mm and 15 percent finer than 0.002 mm) and the remaining 15 percent is clay. From the soil texture triangle (Figure 9.11), Soil A is classified as a *loam*. Soil B is 100 percent sand (by size) and thus it is classified as *sand*. These two soils will have very different engineering properties. You will learn much more about these properties in a geotechnical engineering course. The soil texture is also an important parameter for designing septic systems for rural wastewater management and infiltration basins for urban stormwater management.

MUNICIPAL ENGINEERING ANALYSIS APPLICATION—STORMWATER RUNOFF

One effect of land development projects that engineers must be increasingly aware of is the impact of the development on stormwater runoff. The increased runoff occurs because most land development creates impervious surfaces (e.g., parking lots, roadways, and rooftops). Note that this impervious area is most likely replacing a land cover type that was much more pervious; for example, an agricultural field, a greenway, open fields, or forestland.

One result of the increased imperviousness is that rainwater that falls on the site will now be much less likely to infiltrate. Rainwater that does not infiltrate becomes stormwater runoff. The increased runoff must be managed such that it does not overwhelm storm sewer systems. Additionally, once released, the increased stormwater runoff flow can greatly increase stream velocities, leading to stream bank erosion (Figure 9.12). Moreover, since less water is infiltrating, groundwater aquifers are being recharged less and less.

Although fully analyzing this problem is too advanced for this introductory text, you can still effectively characterize the problem using the Rational Method, which was introduced earlier in this chapter. The Rational Method accounts for this change in land use through a change in the runoff coefficient (see Table 9.1). This change is illustrated in Numerical Example 9.3.

In later chapters, you will learn about design alternatives that can lessen the impact of development on stormwater generation rates.

Figure 9.12 A Double Culvert to Convey Stream (and Storm) Water Under a Bridge. In this picture, construction is taking place to stabilize the stream bank in the photo foreground. The purpose of this project is to reduce stream bank erosion. Note that the culverts are partially blocked at the upstream entrances with steel plates to keep the site dry for ease of construction work. The pipe in the right culvert is attached to an opening in the steel plate and conveys the upstream water past the active construction area, releasing it downstream.

Source: M. Penn.

numerical example 9.3 Effect of Land Development on Stormwater Flowrate

A 20-acre site is currently an undeveloped area with a runoff coefficient, C, of 0.20. The land is sold to a large "big box" retailer. As a result of this land development, the runoff coefficient will increase to 0.65. Note that both values for the runoff coefficient used in this example are estimated from Table 9.1 as middle-range values for similar land uses and surface categories (large retail may be approximated as "light industrial," and undeveloped is assumed to be "lawns, silty/clay soil, moderate slope"). Detailed tables of runoff coefficients for a wide range of specified land uses and soil conditions are used by practicing engineers. Experience provides a basis for professional judgment to determine appropriate values for a given site. For a rainfall intensity of 1 inch/hour, how will the peak flowrate of runoff exiting the site change as a result of the development?

solution

Using the Rational Method (Equation 9.1), the peak flow before development ($Q_{p,\text{pre}}$) can be calculated:

$$Q_{p,\text{pre}} = CiA = 0.20 \cdot 1 \text{ inch/hour} \cdot 20 \text{ acres}$$
$$= 4 \text{ acre} \cdot \text{in} \cdot \text{hr}^{-1} \approx 4 \text{ ft}^3/\text{s}$$
$$= 4 \text{ cfs}$$

After development, the peak flowrate ($Q_{p,\text{post}}$) can be calculated using the same method:

$$Q_{p,\text{post}} = CiA = 0.65 \cdot 1 \text{ inch/hour} \cdot 20 \text{ acres} = 13 \text{ acre} \cdot \text{in} \cdot \text{hr}^{-1} \approx 13 \text{ ft}^3/\text{s}$$
$$= 10 \text{ cfs (reporting to one significant figure)}.$$

Consequently, the peak flowrate has *more than doubled* as a result of development. In practice, an engineer is not likely to report 13 cfs as 10 cfs—the value of 13 cfs would likely be used to ensure a conservative design (i.e., to be certain that storm sewer pipes would be large enough to convey the runoff).

STRUCTURAL ENGINEERING ANALYSIS APPLICATION—TENSION AND COMPRESSION IN BEAMS

In Chapter 3, we discussed tension (pulling) and compression (pushing) forces as they related to bridges. For example, the columns of suspension bridges are under compressive forces while the cables in suspension bridges are under tension. A beam, which is a horizontal structural member, is more complicated, in the sense that portions of a beam will be under compression while other portions will be under tension.

For example, consider Figure 9.13, in which plastic building blocks are used to form a simply supported beam. Note that when the beam is loaded at the center by the force exerted by the finger, that gaps appear between the blocks in the bottom of the beam, illustrating that the bottoms are pulling apart, or that the bottom is under tension. Conversely, the blocks on the top are fitted more closely together (compression) as a result of the loading. This same behavior occurs for a beam in a real structure, when that beam is supported

Figure 9.13 Tension and Compression in Simply Supported Beam.

Source: P. Parker.

on each end. Thus, you can analyze a beam and determine where it is under tension and under compression.

As another example, consider a beam that is cantilevered. A cantilever is a beam that is fixed on only one end. In Figure 9.14, you can see how a beam constructed from building blocks will behave when it is constructed as a cantilever. The top of the beam is under tension, as evidenced by the spaces between the tops of the top row of blocks, whereas the bottom is under compression. Cantilevers are frequently used in structures, including balconies, overhangs, and bridges. A cantilever in a reinforced concrete bridge under construction is shown in Figure 9.15. The bridge is being constructed as a balanced cantilever.

The Firth of Forth bridge, which was introduced in Chapter 8, History and Heritage, is a type of cantilever bridge. Its chief engineer, Benjamin Baker, created a human model of how the cantilever bridge would operate. His idea was photographed (Figure 9.16), and readily demonstrated how the different parts of the bridge would be in compression and tension. That is, the lower members are poles (resting on the chairs) and are in compression, while the upper members are the outer men's arms, and are in tension. The stack of bricks at each end demonstrates the function of the anchorages. This photograph is also an excellent example of effective engineering communication and the importance of knowing one's audience.

Figure 9.14 Tension and Compression Illustrated for a Cantilevered Beam.

Source: P. Parker.

Figure 9.15 Construction of the New "Bay Bridge" Over San Francisco Bay.

Source: Courtesy of Caltrans, District 4 Photography/J. Huseby.

These simple observations about tension and compression in beams help explain the design of reinforced concrete. When forming a concrete beam, steel reinforcing bars (also known as rebar or rerod) can be placed in the portions of the concrete beam that will be in tension. Consequently, for the simply supported concrete beam, reinforcing bars would be placed in the bottom of the beam.

Thus, you have learned to this point some very simple rules that can help you analyze portions of a structure to determine whether members are in compression or tension. Specifically:

- columns are in compression

- cables and ropes are in tension

- arches are in compression

- beams supported at each end are in compression at the top while in tension at the bottom, and

- cantilevered beams are in tension at the top while in compression at the bottom.

Using this information, you can now inspect and analyze many types of structures and begin to surmise which members are under compression and which are under tension.

You may find, as you begin paying more attention to structures and their individual members, that it can be difficult to estimate which forces are at work. For example, it is relatively difficult to determine which members of a truss are in compression and in tension. The software program that you will use in the next chapter will illustrate which members are in compression and tension for a truss bridge, and in future junior- and senior-level courses, you will be able to perform calculations that will allow you to do the same.

TRANSPORTATION ENGINEERING ANALYSIS APPLICATION—PAVEMENT ANALYSIS

Every state has a governmental unit that oversees transportation issues, and in most states this unit is called the Department of Transportation, or DOT. One of the state DOT's responsibilities is to request a budget from the state government to build new roads and renovate existing roads. Recommending which roads are to be resurfaced, widened, straightened, or rerouted has many implications, including effects on an area's natural environment, its economic potential, and its ability to carry on business while the construction is taking place.

Consequently, creating a list of projects to be worked on is a critical task for the DOT. The list of recommended projects, and perhaps more importantly, their prioritization, must be created objectively. The submitted list of proposed projects will come under scrutiny by the legislature, the media, and the general public. Moreover, legislators will many times seek to influence the projects to be carried out in their districts, as such construction projects will benefit their constituents (and help their own chance for re-election).

One aspect to be examined when deciding on how to prioritize road construction projects is the condition of the pavement. There are many ways to evaluate pavement quality, and many states have their own procedures. We present here a method used by the Wisconsin DOT, termed the Pavement Surface Evaluation and Rating (or PASER) process.

Pavement evaluation studies involve physical examination of the roadway. As such, when conducting these studies, safety is of utmost importance. Working in a roadway where vehicles are present is a dangerous situation with significant risks. Engineers must take the appropriate safety precautions before undertaking this type of analysis. If you are planning on conducting such an evaluation, your employer or instructor should educate you on site-specific safety precautions that you are required to take.

Pavement Surface Evaluation and Rating: Asphalt

A roadway consists of pavement (typically either asphalt or concrete) and a base material. Typical cross-sections are provided in Figure 9.17 for concrete and asphalt pavement options. In Figure 9.17, the compacted gravel serves as the base material, or **base course**. Also, the "soil base" is often referred to as the **subgrade**.

To ensure an adequately long pavement life, the pavement needs to be properly constructed and regularly maintained. Maintenance includes filling and sealing cracks, resurfacing, and ensuring adequate drainage.

The PASER manual provides guidance to help evaluators characterize surface defects. Defects may be apparent as loss of pavement material or polishing of the surface; the latter makes the road slippery and dangerous. A second category is surface deformation, which may be caused by rutting (channels created in the wheel paths) or other mechanisms. Various types of cracks are defined by the rating system, and the evaluator also documents the existence of potholes and patches.

As a result of the analysis, each road is given a rating value between 1 and 10, with 1 corresponding to the poorest condition and 10 corresponding to freshly paved surfaces. Photographs corresponding to various scores are shown in Figure 9.18, Figure 9.19 and Figure 9.20.

> **Analysis of Concrete Roads**
> A PASER handbook is also available for concrete roads.
>
> **Pavement Surface Evaluation and Rating: Concrete**

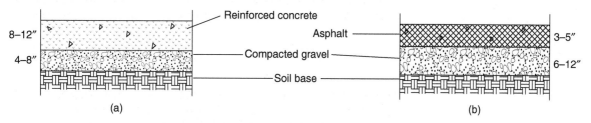

Figure 9.17 (a) Concrete and (b) Asphalt Pavement Cross-Sections.

Figure 9.18 A Resurfaced Road with a PASER Rating of 8.

Source: P. Parker.

Figure 9.19 A Road with a PASER Rating of 6.

Source: P. Parker.

Figure 9.20 A Road with a PASER Rating of 2.

Source: P. Parker.

Outro

We hope that after finishing this chapter, you understand more clearly the analysis process in general, and that by reading the specific applications, you begin to get a sense for what analysis means in practice. A tremendous amount of work is done by engineers in the area of analysis and this type of work may increase as more resources are spent on improving and updating the existing infrastructure as compared to building new infrastructure. Moreover, analysis is often the basis for design—you cannot design a solution unless you understand the problem. In the next chapter, you will explore design in generic terms, and also see how design is used in the various subdiscipline areas of civil and environmental engineering.

chapter Nine Homework Problems

Calculation

9.1 Show mathematically that 1 cfs is nearly the same as 1 acre · in./hour.

9.2 The U.S. Department of Transportation publishes data on the condition of U.S. highway bridges (available at www.wiley.com/college/penn). For this data:

a. Sort the data such that the states are ranked in decreasing order of the number of functionally obsolete bridges (use Excel "help" to find out how to sort data). Sum the number of functionally obsolete bridges in the ten states that have the most functionally obsolete bridges.

b. Calculate the mean number of structurally deficient bridges and the median number of such bridges for the 50 states.

9.3 Create a Gantt chart that shows the courses you need to take to obtain a degree in your chosen area from your university. To construct this chart, the "tasks" will be the individual courses.

9.4 Create a graph of flowrate as a function of channel depth for the concrete channel shown in Figure 9.9, with a slope of 0.005 foot/foot. Use enough data points to construct a smooth graph. Your x-axis should vary between 0 ft and 3 ft and flowrate should be expressed in units of cfs (ft^3/s). Such a graph is termed a **rating curve**.

9.5 An open channel needs to be constructed between two points. The two points are separated by 1,000 feet horizontally and by 15 feet vertically. The channel is to be semicircular in cross-section, constructed of corrugated metal, and is designed to have no freeboard when carrying 8 cfs.

a. What should the diameter be for this situation?

b. If you increase the diameter of your answer to part "a" by 10 percent, how much will this increase the capacity (i.e., flowrate) of the channel?

9.6 Reconsider Numerical Example 9.3. How would the answer to this problem change if:

a. the design storm intensity was 1.5 inches per hour but all other values are the same as stated in the example;

b. the land area was 25 acres but all other values are the same as stated in the example;

c. the pre-development land is a flat lawn with sandy soil but all other values are the same as stated in the example?

9.7 A 10-acre pasture site (moderate slope, sandy soil) is to be developed to a high-density single family subdivision. Calculate the range of the estimated increase in runoff due to development for a 1 inch/hour rainfall. Use a range of values for the pre-development runoff coefficient and a single value for post-development runoff coefficient.

9.8 A sieve analysis was conducted on a 150-gram sample of a coarse-grained soil. The results are provided in the table. Calculate the percent passing for each sieve and create a grain-size distribution plot on a semi-log graph (x-axis as logarithmic scale). This can be done by hand or with a spreadsheet.

Sieve	Size (mm)	Mass Retained (g)
#4	4.75	5
#10	2.00	11
#20	0.85	21
#40	0.425	34
#60	0.250	38
#100	0.150	24
#200	0.075	12
Pan		5

Short Answer

9.9 Why does runoff typically increase from an area of land after it has been developed?

9.10 Where should steel reinforcement be placed in a cantilevered, reinforced concrete beam? Explain your rationale.

9.11 What is the texture of a soil with 50 percent sand, 30 percent clay, and 20 percent silt?

9.12 What is the texture of a soil with 60 percent silt, 10 percent sand, and 30 percent clay?

Discussion/Open Ended

9.13 The American Society of Civil Engineers (ASCE) Code of Ethics states that engineers should not practice outside of their area of expertise. Yet, one component of engineering analysis is conducting background research, which infers that the engineer must step outside their area of expertise. How can you reconcile this apparent paradox?

Research

9.14 The Federal Highway Administration (FHWA) has different ratings for bridges, based on inspections such as the one described in this chapter. Does a rating of "structurally deficient" indicate that the bridge is unsafe? What are some ways that a bridge could earn a "functionally obsolete" rating?

9.15 According to the last inspection of the I-35 W bridge, what was the condition of the bridge? Summarize the results from the section "Inspection Results and Condition Ratings for I-35 W Bridge" in the NTSB Accident Report (available at www.wiley.com/college/penn) for the I-35 W bridge collapse.

9.16 Find three examples of three different types of retaining walls in your community/campus. For each wall,

a. Take a picture of the wall.
b. Note the location of the wall.
c. Measure/estimate as accurately as possible (see Figure 9.21):
    i. the apparent height (H') and length (L) of the wall
    ii. the thickness (T) of the top of the wall
    iii. the angle/slope of the front of the wall (β)
    iv. the angle of the backslope (α)
d. Describe, in about one paragraph, the consequences if each wall were to fail.

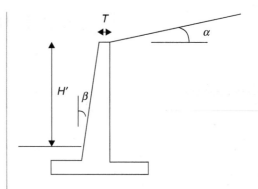

Figure 9.21 Dimensions for Retaining Wall.

9.17 Pavement analysis—Use the PASER manual (available at www.wiley.com/college/penn) to analyze a portion of pavement in your local municipality. Work with your course instructor to select an appropriate and safe location and to review safety procedures. Present your results as a memo, addressed to your instructor, which describes your analysis and addresses all of the components of the engineering analysis process.

Key Terms

- background research
- base course
- citizen engineers
- compression
- data analysis
- data collection
- data organization
- definition of scope
- empirical
- flowrate
- freeboard
- gantt chart
- groundtruthing
- hydraulic radius
- model application
- model development
- open channels
- particle size distribution
- pilot studies
- rational method
- recommendation
- runoff coefficient
- scheduling
- significant figures
- subgrade
- tension

References

ASCE (American Society of Civil Engineers) and WPCF (Water Pollution Control Federation). 1969. *Design and Construction of Sanitary Storm Sewers*. ASCE Manuals and Reports on Engineering Practice no. 37 and WPCF Manual of Practice no. 9.

Soil Survey Division Staff. 1993. *Soil Survey Manual*. Soil Conservation Service. U.S. Department of Agriculture Handbook 18.

Transportation Information Center, University of Wisconsin-Madison. *Pavement Surface Evaluation and Rating: Asphalt Roads*. http://epdfiles.engr.wisc.edu/pdf_web_files/tic/manuals/Asphalt-PASER_02.pdf, accessed on August 8, 2011.

chapter Ten Design Fundamentals

Learning Objectives

After reading this chapter, you should be able to:

1. Describe the steps of the engineering design process.
2. Demonstrate how the design process is applied to meet a need.
3. Explain how constraints, regulations, standards, safety factors, design life, and risk affect the design process.
4. Complete simplified designs for the major subdiscipline areas of infrastructure engineering.

Introduction

When asked the question "What do engineers do?" many children answer, "They build things." A person with a vague understanding of engineering will likely respond, "They *design* things." In many cases, both answers are correct. Without a doubt, the fundamental role of most engineers is design; however, as we have discussed in the previous chapter, design is not the *only* responsibility of an engineer, nor is it something that every engineer will do. Perhaps a better answer to the question is "engineers identify and solve problems." This answer emphasizes that engineering is an all-encompassing discipline that includes much more than design alone.

In this chapter, we will start with a discussion of engineering design in general, and then present civil and environmental engineering design applications.

Introductory Case Study: Rock Lake

Property located along lakes and rivers typically has high market value because of recreational opportunities and aesthetics. These property values may decrease if water quality is impaired. Nutrients such as phosphorus can cause a decrease in water quality due to excessive algal growth (Figure 10.1) turning them pea-soup green in the summer. Residential, commercial, or industrial wastewaters may contain high concentrations of phosphorus. If the amount of phosphorus is high enough, water quality in a river or lake receiving these wastewaters will be impaired. Throughout this chapter, we will repeatedly refer to a hypothetical case study of Rock Lake, where wastewater from homes has caused an algal problem. Engineers solve problems, and before the solutions can be implemented, they must be designed.

Figure 10.1   A Photograph of an Intense Algal Bloom in a Lake.

Source: M. Penn.

The Design Process

ABET, Inc. (formerly the Accreditation Board for Engineering and Technology) uses this definition of design:

Engineering design is the process of devising a system, component, or process to meet desired needs. It is a decision-making process (often iterative), in which the basic sciences, mathematics, and the engineering sciences are applied to convert resources optimally to meet these stated needs.

Let's consider how this definition might apply to the design of a simple everyday item such as a pen. The designer must first determine the desired needs. For example, is the final product to be an inexpensive pen or an executive's pen? If you visit an office supply store, you will see dozens of different pens. Why so many? The answer is that there is a demand for many different types. What is different about them? Shape, material, color, life (disposable versus refillable), line width, and cost are among the major variables. An iterative decision-making *process* comes into play when you consider that many of these variables potentially conflict when designing a specified pen. The number of variables, and thus the challenge of design,

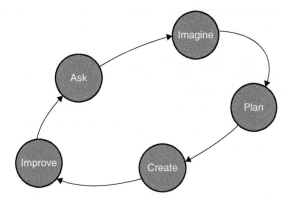

Figure 10.2 A Simplified Conceptualization of the Engineering Process.

Source: Adapted from the Museum of Science, *Engineering Is Elementary* Project.

increases dramatically as the "product" becomes more complex: from pens to alarm clocks to mobile phones to automobiles to skyscrapers.

Figure 10.2 is a conceptualization of the engineering design process, used by the organization *Engineering is Elementary* to introduce children to the field of engineering. We feel it is appropriate to use here, to help you envision engineering as a cyclical process. We start by identifying the problem ("Ask," in Figure 10.2). Then we consider potential solutions ("Imagine"). Next we evaluate the solutions and choose the best one ("Plan"), and then implement the solution ("Create"). Often, further modifications are required as demands change or are re-evaluated ("Improve"), which leads us back to the first step in the process.

More formally, the **design process** is defined to include the following steps:

1. Identify a need (commonly referred to as the **problem statement**)
2. Collect background information
3. Develop alternative designs
4. Evaluate various designs
5. Select the best design
6. Document the design
7. Implement the solution

These seven steps are discussed in the following paragraphs. To help you understand the process, the steps are applied to the introductory case study. In this scenario, the Rock Lake Homeowner's Association approaches a local consulting engineering firm, and asks them to address the fact that the lake often turns green in the summer. The discolored water is caused by algal blooms, which causes the lake to become unsightly and occasionally foul smelling; but perhaps more importantly, the homeowners fear it will lower property values.

PROBLEM (NEED) IDENTIFICATION

Problem identification was previously discussed in Chapter 9, Engineering Analysis. As part of the design process, the need must be clearly specified, for a suitable design cannot be performed to effectively meet economic, social, and environmental needs without a clear objective. Often, design work (and thus product delivery), is prolonged because clarification of, or modification to, the

---

The Need for Design

You may not have thought of this before, but nearly *everything* around you was designed. The popular public radio program "Car Talk" centers around callers asking their car repair questions to two experienced (and humorous) mechanics. One of their favorite sayings about a "bad" car or component is that it was "unencumbered by the engineering design process"—sarcastically referring to the notion that it was not properly designed, in fact, implying that it was not designed at all!

original problem statement occurs. Depending on the project timeline, these changes can be quite problematic, especially if they occur after substantial preliminary design has been conducted.

The engineer studying the Rock Lake problem performed analysis to determine that the cause of the algal blooms is an excess of phosphorus in the lake. Phosphorus is a nutrient that, when present in high concentrations, allows the algae to grow very rapidly. Thus the problem can be *redefined* from preventing "green water" to reducing nutrient input.

Variations of Problem Identification

In many cases, a client has a clearly identified need and a desired approach (e.g., replacing concrete pavement on a city street with asphalt pavement). In other cases, the client may merely have identified a need (e.g., the wastewater treatment facility is not treating sewage to state standards). For some large projects (e.g., bridges, sports stadiums), clients offer a great deal of flexibility in the design approach. For such projects, large, reputable firms are invited to propose conceptual solutions to a panel (the majority of members being non-engineers).

Other Options for Controlling Algae

In addition to reducing nutrient input to the lake, other options exist to control algae. Copper sulfate can be added to the lake, but is expensive because repeated applications are required if the nutrient input is not reduced. In addition, there is concern over copper toxicity to desirable aquatic organisms. One very complex approach is biomanipulation, whereby attempts are made to control the biological assemblage of organisms in the lake (i.e., increasing the amount of organisms that feed upon algae). For the purposes of this chapter, only nutrient reduction approaches will be discussed.

Just Deal With It

In many cases, one solution to a problem is the "do nothing" alternative, sometimes termed "no action." This may be a defensible course of action if the costs (economic, environmental, or social) outweigh the benefits.

COLLECTION OF BACKGROUND INFORMATION

Collection of information was also discussed in Chapter 9, but will be discussed further in this chapter with respect to design. Once a problem has been precisely and objectively defined, it is important to collect background information. What solutions have been implemented in the past to solve similar problems? "Starting from scratch" is often inefficient. Equally problematic is assuming that any solution *must* be based on previous solutions.

Many routine design aspects of engineering projects are simply adapted with minor changes from previous designs. In fact, it is common for entry-level engineers to be involved in this type of work to "learn the ropes." As an engineer's skills and experience accumulate, he or she should be given increasingly challenging responsibilities. Young engineers can learn much from the collective experience of coworkers and from reviewing in-place solutions that were successfully implemented.

For the Rock Lake scenario, the engineer must determine the actual sources of the nutrient. Possible sources of nutrients include discharges of domestic wastewater, runoff of excess fertilizer, bird droppings, atmospheric deposition from rain and dust. By collecting and analyzing water samples from in and around the lake, it was determined that the primary source of nutrients is from the wastewater from individual homes. The homes surrounding the lake are currently served by individual septic systems, a very common method for wastewater treatment in rural areas. Septic systems consist of a tank in which solid particles settle out from the wastewater and degrade; the tank is followed by a **drainfield** from which the partly treated wastewater is allowed to seep into the soil, ultimately mixing with groundwater that flows into the lake. Research revealed that many of the septic systems are several decades old and while they are functioning to original specifications, the standards 30 years ago were not as stringent as those of today. The design of septic systems will be illustrated in the Environmental Engineering Design Application at the end of this chapter.

Note how the problem identification has progressed for this case study, from removing the green color from the lake to decreasing nutrient discharges into the lake to eliminating the domestic wastewater discharges. This refining of the problem is typical of many engineering projects.

DEVELOPMENT OF ALTERNATIVE DESIGNS

Engineers must think in terms of designs (plural) rather than design (singular). Rarely is there ever one correct solution. Typically, the client will meet regularly with the engineer initially to ensure that the problem is properly identified. But there is always potential for misinterpretation; thus, follow-up meetings are recommended. Engineers then develop several alternative designs to meet the stated objectives. To many engineers, this step is the most exciting, because it provides them with the opportunity to "pull all of the tools out of their toolbox." Some of the designs might work, but can be ruled out for reasons such as cost—either initial cost (including capital cost) or lifetime costs (operation and maintenance, or O&M). In order to estimate the costs associated with an alternative design or to assess its feasibility, engineers must undertake preliminary engineering design.

After researching options, the engineer worked closely with the Rock Lake Homeowner's Association and proposed three possible alternatives to minimize the wastewater discharge to the lake. The three alternatives are illustrated in Figure 10.3:

Alternative A: replace all of the septic systems with new septic tanks and enlarged drainfields.

Alternative B: replace the septic systems with a sanitary sewer system and build a centralized wastewater treatment facility to treat the collected wastewater.

Alternative C: collect and pump the wastewater to a wastewater treatment facility at the nearby city.

Note that each of these alternatives will solve the problem—the next step is to select the "best" alternative.

EVALUATION OF DESIGNS

Roadway Design Alternatives Case Study

A proficient engineer not only can identify several possible solutions, but he or she has the knowledge to work with the client to determine the design that best meets their needs. Evaluation can take many forms. In some cases, especially with new technologies or new applications of existing technologies, prototypes (or pilots) are built and tested. In other cases, especially when pilots are prohibited by cost or short project timelines, computer modeling will be performed. When testing a design, all anticipated conditions should be considered. Many engineering failures result from improper testing or lack of consideration of potential situations in which the product will be used.

Personal experience with automobiles may illustrate such "failures." An engine design may be such that it is very compact, thus minimizing body size and maximizing gas mileage. But if the oil filter is surrounded by other components such that it cannot be readily accessed for changing the oil, the car's owner is not likely to be happy with the design.

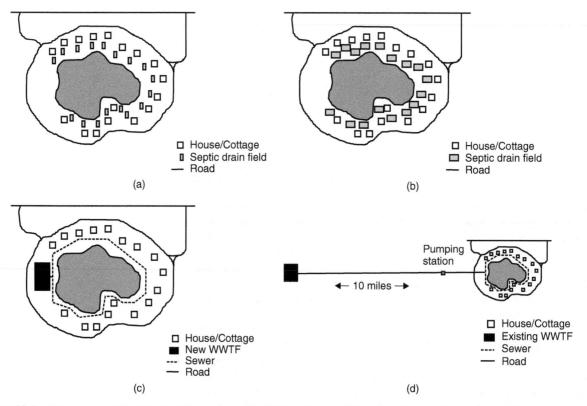

Figure 10.3 Illustration of Rock Lake Alternatives. (a) Original layout (b) Alternative A with improved (larger) drain fields. (c) Alternative B with new WWTF (d) Alternative C with conveyance to existing city WWTF (note change of scale).

Table 10.1

Decision Matrix for Rock Lake Alternatives Based on Ranking

Alternative	Installation Costs	O&M	Ease of Operation	Public Acceptance	Client Acceptance	Ease of Maintenance	Odors and/or Noise	Land Requirements	Construction Disruption
A	1	1	1	1	3	1	3	2	1
B	2	3	3	3	2	3	2	3	3
C	3	2	2	2	1	2	1	1	2

When comparing the three alternatives for Rock Lake, there are many competing variables. Importantly, initial ("upfront") cost is not the only variable, nor is it necessarily the most important variable (although some clients may mistakenly think that it is). For example, the alternative that is the least expensive to install might have the highest operating cost; or, the alternative that has the lowest total cost may be the alternative that is the least likely to be readily accepted by the public. Additional expenses are often accepted by the client if the results are more aesthetically pleasing, safer, longer-lasting, or easier to operate.

A common method of comparing alternatives is to list all of the competing variables and arrange them in a matrix. For the Rock Lake project, variables include installation costs; operations and maintenance (O&M) costs; ease of operations; public acceptance; client acceptance; ease of maintenance; possibility of odors and noise; land requirements; and extent of disruption caused by construction. A spreadsheet was used to compile the matrix, and the results are shown in Table 10.1. In this matrix, each alternative was ranked for each variable; a score of 1 was assigned to the most favorable alternative, and a score of 3 was assigned to the least favorable alternative for a given criterion. Rankings may be objective, based on quantitative information (e.g., cost), or may be subjective (e.g., public acceptance).

SELECTION OF BEST DESIGN

Communication between the engineer and the client is critical to making sure that the best design is selected. For example, when presented with a real-life engineering problem, engineering students are typically very eager to apply their technical knowledge to solve the client's problem. Sometimes the students are "too eager" to solve the problem—they are convinced that their proposed solution is the best for the client, without allowing for input from the client.

Many engineering companies have acted in a similar manner. Especially in the field of consulting engineering, where the company is contracted by a client to solve a problem, the very nature of the client–engineer relationship is such that the engineer is more knowledgeable about the topic than the client. The client may underestimate the need for direct input, assuming the engineers are the experts. However, any engineered solution ultimately gets "handed over" to the client, and thus it is imperative that the client understands and accepts the solution, and that the solution meets current and future needs.

When presented to the Rock Lake Homeowner's Association, the decision matrix makes it relatively easy for the engineer and the non-engineering client to discuss the advantages and disadvantages of the different alternatives.

Table 10.2

Improved Matrix for Rock Lake Alternatives Based on Scoring (5 = Highest, 1 = Lowest)

Alternative	Installation Costs	O&M	Ease of Operation	Public Acceptance	Client Acceptance	Ease of Maintenance	Odors and/or Noise	Land Requirements	Construction Disruption
A	5	5	5	5	1	5	2	1	5
B	2	1	1	1	2	1	1	3	1
C	1	3	4	4	5	4	5	5	2

Although in Table 10.1, Alternative A (replacing the existing septic tanks) has the most first place rankings, the client does not have a high opinion of continuing with a treatment practice that has failed in the past. Alternative C, although more expensive, has the advantage of being the "out of sight, out of mind" alternative, as witnessed by the favorable scores for Client Acceptance and Odors and Noise.

An improved matrix would include a *scoring* of each alternative for each variable (Table 10.2) rather than a ranking. This provides a more quantifiable basis for selection as it provides a degree of separation between alternatives. By ranking alone (e.g., 1, 2, 3) the *relative difference* between alternatives is not distinguishable, whereas a scoring of alternatives (e.g., 1 to 5, with 5 being the highest, most preferred rating) demonstrates that one alternative is not only preferred, but that it is perhaps much more preferred.

An additional improvement to the matrix would be to add a weighting factor to each variable (e.g., 0 to 1). The engineer and the client should work together to establish the weighting factors. This prevents incorrect assumptions by the engineer of the value that the client places on a variable, and prevents incorrect priorities by the client based on technical ignorance or misperceptions. By summing the value of each variable multiplied by its weighting factor, a total score for each alternate can be obtained (Table 10.3). When entered into a spreadsheet, weighting factors can be easily varied to see the impact on the total scores.

Table 10.3

Further Improvement of the Decision Matrix for Rock Lake Alternatives with Weighting Factors

Alternative	Installation Costs	WF	O & M	WF	Ease of Operation	WF	Public Acceptance	WF	Client Acceptance	WF	Ease of Maintenance	WF	Odors and/or Noise	WF	Land Requirements	WF	Construction Disruption	WF	Total score
A	5	1.0	5	0.5	5	0.2	5	0.3	1	1.0	5	0.2	2	1.0	1	0.2	5	0.5	16.7
B	2	1.0	1	0.5	1	0.2	1	0.3	2	1.0	1	0.2	1	1.0	3	0.2	1	0.5	7.3
C	1	1.0	3	0.5	4	0.2	4	0.3	5	1.0	4	0.2	5	1.0	5	0.2	2	0.5	17.8

Note: Weighting factors are abbreviated as "WF" in this table.

From this decision matrix, Alternative C (highest score) can be established as the best design. Alternative B (lowest score) is clearly out of consideration. However, Alternative A has a score that is nearly as high as Alternative C. In such a case, when two alternatives have similar scores, revisiting the weighting factors may have merit.

DOCUMENTING THE DESIGN

Upon reaching an agreement with the client for the best design, the engineer must document the design in a final report. Additionally, intermediate reports are often prepared for clients to assist in the evaluation of alternatives. Details of the final report will vary depending on client requirements. Engineers also document the design through the specification of equipment and materials and design drawings needed to send the project out for construction bidding. Details must be accurate and easily interpreted such that the documents can be "handed over" to construction contractors, ideally without any need for clarification. Designs are typically reviewed and "stamped" by a licensed professional engineer. Documentation is also critical from a liability standpoint; if something "goes wrong" later, the question will arise as to whether the designer or the contractor is at fault. Final documentation also includes "as-built" drawings that incorporate changes that occurred during construction. Further documentation for the client includes operations and maintenance manuals.

Examples of Engineering Design Drawings

IMPLEMENTATION OF SOLUTION

In many ways, the implementation step of design is the most often overlooked yet most important step of the process. The best design is meaningless if it is not implemented in a manner so that the client ultimately uses it with satisfaction.

Implementation is multi-faceted. It may involve any of the following: production, delivery, installation, training, operation, and maintenance. Each of these steps may vary in complexity depending on the scope of the project. An engineer's ability to foresee complications in any of these steps will lessen the likelihood of project failure or delay. Many of the implementation steps may require work with other companies (suppliers, subcontractors, etc.). Engineering firms work diligently to establish long-lasting and trusting relationships with suppliers and contractors.

Successful scheduling of implementation to ensure delivery to meet deadlines requires consideration of how each step of the process is interrelated and identification of critical components. As discussed in the Construction Engineering Analysis Application in the previous chapter, prioritization is very important. For example, ten parts from different suppliers may be needed to assemble a component, but if two of the ten are absolutely required before assembly can even begin, the scheduling must ensure their prioritized delivery. Flexibility is also important. In many civil and environmental construction projects, weather becomes a limiting factor, and must be accounted for when developing schedules.

Proper implementation of a design is the signature of its success. Only when a design is properly implemented will a client be fully satisfied. A satisfied client is obviously much more likely to hire an engineer in the future, and equally important, they are likely to speak favorably of the engineer to their peers, potentially leading to new clients.

> **Don't Forget the End User(s)**
> A new multi-million dollar ice arena was constructed 15 miles from the authors' campus. A student that plays hockey was asked how he liked the new facility. "It's beautiful, but it was NOT designed for hockey players," was his response. He is a goalie and has to carry two large bags of equipment. Apparently he has to open *five* different doors to get from the parking lot to the locker room! Spectators have to go through only two doors to get to their seats. Obviously, the designers were not hockey players.

Design Factors

Now that we have discussed the major steps of the design process, we will address some of the key factors that engineers must consider throughout the process.

DESIGN CONSTRAINTS

Without a doubt, life-cycle cost (capital and O&M over the expected life of the product) is typically the major constraint in any design. The client will be willing and able to spend a certain amount of money. For large projects, the money is often obtained as a loan, so additional costs are incurred due to interest payments over the borrowing period and the initial capital cost is indeed a major driver for decisions. For larger projects, preliminary engineering is performed to determine an estimate of the project cost, so that the client may determine whether or not undertaking the project is economically feasible. It would be short-sighted to begin a multi-million dollar project and proceed to the presentation of the best design, only to find out that the client cannot afford it.

In infrastructure engineering, the project site often provides many constraints. Listed below are several common examples:

- Size of the project site
- Need for buffer areas around the project
- Access to the site for both construction equipment and client requirements (e.g., a railway)
- Location of existing utilities (sewers, natural gas lines, water mains, electricity, etc.)
- Environmental factors (presence of wetlands, old dump sites, endangered species, etc.)
- Geologic setting (topography, soil depth, type of bedrock, etc.)

The engineer should conduct an extensive review of the site and determine which of the above (or other) constraints may impact design.

REGULATIONS

Engineering projects are required to meet many regulations established by federal, state, or local governments. The degree of regulatory requirements can vary substantially based on the type of project. Numerous permits are typically required and engineers spend a significant amount of time preparing applications for such permits. In many cases, construction work may not begin until permits have been granted, thus making this a critical step in the project timeline.

State-level regulations are referred to as codes, and can be thought of as rules that must be followed. State codes often specify both design criteria and performance criteria. **Design criteria** are objective requirements (e.g., pumps must be designed for a maximum flow of 10 gallons per minute). **Performance criteria** do not specify the type, size or shape of the material/components to be used, but rather set a target for performance (e.g., pumps must be designed to ensure that maximum expected flows can be conveyed). Design criteria significantly limit engineering options, whereas performance criteria allow engineers flexibility in design to meet the criteria. Engineers must be aware of all applicable codes on any project. While failure to meet code requirements does not guarantee that a design will not perform to meet client needs, it can result in penalties, delays, and loss of professional stature.

Many local governments establish ordinances that are more stringent than state or federal requirements. Again, it is critical that the engineer be aware of and understand all applicable local regulations. Examples of local ordinances include maximum building height, buffer requirements, setbacks from roadways or streams, lighting requirements, lot sizes, and parking provisions.

STANDARDS

In addition to federal, state and local regulations, many engineering subdisciplines have developed standards of practice. While there may be no direct legal requirement to follow these standards, they are developed by practicing engineers to serve as guidelines for design. Examples include *Building Code Requirements for Structural Concrete* (American Concrete Institute), *Manual for Steel Construction* (American Institute of Steel Construction), *A Policy on Geometric Design of Highways and Streets* (American Association of State Highway and Transportation Officials) and *Recommended Standards for Wastewater Facilities* (Great Lakes-Upper Mississippi River Board of State Public Health and Environmental Managers).

SAFETY FACTORS

Engineering work is inherently risky. The failure of a bridge or a drinking water treatment facility can lead to loss of human life, injury, and suffering. The conditions under which designed projects must function are highly variable. The very nature of civil engineering materials (steel, soil, pavement) contributes to the uncertainty. As a result, safety factors are often included in final designs. For example, the volume of a drinking water treatment process tank might be designed to be 100 cubic meters to meet state code. Common engineering practice might include a safety factor of 1.2, leading to a final designed volume of 120 cubic meters. By oversizing the tank, the engineer has provided a factor of safety so that it should meet treatment objectives in the event of unforeseen circumstances. Also, consider that bridges crossing waterways are required to have a certain amount of *freeboard* (the vertical distance between the bottom of the bridge deck and the top of the anticipated high water level); this freeboard is a type of safety factor which allows for the passing of ice and/or debris (e.g., fallen trees) and minimizes the risk of the bridge being overtopped by floodwaters. Geotechnical engineers use very large safety factors given the uncertainty inherent in the material (soil) with which they work. You will learn much more about safety factors in subsequent engineering courses.

In many cases, safety factors have been incorporated into codes or standards; it is important for the engineer to know if this is the case, because applying an *additional* safety factor may lead to overdesign and an unnecessary increase in cost. Safety factors also may vary depending on risk, where

Situational Challenges

The Chicago River turns green once a year, but not because of algae blooms. The river is dyed green for the annual St. Patrick's Day Parade, an event that hosts over 15,000 marchers and draws in excess of 250,000 spectators. These large crowds create a temporary, but significant, stress on the infrastructure (especially with respect to street traffic).

St. Patrick's Day and the Chicago River

a higher factor will be applied if data to completely define the problem is lacking or of questionable quality.

ACCEPTABLE RISK

In many cases, designs are governed by an acceptable level of risk. For example, storm sewers that collect and convey stormwater are sized for a maximum design flow of water. If the runoff entering the sewers exceeds capacity, street flooding will occur. Engineers could size the pipes to convey any conceivable amount of water, but costs escalate as pipe size increases. Sewer pipes are often sized to handle storms with a 5 to 10 percent probability of occurring in a given year (in other words, there is a 90 to 95 percent probability that no street flooding will occur). The impact of street flooding is often minimal, isolated, and temporary.

On the other hand, levees and floodwalls are built to protect *entire cities* from floodwaters of rivers or storm surges of hurricanes. In these cases, the potential damage can be severe and tremendously expensive, which warrants greater expenditures to minimize risks. Floodwalls and levees are often designed to protect at the 0.5 to 1 percent probability level of overtopping. The costs to reduce this risk to an even lower level have been considered excessive in the past, but this is an issue of current debate in the profession.

It is very important that engineers communicate risk to the public. Citizens often assume that if a levee is in place that they are totally safe, when in fact, they are not. Billions of dollars are being spent redesigning and reconstructing the levees/floodwalls in New Orleans after the Hurricane Katrina disaster. The new system will protect at the 1 percent probability level. The Army Corps of Engineers is making great efforts to fully communicate this risk to the public.

DESIGN LIFE

The **design life** is the anticipated useful period (life) of a product or system. If properly designed and maintained and subjected to anticipated uses and conditions, the product/system should continue to meet client needs for at least the design life (and hopefully longer). The engineering extremes of design life range from temporary road diversions of a few weeks duration to 10,000 years for the Yucca Mountain Radioactive Waste Repository in Nevada. Design lives of civil and environmental projects are typically in the 20- to 100-year range (see Table 10.4).

As the design life increases, so does the uncertainty (and thus the risk) in predicting future conditions to which the design will be subjected over its lifetime. Inevitably, selecting a longer design life results in greater cost. Imagine designing a highway to last 100 years. An advantage of such a design would be that the pavement theoretically wouldn't have to be replaced for a century (which is three times longer than the 30-year design life of common pavements). However, the cost of constructing and designing a pavement capable of lasting this long would be extremely expensive, requiring current taxpayers to bear the entire financial burden of several generations in the future. Perhaps most importantly, we cannot adequately predict traffic volumes that will occur in 100 years. How many lanes would be required: two, four...six? If we underpredict the future demand, the highway would not meet requirements at its design life (imagine horrendous traffic jams) and

Table 10.4

Typical Design Lives of Various Components of Infrastructure Projects

Project	Design life (years)
Commercial/Industrial buildings	30–60
Skyscrapers/Civic buildings	120+
Bridges	30–100
Dams	50–100
Drinking water treatment facilities —Concrete structures —Mechanical/Electrical	 60–70 15–25
PVC pipelines	75–100
Pavements	20–40
Culverts	20–75
Retaining walls	30–60

it would have to be widened. If we overpredict future demand, we build an unnecessarily large and costly highway.

Proper maintenance is paramount to meeting and exceeding design life. It is very unlikely that a car engine will achieve 200,000 miles of use if the oil is not changed regularly. Likewise, civil and environmental engineering systems must be maintained. If a steel bridge is not painted at appropriate intervals, it will rust and weaken. If sewer pipes are not cleaned of debris and intruding tree roots, their capacity will greatly decrease and they may clog. The examples are endless. The engineer must work with the client and communicate not only the need for maintenance, but also the costs and implications of neglecting it.

Case Study: Chicago River Reversal

In the mid-1800s, much of Chicago's domestic sewage was discharged to the Chicago River. The Chicago River drained to Lake Michigan, which was the source of drinking water for the city. You may find this alarming, to imagine that a city would discharge its untreated waste into its drinking water source. However, the engineers of the day may have believed that enough dilution would occur to reduce the risk of disease, or perhaps they did not even consider the disease risk. Although you may consider this lack of consideration to be negligence on behalf of the engineers, you must remember that at the time, the link between wastewater-borne pathogens and drinking water related diseases such as cholera was only beginning to be understood.

With increasing awareness of the link between wastewater and water-borne diseases, engineers and city planners had identified a problem—a need to separate the wastewater discharge from the drinking water intake. Several alternatives existed, such as extending the intake for the drinking water treatment plant farther out into Lake Michigan. Another alternative was to discharge the wastewater far to the south and build a new drinking

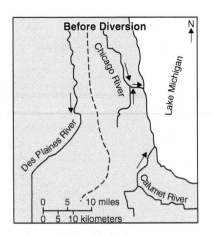

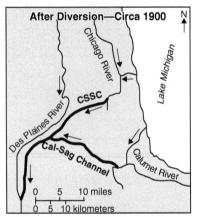

Figure 10.4 **The Natural River System of the Chicago Area (Above) and the Engineered Canal System (Below) Including the Chicago Sanitary and Shipping Canal (CSSC) and the Cal-Sag Channel.** The subcontinental divide (see Figure 10.5) is shown as dashed north-south line. Note the reversal of flow direction for the Chicago and Calumet Rivers.

Source: Adapted from U.S. Geological Survey, 1999.

water intake farther to the north. Also considered was disposal of the sewage on land. Perhaps engineers considered building a wastewater treatment facility, but these were not common at that time. Another method considered was to reverse the flow of the Chicago River, such that it would flow out of Lake Michigan and empty into the Des Plaines River (see Figure 10.4 and Figure 10.5). Note that the Des Plaines River eventually flows into the Mississippi River, and thus this option would result in Chicago's untreated sewage being carried to the Gulf of Mexico rather than to Lake Michigan.

You can imagine the many competing variables: cost, effectiveness, public acceptance, ability to meet regulatory requirements (of which there were very few at that time in history), etc. Technical feasibility was another variable. The reversal of a river surely would have appeared to many people as technically infeasible, but the river could be readily reversed by building a canal to connect the river to the Des Plaines River.

The river reversal option was chosen. It took 8 years and 8,500 workers to dig the 28-mile long canal from 1892 to 1900, without the aid of the heavy construction equipment that is available today. It is one of the largest public works project in history.

You might be surprised that the river reversal was the alternative that was chosen, as it seems so obvious today that this alternative was simply passing the problem onto the unfortunate people downstream, and in no way was protecting the environment. However, the reversal of the river was very effective, and combined with other technological advances of the time, helped dramatically decrease the incidences of water-borne diseases such as cholera. It is important to note, however, that the reversal of a river would very likely never be approved by regulatory agencies or accepted by the public today. In fact, there was substantial controversy surrounding the original plan, and additional controversy (and several lawsuits) regarding control of the amount of water flowing out of the lake. Note that such controversy is not uncommon for public projects. At any scale, each individual project has advantages and disadvantages, and as a result there are people who are in favor, and those that are opposed to the project.

This case study highlights many of the steps of the engineering design process, and emphasizes that "engineers identify and solve problems." Moreover, the case study shows that engineers have had a dramatic influence on public health. Some have suggested that the 20-year addition to life expectancy

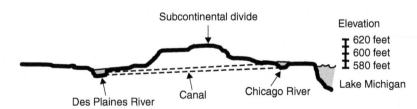

Figure 10.5 **An East-West Cross-Section of the Area Shown in Figure 10.4.** Vertical scale is exaggerated. The subcontinental divide is the ridge of highest elevation running north to south. Rivers to the east of the subcontinental divide flow into Lake Michigan and ultimately to the Atlantic Ocean via the St. Lawrence River. Rivers to the west of the divide flow to the Mississippi River and ultimately to the Gulf of Mexico. The canal diverts water across the natural divide.

during the 20th century can nearly all be attributed to the work of civil and environmental engineers. This dramatic increase in life expectancy is due to the widespread installation of sewage collection pipes, wastewater treatment, and drinking water treatment.

Unfortunately, engineering solutions sometimes lead to future problems, and this is illustrated by the Chicago River reversal. For example, the placement of the canal established a direct hydrologic connection between the Great Lakes and the Mississippi River that did not previously exist. As a result, invasive species such as the Asian Carp (Figure 10.6) are free to travel up the river to the lakes and potentially disrupt the ecology; this is currently a major concern. In January 2010, the U.S. Supreme Court denied a request from several Great Lakes states to force the Army Corps of Engineers to shut navigational locks to prevent the carp from entering Lake Michigan. Scientists and engineers have installed electrical barriers in the canal to deter the carp from moving upriver, but the effectiveness of these devices is questioned.

Further concerns have arisen with respect to very strict laws for water withdrawals from Lake Michigan. As water demand increases with population, there are more attempts to take water from Lake Michigan. Uncontrolled withdrawals could ultimately lead to a lowering of the lake's water level. Water withdrawals from all of the Great Lakes is strictly regulated by a commission representing the bordering states and provinces (Illinois, Indiana, Michigan, Minnesota, New York, Ohio, Pennsylvania, Wisconsin, and Ontario).

Figure 10.6 An Invasive Species, the Asian Carp, is a Potential Threat to the Lake Michigan Ecosystem.

Source: U.S. Fish and Wildlife Service.

Design Applications

For the remainder of this chapter, we will present sample design tasks undertaken by five subdiscipline areas of civil and environmental engineering: construction engineering, environmental/water resources engineering, geotechnical engineering, structural engineering, and transportation engineering.

CONSTRUCTION ENGINEERING DESIGN APPLICATION—CUT AND FILL

Nearly all sites for which infrastructure components are designed require some amount of earthwork. The earthwork may take the form of excavating trenches or smoothing out irregularities in the land surface. In many instances, extensive earthwork is required to provide: a level site for a building pad, a corridor in which a roadway will be constructed, a level site for a parking lot, etc. Soil material may need to be excavated from a certain portion of the site (termed **cut**) and/or may need to be added to another portion of the site (termed **fill**).

For example, consider the building site shown in Figure 10.7. A level building pad needs to be created with the use of earth moving equipment for the proposed building. Since the site is not flat, a certain amount of cut and fill will be required. In other words, the low areas will need to be filled in and the high areas will need to be "dug out." This can be understood by comparing two cross-sections.

Figure 10.7 Building Site.

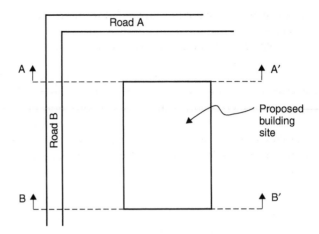

Figure 10.8 Cross-Section View, with Exaggerated Vertical Scale.

Figure 10.9 Cut and Fill Regions.

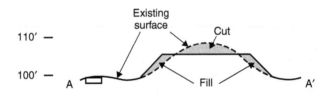

First, consider a profile (cross-section) of the land along the line labeled A-A′, as shown in Figure 10.8. The land elevations are shown using tic marks to the left of the drawing, and are expressed in feet (above some datum, often mean seal level, or MSL). Also shown in Figure 10.8 is the proposed building pad, to be constructed at an elevation of 105 feet. At the center of the building pad, earth will need to be excavated down to an elevation of 105 feet. At the edges of the building pad, backfill material will be needed to fill in the areas beneath the building pad but above the existing ground. Additionally, the soil will need to be sloped at a uniform slope away from the edge of the building pad. The proposed cut and fill plan is illustrated in Figure 10.9.

To estimate the quantity of cut and fill (a volume) for this site, we will introduce a method called the **average end area** method. First, consider that the shaded cut and fill regions in Figure 10.9 have an associated area, say in square feet.

Given a second cross-section, along B-B′, the volume of cut and fill between the two cross-sections can be estimated. Using the average end area method, the volume of cut (or fill) is calculated using Equation 10.1:

$$V = \frac{A_1 + A_2}{2} \cdot L \qquad (10.1)$$

where

$V =$ volume of cut (or fill)

$A_1 =$ the cross-sectional area of the cut (or fill) at the first cross-section

$A_2 =$ the cross-sectional area of the cut (or fill) at the second cross-section

$L =$ the distance between the two cross-sections.

---

numerical example 10.1 Cut and Fill Volume

Assume that the area of cut in Figure 10.9 is 500 square feet and that a cross-section along B-B' shows a cut area of 50 square feet. If the two cross-sections are separated by 100 feet, estimate the volume of cut.

solution

The volume of cut can be estimated by using Equation 10.1.

$$V = \frac{500 \text{ ft}^2 + 50 \text{ ft}^2}{2} \cdot 100 \text{ ft} = 27,500 \text{ ft}^3$$

Just as engineers need to quantify the volume of cut and fill, they are also interested in whether the volumes of cut and fill **balance**. It is desirable that the volume of cut is approximately equal to the volume of fill for a given project; in such a case, the volume of cut and fill are said to be balanced. If a site has excess cut, the contractor must find a place to dispose of this cut material. If the site has excess fill, additional fill material must be trucked in from off site. Thus in cases of excess fill or cut, additional expenses are incurred.

Consider the case of building a new 5-mile road through existing irregular terrain. Imagine that to ensure accuracy in the volumes of cut and fill, you would need to obtain cross-sections every 50 feet and perform volume calculations between them. Clearly, this is a tedious and time-consuming process, as the roadway consists of more than 500 cross-sections. Computer software is used to perform these calculations, but the user should always verify the calculations by hand, using a simple method such as the end-area method, to ensure that the computer-generated values are realistic (i.e., to ensure that mistakes did not occur with data entry).

Cut and Fill Video

ENVIRONMENTAL ENGINEERING DESIGN APPLICATION—SEPTIC SYSTEMS

Municipalities collect wastewater from homes, industries, stores, etc., using sewers that convey the wastewater to treatment facilities. In rural areas, because of low population densities, it is not feasible to install sewer systems. As a result, rural homes, schools, and other facilities typically have on-site treatment of wastewater using septic systems. The Environmental Protection Agency (EPA) estimates that 25 percent of all homes in the nation use septic systems. A typical residential septic system is shown in Figure 10.10. The typical septic system includes a septic tank for collecting settling solids and a **drainfield**.

Treatment occurs in the drainfield by two primary mechanisms: decomposition by bacteria and adsorption (attaching) of contaminants to soil particles.

A **drainfield** is a distribution system of pipes that releases the partially treated wastewater from the septic tank into the soil.

Figure 10.10 Cut-Away View of a Typical Residential Septic System.

Source: EPA, 2005.

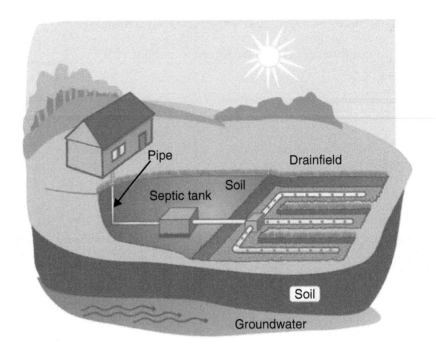

A site assessment is required to determine the type of soil at the site. If the soil is too sandy, or if the groundwater table is close to the land surface, the septic tank **effluent** may percolate too quickly into the groundwater below.

Without sufficient natural soil treatment, neighboring drinking water wells may be at risk of contamination. If the soil contains too many fines (silt and clay), the septic tank effluent will not percolate down through the soil and can accumulate in the drainfield, potentially pooling at the ground surface and producing a health hazard. In the Geotechnical Engineering Analysis Application in Chapter 9, you learned about soil size distributions and soil textures. This information may be used to determine whether or not a site is appropriate for a septic system. In some cases, sites are not appropriate. When purchasing rural land for a future home site, the buyer should always confirm that the site is suitable for a septic system. If it is not, more costly options such as a **holding tank** may be required.

Typical design considerations for septic systems are:

1. What volume is required for the septic tank to achieve proper settling?

2. What are the dimensions of the drainfield?

Wastewater that is discharged after treatment is termed **effluent**.

A **holding tank** is a storage tank for wastewater. It must be frequently pumped out at significant expense.

SEPTIC TANK DESIGN A typical septic tank is shown in Figure 10.11. While the primary treatment objective is to collect settleable solids, additional treatment also occurs. Fats, oils, and greases (collectively termed **scum**) float on water, and the PVC tees at the inlet and outlet of the tank are designed to trap these substances. Biological decomposition of organic wastes also occurs in the tank. All three of these treatment processes are fundamental to municipal wastewater treatment, and are performed in more highly engineered systems at larger wastewater treatment facilities. Proper "operation" of a septic tank requires that accumulated sludge must be periodically removed to avoid septic system failure.

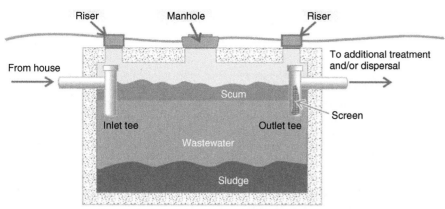

Figure 10.11 Cross-Section of a Septic Tank.

Source: EPA, 2005.

The size of a septic tank is based on the concept of **detention time**, τ, which is the average amount of time that water remains in the tank. Detention time is calculated using Equation 10.2:

$$\tau = V/Q \qquad \text{(10.2)}$$

where

  V = tank volume
  Q = flowrate

Equation 10.2 can be rearranged to solve for the necessary tank volume:

$$V = \tau \cdot Q \qquad \text{(10.3)}$$

Requirements for septic tank detention time vary from state to state, depending on design codes. A typical value is 24 hours. The flowrate of wastewater depends on water usage in a home. Typically, state codes provide a value for wastewater flowrate per person (or per capita), which can be multiplied by the number of bedrooms (assuming two parents in one bedroom and one child per additional bedroom) to get the total flowrate.

Typical residential wastewater generation is 50 to 60 gallons per person per day. However, if we designed for the average flow, we would be underestimating the flow of approximately half of all homes. However, if we design for extreme flows (e.g., 200 gallons per person per day), most homeowners would end up with tanks that were larger (and more costlier) than necessary. Another consideration is that new homes often have more water efficient appliances, toilets, etc. A value of 70 gallons per capita per day is often recommended, but state or local codes may provide specific requirements.

numerical example 10.2 Septic Tank Sizing

Determine the tank size for a three-bedroom home.

solution

For design purposes, a three-bedroom home (for four residents) is expected to generate 280 gallons per day. For a 24-hour (1-day) detention time, a 280-gallon tank would be required, as determined by Equation 10.3:

$$V = \tau \cdot Q = 1 \text{ day} \cdot 280 \text{ gal/day} = 280 \text{ gal}$$

However, as mentioned previously, solids (or sludge) settle in the bottom of the tank. This effectively decreases the storage volume of the tank for water above the sludge. As the effective volume decreases, the detention time (and thus treatment efficiency) also decreases. It is common to double the size of the tank to account for sludge accumulation. Thus our required tank size will be 560 gallons.

Septic tanks are supplied by manufacturers in incremental sizes (e.g., 500, 750, 1,000 gallons). To special-order a 560-gallon tank would cost more than a stocked 750-gallon tank; therefore, we will round up to the next largest size and choose the 750-gallon tank. (Note that this practice of altering designs to meet supplier specifications is routine in engineering.)

DRAINFIELD DESIGN   A drainfield has two key design criteria: total area (square feet) and shape (length × width). The area is based on the soils at the site. The shape is determined by lot size, location relative to the home, requirements for separation (or **setbacks**) from lot lines, and other factors. Many state codes have specific requirements for drainfield area based on soil type. One such design standard is the **application rate** (AR), the maximum flowrate of septic tank effluent to seep through a given area of soil. Typical values for application rates of septic tank effluent range from 0.2 to 1.0 gallons per square feet per day. Drainfields are often constructed of parallel pipes in trenches. Total drainfield area is a function of trench area and the spacing between parallel trenches (often 8 to 10 feet from center to center of parallel pipes). Trench width is often 5 feet, and pipe lengths are often required to be less than or equal to 100 feet (to minimize maintenance problems). Trenches are typically several feet deep and are dug parallel to topographic contours (i.e., they are "level"). Trenches are backfilled with gravel once the slotted drainpipes are installed, and covered with topsoil for seeding. Filter fabric is placed between the gravel and the topsoil to prevent fine particles from the soil from washing down into the gravel and clogging the drainpipes.

The trench area is calculated using Equation 10.4:

$$A = \frac{Q}{AR} \tag{10.4}$$

where

A = trench area
Q = flowrate
AR = application rate

Trench length is calculated by dividing the trench area by the trench width.

Determine the size for a drainfield to follow the septic tank sized in the previous example, given that $AR = 0.4$ gallons per square feet per day. Also, given a trench width of 5 feet, a maximum trench length of 100 feet, and the fact that trenches are to be spaced 10 feet on centers, provide a possible layout for the trenches.

numerical example 10.3 Drainfield Sizing (Continued)

solution

The trench area can be obtained from Equation 10.4:

$$A = \frac{Q}{AR} = \frac{280 \dfrac{\text{gal}}{\text{day}}}{0.4 \dfrac{\text{gal}}{\text{day} \cdot \text{ft}^2}} = 700\,\text{ft}^2$$

By dividing the total area by the trench width (5 feet), we obtain a total trench length of 140 feet. Since the maximum trench length is 100 feet, the system would likely be designed as two 70-foot trenches, or three 47-foot trenches (or perhaps even four 35-foot trenches depending on site layout).

The drainfield layout for three 47-foot trenches is shown in Figure 10.12.

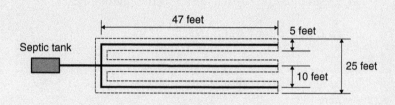

Figure 10.12 Plan View of Drainfield Layout.

GEOTECHNICAL ENGINEERING DESIGN APPLICATION—FOOTINGS

Footings are used to distribute the weight of the foundation and its overlying structure to the underlying soil or rock. A spread footing, on which the foundation wall rests, is illustrated in Figure 10.13. Various types of soil and rock have differing abilities to support a load. If too much weight is applied to a unit area of soil, the soil will compress, resulting in unwanted settlement. Evenly distributed settlement will result in a lowering of the elevation of the structure. Uneven, or differential, settlement may result in the tilting of a structure (e.g., the Leaning Tower of Pisa in Italy, or the bridge in Figure 10.14) or the formation of cracks in foundation walls (Figure 10.15), which compromise structural integrity. Properly designed and constructed footings prevent settlement.

The strength of a soil is characterized by its **bearing capacity** (also referred to as its **load capacity**) and is expressed as a force per unit area. The bearing capacities (in pounds per square foot, psf) of various geologic materials are shown in Table 10.5. Actual values may be determined by geologic testing of the material at a building site to ensure proper design. Note that the bearing capacity is significantly dependent on how much the soil is compacted.

The lower the bearing capacity, the more susceptible a soil is to compaction, and thus settlement, and the larger the footing area requirement to spread the weight of the structure and its foundation.

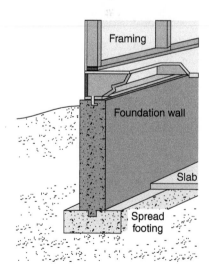

Figure 10.13 Diagram of a Spread Footing.

Figure 10.14 Differential Settlement of the Foundation (Abutment) of a Bridge. Note that the right side of the bridge entrance is lower than the left.

Source: U.S. Geological Survey.

A simplified approach to footing design follows.

Step 1: Determine footing area.

$$A_{\text{footing}} = \frac{W}{q}$$ **(10.5)**

where

$A_{\text{footing}} =$ footing area
$W =$ structure weight
$q =$ bearing capacity

Figure 10.15 A Cracked Foundation Wall Caused by Differential Settlement. The crack extends from the bottom of the concrete wall to the top, and is wider at the top. The right side of the wall settled more than the left, placing the wall in tension. As a result of the crack, the structural integrity of the wall is compromised. The portion of the wall to the right of the crack is tilting (rotating at the base of the wall) toward the viewer.

Source: M. Penn.

Table 10.5

Bearing Capacities of Soils and Rocks

Load-Bearing Material	In-Place Consistency	Nominal Allowable Bearing Capacity (psf)
Crystalline rock (e.g., granite)	Hard, sound rock	160,000
Sedimentary rock (e.g., sandstone)	Medium-hard, sound rock	40,000
Gravel	Compact Loose	14,000 6,000
Coarse sand	Very compact Medium to compact Loose	8,000 6,000 3,000
Fine sand or silty sand	Very compact Medium to compact Loose	6,000 5,000 3,000
Silts	Very stiff to hard Medium to stiff Soft	6,000 3,000 1,000
Clays	Very stiff to hard Medium to stiff Soft	8,000 4,000 1,000

Source: USACE, 1992.

Step 2: Determine footing width:

$$B = \frac{A_{\text{footing}}}{P}$$ **(10.6)**

where $B = $ footing width
$P = $ perimeter length of foundation wall

The weight of the structure is the sum of the **dead load** (structural components such as foundation wall, floors, joists, roofing), the **occupancy live load** (furnishings, appliances), and the **snow load** on the roof, if applicable, which can be quite substantial in areas receiving heavy snowfalls (e.g., 40 psf in Minnesota). Some "rules of thumb" values that can be used for preliminary calculations are found in Table 10.6.

Table 10.6	
Typical Values for Footing Calculations	Load of a conventional wood-frame residential structure (dead + live, excluding snow load): 1st floor (with roof) = 60 psf 2nd floor additional increment = 40 psf
	Snow load= 0–50 psf depending on local snowfall accumulation
	Footprint area of the building is determined from the floor plan (footprint = length × width)
	Weight of foundation wall = height (feet) × wall width (feet) × perimeter length (feet) × density of concrete (150 pounds per cubic foot, pcf)
	Structural weight (pounds) = [(Structural load + snow load, psf) × footprint (square feet)] + foundation wall weight

numerical example 10.4 Footing Width Determination

Determine the footing width for a two-story 3,200 square feet home to be built on a silty sand site in Minnesota. Foundation walls are made of 8 inch-thick concrete and are 8 feet tall. Assume a snow load of 40 psf.

solution

The building footprint is obtained by assuming that each story is of equal area.

$$3200 \, \text{ft}^2/2 \, \text{stories} = 1,600 \, \text{ft}^2 \, \text{footprint}$$

The dimensions of the floor plan can be obtained by assuming a square floor plan. Consequently, the dimensions are 40 feet × 40 feet. Given these dimensions, the perimeter length of the foundation walls is determined to be 160 feet (=4 sides × 40 feet per side).

Given this information and the rules of thumb in Table 10.6, the weight of the structure can be calculated. We will assume that the roof area is equal to the footprint area.

Weight of foundation wall = 160 ft × 8 ft × 0.67 ft × 150 pcf = 129,000 lbs
Weight of house = 1,600 ft² × 60 psf + 1,600 ft² × 40 psf = 160,000 lbs
Weight of snow = 1,600 ft² × 40 psf = 64,000 lbs

Thus, the total weight to be supported by footings equals 353,000 pounds. Given this weight, the footing area and width can be calculated using Equation 10.5 and Equation 10.6, respectively. From Table 10.5, the bearing capacity of the soil is 4,000 psf.

$$A_{footing} = \frac{W}{q} = \frac{353,000\,lb}{4,000\,psf} = 88.3\,ft^2$$

$$B = \frac{A_{footing}}{P} = \frac{88.3\,ft^2}{160\,ft} = 0.55\,ft = 7\,in.$$

Note that this width is less than the wall width (8 inches), and thus, for this application, no footing may be required. However, if the soil had a load bearing capacity of only 2,000 psf, the footing width would double to 14 inches. Thus the footing width required for a given structural load is inversely proportional to the load bearing capacity of the soil. Often, safety factors in the range of 1.5 to 2.5 are applied, so that the design footing width will be larger than the calculated footing width.

The design of footings must also take into account the strength of the concrete itself; that is, the concrete footing must be strong enough to support the overlying load. You will learn about the strength of concrete and the reinforcement of concrete with steel in subsequent civil engineering courses. Depending on the type and scale of the structure and the soil at a site, footing designs may be much more complex and require additional geotechnical and structural knowledge.

MUNICIPAL ENGINEERING DESIGN APPLICATION—LAND DEVELOPMENT

One common design task for engineers involved in land development is to divide up a piece of land into parcels. The parcels may be developed for any number of clients, including commercial, business, retail, and residential. For this application, we will consider the dividing of land into residential building lots.

Consider a site to be developed as shown in Figure 10.16. It is nearly 28 acres in size, and is bounded by a road to the north and a stream to the south. The site slopes down to a creek, and the 100-year floodplain is identified in Figure 10.16. You should be able to verify this size using an engineering scale.

How to Use an Engineering Scale

The manner in which land is divided into parcels will depend on many site constraints. For this site, specific constraints are defined as follows:

- Most parcels are to be sized between one-fourth and one-third of an acre.

- Roads are to be 40 feet wide.

- Cul-de-sacs, if used, are to have a diameter of 90 feet.

- The newly created subdivision should provide no more than two access points to the existing Mitchell Hollow Road.

- Houses will have a footprint of about 2,000 square feet.

- Each parcel must have 60 feet of **frontage** on the road. Frontage is defined as the full length of a parcel measured alongside the road on which it borders.

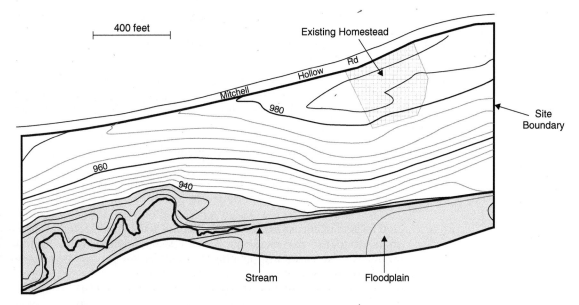

Figure 10.16 Proposed Land to be Developed. Topographic contour interval is 4 feet.

- Portions of the floodplain can be included in parcels, but houses cannot be built within the floodplain.

- 5 percent of the developable land must be left open for parks/recreation/nature areas.

- In addition, the subdivision of the land into parcels should minimize the number of odd-shaped lots.

There are many feasible ways of dividing the land into parcels. One possible solution is shown in Figure 10.17, which contains 65 lots. For this solution, the developer chose to not include the floodplain in any of the parcels; rather, this area will be set aside as a nature area/parkland.

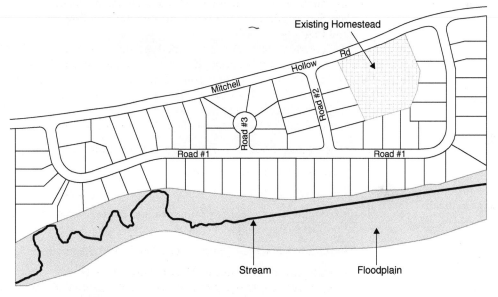

Figure 10.17 Sample Residential Layout.

West Point Bridge Design
Software

STRUCTURAL ENGINEERING DESIGN APPLICATION—BRIDGES

Many bridges are considered among the greatest achievements of civil engineers (e.g., the Brooklyn Bridge in New York City shown in Figure 10.18, which was built in 1883). The structural design of bridges is very complex, but a simplified process (for educational purposes only) for truss bridges has been incorporated into the West Point Bridge Design software that is available for download.

Truss bridges serve as a classic example to demonstrate how bridge structures balance forces. Structural members of a truss, depending on design, may be under tension or compression. Depending on the distribution of the load being supported, the size of the structural members (and their connections) must be appropriately designed.

The West Point Bridge Design software allows you to analyze compression and tension of a truss bridge using different sizes, cross-sectional shapes, and materials for the truss members. The best design provides a structurally sound bridge for the least cost (including not only materials, but also construction and maintenance).

In addition to functionality (span length, number of traffic lanes, clearance height for boat traffic above high water level, etc.), aesthetics often plays a key role in bridge design. Large bridges are icons of a city or region, and often significant additional expenditures are made for aesthetic reasons (Figure 10.19).

TRANSPORTATION ENGINEERING DESIGN APPLICATION—INLET SPACING

One task facing transportation engineers is to ensure that rainwater is effectively and safely conveyed out of the traffic lanes. For this reason, most roads are **crowned** in the middle; that is, the highest point of a road's cross-section

Figure 10.18 Brooklyn Bridge.

Source: Copyright © Klaas Lingbeek-van Kranen/iStockphoto.

Figure 10.19 The Juscelino Kubitschek Bridge in Brazil.

Source: Copyright © OSTILL/iStockphoto.

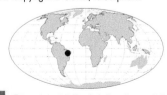

is in the center of the roadway. The most typical cross slope is 2 percent (i.e., 2 inches per 100 inches).

After water is conveyed to the edge of the road by the road's crown, the water is conveyed in gutters along the road's edge. A curb serves to keep the water in the gutter's channel. As the water travels in the gutter, the flow increases as additional water from the roadway is collected. Consequently, the gutter flow of water will begin to spread out into the roadway. An **inlet** is installed in the gutter to intercept this flow and to convey the water to a sub-surface storm sewer network or into a receiving stream or ditch.

The most common type of an inlet is a **combination inlet** (Figure 10.20), a combination of a grate set into the bottom of the gutter and an opening in the side of the curb. If the grate does not clog with debris, then stormwater will be captured by the grate as long as the flowrate does not exceed the grate capacity. If the grate clogs, the curb opening will capture less of the stormwater and the excess stormwater will bypass to the next inlet.

Inlets must be spaced such that the **spread** of water does not become too large. One common standard for stormwater inlet spacing is to ensure that the width of one travel lane (e.g., 12 feet) is open to allow for emergency vehicles to pass. This standard applies to a storm of a certain rainfall intensity (often expressed in inches per hour). This intensity is typically set equal to the known intensity of a 10-year storm for a given locale.

To determine the spacing of inlets, an inlet must be located such that it prevents the spread from becoming too large. The flowrate that will cause this allowable spread to occur can be estimated using Manning's Equation (Equation 9.2 in Chapter 9). Once this flow is known, the amount of contributing area (watershed) that will generate this amount of flow can be estimated using the Rational Method (Equation 9.1 in Chapter 9). It is this contributing area that dictates the spacing. These concepts are best illustrated with an example.

Figure 10.20 Stormwater Entering a Combination Inlet.

Source: M. Penn.

Spread is the distance from the curb to the edge of the flowing water in the street.

numerical example 10.5 Inlet Location

Consider a 36-foot wide roadway that is crowned in the center with a 2.0 percent slope. The roadway, along with a 6.0-inch high curb, is shown schematically in Figure 10.21. Also note that the roadway has a longitudinal slope (along the roadway centerline, into the page) of 1.0 percent (0.010 foot/foot). A ridge runs parallel to each side of the roadway, such that 100 feet of land on each side of the road contributes runoff to the road. Determine the location of the "first" inlet (i.e., the first inlet from the top, or **crest**, of the 1.0 percent road centerline). We will use a rainfall intensity of 4.0 inches/hour for our analysis.

solution

The spread for the cross-section in Figure 10.21 is 12 feet, which allows for a 12-foot lane to remain open during the design rain event. Given this spread and the cross-section geometry, we can use Manning's Equation to estimate the flow that this cross-section can convey.

Recall Manning's Equation (Equation 9.2: $Q = \frac{\alpha \cdot A}{n} R_h^{\frac{2}{3}} S_o^{\frac{1}{2}}$). The area ($A$) and hydraulic radius ($R_h$) for the channel can be obtained from Figure 10.22, which is an exaggerated drawing of the flow cross-section. The length b is equal to the allowable spread (12 feet). The length a is equal to the height of water at the curb, and can be calculated knowing the cross slope (2.0 percent, or 2 feet of rise per 100 feet of run). Thus, a is equal to 0.24 feet (12 feet × 0.02 foot/foot). From this information, the cross-sectional area of the triangle in Figure 10.22 is the area of a triangle ($0.5 \cdot b \cdot h = 1.44\,\mathrm{ft}^2$). The wetted perimeter is equal to the sum of a and c, and given the small angle formed by b and c, the length of c is essentially equal to b. Thus, the wetted perimeter is equal to 12.24 feet[†] and the hydraulic radius (A/P) is found to be 0.12 feet. By inserting the values of A, R_h, S_o (0.010 foot/foot), and n (0.016) into Manning's Equation, we calculate the flowrate to be 3.3 cfs. This is the flow conveyed by one of the flow cross-sections depicted in Figure 10.21.

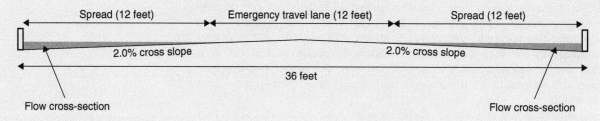

Figure 10.21 on schematic with labels: Spread (12 feet), Emergency travel lane (12 feet), Spread (12 feet), 2.0% cross slope, 2.0% cross slope, 36 feet, Flow cross-section, Flow cross-section

Figure 10.21 Schematic of Roadway Cross Slope (Exaggerated Vertical Scale) and Spread of Accumulated Stormwater.

Figure 10.22 Channel Cross-Section.

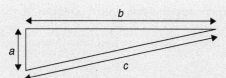

To determine the inlet spacing that will preclude the spread from being greater than 12 feet, the length of roadway that will produce a flow of 3.3 cfs must be determined. This can be calculated by solving the Rational Method (Equation 9.1: $Q_p = CiA$) for area (A), and then calculating the length of roadway that corresponds to this area. The intensity, i, is given as 4.0 inches/hour and we will assume that C for the roadway and adjoining land area is 0.50 (corresponding to the high end of "single family residential" values in Table 9.1 in Chapter 9). Therefore, the Rational Method equation can be solved for area, A.

$$A = \frac{Q}{Ci} = \frac{3.3\,\mathrm{cfs}}{0.50 \cdot 4.0\frac{\mathrm{in.}}{\mathrm{hr}}} = 1.65\,\mathrm{acres} = 72,000\,\mathrm{ft}^2$$

This value represents a rectangular area of land and roadway that contributes runoff to an inlet. This rectangle has a width of 118 feet (corresponding to one-half of the roadway plus the distance from the back of the curb to the ridge line). Consequently, the length of the rectangle is 609 feet (610 feet expressed to the correct number of significant figures), and this length represents the distance from the crest of the roadway to the first inlet. Note that inlets are required on both sides of a crowned roadway.

[†]Note that in the case of intermediate calculations, an "incorrect" number of significant figures should be used in order for the final answer to be correctly calculated. Otherwise, if intermediate calculations are rounded off, this could result in error being propagated through the calculations.

This example illustrates the concepts involved in spacing inlets and has necessarily made many simplifying assumptions. For example, the longitudinal slope typically changes along the length of the roadway depending on topography. You might be wondering what the spacing of the next inlet should be. Most likely it will be closer, given that the first inlet will not collect all of the flow, but will bypass a certain amount of it because the spread of water extends into the roadway beyond the capture zone of the inlet.

Outro

For most engineers, their primary responsibility is, in fact, design work. The design process allows the engineer to methodically develop and evaluate solutions to problems while accounting for design constraints. The details of final designs require technical skills that you will develop in subsequent courses, and throughout your career. Bringing the designs to fruition requires excellent communication skills and attention to detail during implementation.

chapter Ten Homework Problems

Calculation

10.1 For the area of land shown in Figure 10.16, create an alternative lot layout. Use lots that are approximately 1 acre in size. Your final design should meet all other constraints used for the lot layout in Figure 10.17. What is the roadway length for your layout? How does it compare to the roadway length in Figure 10.17? (One of the common goals of residential land development is to minimize the ratio of roadway length to the number of parcels.)

10.2 Reconsider the road cross-section in Figure 10.21. How much more flow could be conveyed by that cross-section if the allowable spread was 14 feet instead of 12 feet?

10.3 Reconsider Numerical Example 10.5. How would the distance to the first inlet vary if the longitudinal slope was only 0.0010 foot/foot?

10.4 A site is to be graded such that a level building pad will be installed at an elevation of 110 feet. Cross-sections of the existing ground are provided in Figure 10.23. Section A and Section B are 150 feet apart. Note that the vertical scale is different from the horizontal scale; this is often done in engineering to exaggerate the variation in the vertical direction. How much fill is needed between Section A and Section B?

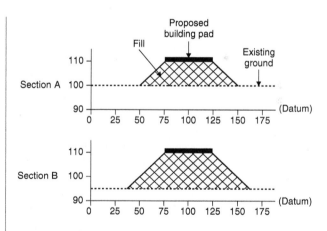

Figure 10.23 Cross-Sections for Homework Problem 10.4. Elevations and distances are in feet.

10.5 Repeat Homework Problem 10.4, except consider that the building pad is at an elevation of 105 feet. Assume that the base widths for the fill area are the same as for Homework Problem 10.4.

10.6 At what elevation should the building pad of Homework Problem 10.4 be placed in order to balance the cut and fill volume?

10.7 Determine the footing width for a one-story 800 square feet conventional wood-frame house with a 2:1 length-to-width ratio to be constructed on a site in a warm climate (i.e., no snow load) with the following soil:

a. Compact coarse sand
b. Loose fine sand
c. Stiff clay
d. Soft silt

10.8 A rectangular concrete septic tank is 3 feet wide × 5 feet long × 3 feet high. The outlet tee is located 0.5 feet below the top of the tank. Sludge solids accumulate in the bottom of the tank at a rate of 4 inches/year. Wastewater enters the tank with a flowrate of 200 gallons/day.

a. What is the detention time of wastewater in the tank without sludge accumulation?
b. What is the detention time of wastewater in the tank after two years of sludge accumulation?

10.9 Size a septic tank for a house with a maximum occupancy of six people. State all assumptions.

10.10 Design a drainfield layout for the previous problem for a soil with an application rate of:

a. 1.0 gallons/square foot/day
b. 0.5 gallons/square foot/day
c. 0.25 gallons/square foot/day

Short Answer

10.11 Why is it important for an infrastructure engineer to communicate the importance of maintenance to a client?

10.12 What are the steps of the design process?

10.13 How does a decision matrix help the engineer and client choose the best design?

10.14 What is the basic engineering principle upon which foundation footing designs are based?

Discussion/Open-Ended

10.15 Document an example on your campus where a decision was made to spend more money than otherwise necessary in order to incorporate aesthetics in the design of an infrastructure component.

10.16 Suggest an aesthetic improvement to an infrastructure component of your campus. What additional design considerations and potential cost increases would be necessary for this improvement?

Research

10.17 The Hoover Dam Bypass project has recently been completed. Read the Executive Summary (available at www.wiley.com/college/penn) of the Environmental Impact Statement. Of the three bypass alternatives proposed, was the least cost alternative the preferred alternative? What other criteria in addition to cost were considered?

10.18 Consider one of the following components located on your campus, as selected by your instructor. List the competing variables that may have been considered in the design of that item. What were the design constraints? What regulations apply to the component? What is a typical design life? How would you define "failure" of the component? What risks are involved if the component fails? Use the Internet to conduct your research unless your instructor suggests alternative sources (e.g., your university library or a textbook from a subdiscipline area).

a. Environmental/water resources
   i. culvert
   ii. detention pond
b. Structural
   i. bridge
c. Transportation
   i. intersection
   ii. parking lot
d. Geotechnical
   i. retaining wall

10.19 For one or more items (as selected by your instructor) in Homework Problem 10.18, list several alternative designs that engineers might have considered before implementing the final design. Make sure to consider some non-technical solutions.

10.20 Use Internet resources to investigate the concept of LID (low impact design) with regard to residential and commercial land development. Write a one-page paper in which you describe the pros and cons of LID and state whether you think the pros outweigh the cons, or vice versa. List all sources.

10.21 Choose a national civil engineering landmark (list available at www.wiley.com/college/penn) and perform research to identify the design constraints associated with the project.

Key Terms

- application rate
- average end area
- bearing capacity
- combination inlet
- crest
- crowned
- cut
- dead load
- design criteria
- design life
- design process
- detention time
- drainfield
- effluent
- fill
- footing
- frontage
- holding tank
- inlet
- load capacity
- occupancy live load
- performance criteria
- problem statement
- scum
- setbacks
- snow load
- spread

References

Environmental Protection Agency (EPA). 2005. *A Homeowner's Guide to Septic Systems*. EPA-832-B-02-005.

Museum of Science. *Engineering is Elementary*. http://www.eie.org/content/engineering-design-process, accessed August 8, 2011.

United States Geological Survey (USGS). 1999. *Illinois*. Fact Sheet 014-99.

United States Army Corps of Engineers (USACE). 1992. *Bearing Capacity of Soils, Engineering Manual*. 1110-1-1905.

chapter Eleven Planning Considerations

Learning Objectives

After reading this chapter, you should be able to:

1. Define planning and explain why it is necessary.
2. Explain how a Capital Improvement Program is used for infrastructure planning.
3. Define context-sensitive design and describe how it is related to transportation planning.
4. Explain how effective land use planning can lead to more efficient infrastructure.
5. Explain the importance of emergency planning.
6. Predict future populations given historical population data.
7. Compare and contrast infrastructure, transportation, land use, regional, and emergency planning.

Introduction

Planning is the act of formulating a course of action, of creating a roadmap to what the future should be. As a student, you should plan when you will take specific courses in order to graduate on time. Once you graduate, you will hopefully begin to plan for a financially secure retirement. Planning, as it relates to the infrastructure, is the act of formulating a specific course of action for future roadway expansions, wastewater treatment facility upgrades, airport runway expansions, and the like.

Planning is by nature a multidisciplinary endeavor, is always carried out by teams, and is a very people-centered profession. Indeed, to be effective, planning *must* be multidisciplinary as many stakeholders are involved. Many of these stakeholders have competing interests. For example, imagine the stakeholders involved in choosing the route for a highway bypass around a municipality. Landowners will be affected as the bypass will require their land to be purchased; some landowners will welcome this source of income while others will resist based on a decrease in privacy, increase in noise, or diminished economic viability. Some farmers may be affected as their fields are no longer contiguous or accessible. Elected officials will have a stake in that they want the bypass to succeed in order to attract businesses to their community; difficulties that arise during the project may lead to their inability to be re-elected. Some business owners (e.g., hotel owners) will benefit from the increased economic activity associated with construction while others will suffer as traffic is moved away (either temporarily during construction or permanently due to the bypass). In addition, a number of citizen groups exist that will be concerned about a wide variety of potential impacts, including environmental degradation, loss of wildlife habitat, and increased noise.

The Colors of Infrastructure

Some people distinguish between gray infrastructure and green infrastructure. The former refers to the physical, built infrastructure including roads, bridges, etc. On the other hand, green infrastructure refers to the interconnected network of open spaces and natural areas, and includes greenways, wetlands, parks, and forest preserves. Benefits of the green infrastructure are obvious in terms of wildlife habitat and native species, but a healthy green infrastructure can benefit a community by increasing stormwater infiltration (and thereby recharging groundwater sources and decreasing the risk of flooding), reducing the urban **heat island effect**, and enhancing community interactions.

Proper planning is needed to ensure that gray infrastructure and green infrastructure can coexist. Ideally, both types of infrastructure should be planned simultaneously and complementarily.

Introductory Case Study: Las Vegas, Nevada Population Growth

The historical populations of the city of Las Vegas and Las Vegas metropolitan area are presented in Figure 11.1; this area is one of the fastest growing areas in

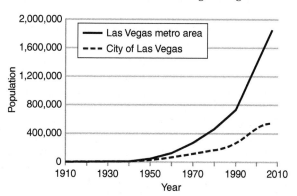

Figure 11.1 Population Growth of the City of Las Vegas and the Metropolitan Area. The metropolitan area includes the city in addition to the surrounding suburbs and smaller cities and towns.

Figure 11.2 Expanding Growth of Las Vegas.

Source: Southern Nevada Regional Planning Coalition.

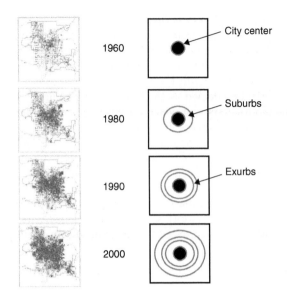

1960 — City center

1980 — Suburbs

1990 — Exurbs

2000

Exurbs are the primarily residential areas that lie outside a city and usually beyond its suburbs.

Impact of Las Vegas Population Growth on Groundwater Resources

We recommend that you read the article "Las Vegas, Nevada: Gambling with Water in the Desert," which is part of U.S. Geological Survey Circular 1182, *Land Subsidence in the United States*.

Gambling with Water in the Desert

the United States and has doubled in population in the last 20 years. Note from Figure 11.1 that the suburban areas collectively have a much greater population than the city and two or three times the rate of growth.

What does this rapid growth have to do with planning? In one word, everything. The city's infrastructure must be able to serve this growing population. In order to ensure that this expanding infrastructure is built in an orderly and efficient manner, proper planning is essential.

The dramatic growth that has occurred outside the city limits is shown in Figure 11.2. The four maps in the left-hand column of Figure 11.2 represent the increase in developed area between 1960 and 2000. The four representations in the right-hand column illustrate how the increase in developed area has occurred similar to the adding of concentric rings, with the bull's eye representing the city center, the first ring representing the suburbs, and the outer rings representing the **exurbs**.

The Need for Planning

Crazy India Intersection

The need for planning might be best understood by considering what happens when an infrastructure component or system is constructed without being carefully planned. Consider the roadway shown in Figure 11.3 and the intersection recorded in the video on the textbook website. These were most likely not planned for the type and extent of traffic that moves through the intersections daily. As a result, the vehicle, bicycle, and pedestrian traffic are dramatically chaotic and very unsafe.

Most large urban areas in the northeastern United States were established long before the use of automobiles. As a result, many streets in city centers are relatively narrow, and no provision for right-of-ways has been made. Adding underground utilities such as water mains, wastewater conveyance pipes, telephone lines, and electricity lines, can be very complicated given that no provision was made for these utilities when the cities were built. Moreover, narrow streets also limit the availability of on-street parking and sometimes prevent two-way traffic flow as an option. This failure to plan for unforeseen infrastructure cannot be faulted, but it does illustrate the effect that a lack of planning can have. The problem is even more dramatic outside of

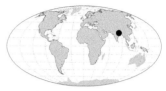

Figure 11.3 An Urban Roadway in Bangladesh.

Source: Courtesy of J. Jordens.

North America, where many urban areas were initially developed several centuries ago (Figure 11.4).

A failure to plan makes future reconstruction and renovation more difficult. An example of this is the reliance on the automobile as the prime means of moving people around in urban areas for much of the 20th century. Now that many urban areas are becoming increasingly congested with traffic and highway expansion is infeasible due to space or cost constraints, elected officials are beginning to see the wisdom in mass transit and non-motorized transport (pedestrian and bicycle). Yet transitioning to mass transportation is hindered and is much more expensive than it would be if mass transit had been part of the original plan.

The effectiveness of planning directly influences the "livability" of a city, or how desirable a city is to potential residents. This influence can readily be seen by viewing the set of criteria (Table 11.1) that *Forbes* magazine uses in their annual ranking of "most livable" and "most miserable cities." Planning affects many of the criteria listed in Table 11.1. The most obvious criterion affected by effective (or ineffective) planning is traffic congestion.

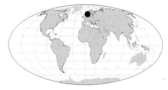

Figure 11.4 A Very Narrow, Centuries-Old Street in Norway.

Source: Courtesy of R. Bonell.

Table 11.1	Most Livable	Most Miserable
Criteria Used by *Forbes* Magazine in Ranking Most Livable and Most Miserable Cities	High income growth per household	Long commute times (traffic congestion)
	Low cost of living	High income taxes
	Low crime levels	High unemployment
	High culture and leisure index (e.g. museums, parks, recreation)	High crime rates
	Low unemployment	High corruption
		High sales taxes
		Many "Superfund" (highly contaminated) sites
		Poor professional sports team performance
		Bad weather

The amount of employment and the cost of living are also related to planning effectiveness. Although some of these criteria may be misleading or unimportant to many people (e.g., professional sports), these rankings greatly influence people's perceptions and decisions.

The need for planning can also be illustrated by Boston's Central Artery/Tunnel Project (the "Big Dig"). The Big Dig is considered one of the most ambitious engineering projects undertaken, and consisted of several large projects such as moving 3.5 miles of Interstate 93 underground. The total cost of the project was over $20 billion.

As mentioned previously, the streets in older cities were not designed (nor planned) with automobile use in mind, and Boston's street layout is notoriously tangled and inefficient. In the 1950s, in an attempt to alleviate this problem, an expressway was built (the Central Artery). However, in retrospect, the planning for this project appears to have been relatively near-sighted. The expressway in essence put up a barrier between the central city and the waterfront, which can be observed in Figure 11.5. The residences of nearly 20,000 people and many businesses were permanently displaced as a result of its construction. The expressway quickly became unable to carry the necessary

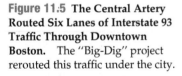

PBS Video on the "Big Dig"

Figure 11.5 The Central Artery Routed Six Lanes of Interstate 93 Traffic Through Downtown Boston. The "Big-Dig" project rerouted this traffic under the city.

Source: Courtesy of the Massachusetts Department of Transportation.

traffic volumes. Designed for 75,000 vehicles per day, the expressway was carrying 200,000 vehicles per day by the mid-1990s. By 2010, transportation models predicted that stop and go traffic could be expected for 16 hours per day! Part of the problem is due to the simple fact that the population of Boston and its surrounding suburbs was rapidly growing between the 1950s and the 1990s, and therefore traffic volumes were increasing. However, better regional planning, land use planning, and transportation planning could have eliminated or greatly alleviated these problems.

This example illustrates how the improper planning of the infrastructure affects many aspects of our world, including economic, environmental, social, and safety aspects. (Each of these aspects will be covered in individual chapters in this text.) For example, the Central Artery affected the economy of the central city by dividing it from profit centers along the waterfront. Traffic congestion also negatively affected the economy due to decreased economic activity (there is minimal economic production in idling vehicles) and increased fuel usage. Increased traffic congestion also contributed to decreased air quality. Social impacts were evident by the displacement of thousands of individuals from their homes and businesses and by the reduced enjoyment of the waterfront amenities. The effect on safety is demonstrated by the fact that the accident rate on the Central Artery was four times the national average when compared to similar types of roadways.

To alleviate these problems, the project eventually known as the "Big Dig" was envisioned to route the Central Artery traffic *beneath* the city (Figure 11.6). Planning for the Big Dig began in the 1970s but ground was not broken until 1991. The intervening time (nearly two decades) between initial planning and the start of construction is typical of very large projects. Much of this time was spent in non-design activities, the most important of which was finding ways to finance the project.

Given the huge scale of the project (Figure 11.7 and Figure 11.8), which included two major tunnels and a cable-stayed bridge, it was imperative that the Big Dig was properly planned. For example, a tunnel was bored under the existing expressway, which meant that engineers had to plan and design how to keep the expressway open while a tunnel was constructed beneath it. A tangled web of utilities ran beneath the construction site (Figure 11.9), and at its deepest (120 feet below ground surface), the tunnel had to pass under a subway line (Figure 11.10). Moreover, engineers and planners had to decide what to do with the 16 million yards of excavated soil. Finally, there were environmental concerns related to excavation of toxic materials, an increase in noise and dust, and disturbance to harbor life, including lobster migration.

Figure 11.6 Northbound Rush Hour Traffic Entering the Central Artery Tunnel (Arrow). Prior to the Big Dig project, this traffic was routed through, rather than under, downtown Boston (see Figure 11.5).

Source: Courtesy of the Massachusetts Department of Transportation.

Remember that full-sized versions of textbook photos can be viewed in color at www.wiley.com/college/penn.

Figure 11.7 Big Dig Project Map.

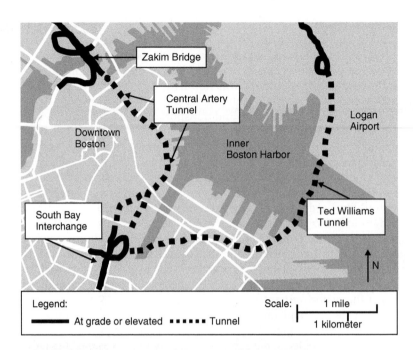

Figure 11.8 Construction of the South Bay Interchange, One Portion of the Big Dig Project in Boston.

Source: Courtesy of the Massachusetts Department of Transportation.

Boston's Missing North/South Rail Link

Boston's commuter rail lines terminate at two separate stations, the North Station and the South Station. These two stations are separated by only one mile, but travel between them is accomplished by walking, taking two separate light rail trains, or by taxi, which exacerbates the automobile traffic congestion. In retrospect, the planning for the Big Dig might have addressed this missing link by adding a rail line in the same tunnel in which the Central Artery was rerouted.

The following sections of this chapter will focus on five types of planning: infrastructure planning, transportation planning, land use planning, regional planning, and emergency planning. As you will see, these five types of planning are interrelated and share many of the same principles.

Infrastructure Planning

The infrastructure supporting our country is vast, and must be maintained, retrofitted, and renovated to continue to meet its intended purpose, often for a larger population than originally planned. If existing components or sectors can no longer meet their intended purpose, they must be expanded or new components must be designed and constructed.

A municipality must decide how best to spend its limited funds. For example, given a finite amount of money, the civic leaders may need to choose between spending money on a new police station or on an upgrade to the wastewater treatment plant, or between drilling a new water supply well, or resurfacing several street segments.

Figure 11.9 Nine Different "Live" Utilities Run Through This Excavated Area, Some of Which are Not Visible in the Picture.

Source: Courtesy of the Massachusetts Department of Transportation.

Figure 11.10 Layers of Infrastructure in Downtown Boston. The "Big Dig" was so named for the lower-most "layer," the tunnel replacing the Central Artery.

Source: Redrawn with permission from the Massachusetts Department of Transportation.

Perhaps it is evident that the planning process can be contentious as there are many competing variables. Public safety, public benefit, and public security should be at the forefront, but in reality must be balanced with public opinion, political realities, and the availability of funding. Also, federal and state regulations are constantly changing, and a new and/or stricter regulation may very well *require* a certain project to be completed if fines and other penalties are to be avoided. Priority will be given to those projects for which funding is available (e.g., a clean air fund that provides tax-free loans for implementation of certain types of abatement projects). Also, the city will want to allocate funds for projects that enhance the ability of the city to grow economically (e.g., creation of an industrial park or improvements to the schools that will attract more young professionals and their families). At the same time, the "general public" values improvements and additions to the parks, decreases in commuting time, and convenient parking.

Another variable that must be assessed is the *condition* of the infrastructure component. ASCE's *Report Card for America's Infrastructure*, introduced in Chapter 1, assigns a grade to different areas of the infrastructure to inform congressional decisions on which areas are most in need of funding. A similar exercise should be carried out by municipalities in order to help prioritize future projects.

The "triple bottom line" is one paradigm by which plans are increasingly being shaped. This paradigm was first applied to the business world but is pertinent and is increasingly applicable to government entities. The triple bottom line considers "people, planet, and profit." Thus, with respect to the infrastructure, decisions on spending should be made with respect not

Was it worth it?
While the benefits of the Big Dig are numerous, the project was plagued by delays, substandard materials and construction, and cost overruns. The final project cost of approximately $20 billion is nearly *three* times that of the estimated cost when construction began. And countless lawsuits have arisen. In 2006, only months after project completion, a concrete ceiling panel collapsed on a car, killing the passenger and injuring the driver. In order to conduct safety inspections in the aftermath of the tragedy, the tunnel was closed for several months, temporarily recreating the traffic jams that the tunnel was designed to alleviate.

just to the traditional "bottom line" of making money (in the private sector) or of balancing the budget (in the public sector), but should also consider the impact on people ("human capital") and on the environment ("natural capital"). For example, infrastructure projects should treat *people* fairly regardless of economic level or race (see discussion on "environmental justice" in Chapter 16, Social Considerations). Moreover, infrastructure projects should seek to protect the *planet*, perhaps by selecting projects that will enhance or create wildlife habitats or increase stormwater infiltration. Boston's Big Dig project, for example, addressed people and planet by creating an extensive and beautiful park space and by reuniting portions of the city that had been separated by the Central Artery.

Case in Point: Planning in San Antonio, Texas

San Antonio, Texas, has established a 50-year water management plan. It includes projected water demand, identification of potential additional water sources, drought risk management, and conservation efforts. In areas with increasing demand and steady or decreasing water availability, such long-term plans are necessary because the political, technical, and regulatory hurdles associated with securing water sources are significant. Anticipated water sources and demands for San Antonio through 2060 are shown in Figure 11.11. Given the uncertainty of future water demand, three water use scenarios are presented: the high demand scenario corresponds to a daily water use of 126 gallons per capita per day (gpcd); normal demand corresponds to 116 gpcd; and low demand corresponds to 106 gpcd. From Figure 11.11, it is apparent that existing water supplies will likely not meet future demand; for example, in 2060, demand will be greater than the available water supply for the normal demand and high demand scenarios. This information is used by water planners to evaluate the costs and benefits of various alternatives.

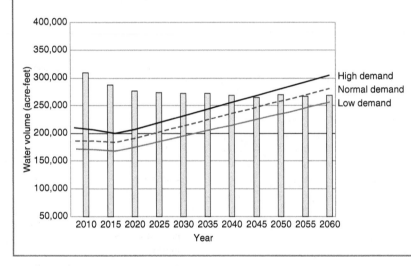

Figure 11.11 San Antonio Water Systems.   Water Sources and Demands for San Antonio Water System. The shaded bars represent annual volume of available water supply in a non-drought year (during droughts, less water supply is available). The three lines represent volumes of water demand for three different scenarios. The water supply volumes shown in the graph are based on existing sources in 2009 and do not include proposed sources of additional water. The San Antonio water utility continually evaluates supply and demand, updating projections to guide water management.

Source: San Antonio Water System, 2009.

A 100-Year Plan

The city of Calgary, Alberta, is experiencing rapid growth, fueled (pun intended) by a wealth of regional fossil fuels. Some might find it ironic that an "oil city" is also a "green city" (increasingly dependent on renewable energy). In 2005, a "visioning" process began for a long-term sustainability plan for the city in which an estimated 18,000 residents participated.

A 200-plus-page document, the Imagine Calgary Plan (ImagineCalgary, 2006), was published in 2006 detailing a 100-year vision and specific targets to be achieved in 30 years in the following areas: infrastructure (communications, energy, food, goods and services, housing, transportation, waste management), economic well-being, governance, natural environment, and social well-being. The plan is supported by many local organizations and an implementation team is working to solidify collaboration with provincial and federal agencies.

Managing and prioritizing ongoing projects is the purpose of infrastructure planning. Given that projects are carried out continuously, states and municipalities must also continuously plan. One annual outcome of infrastructure planning for many cities is a document that prioritizes the infrastructure projects to be completed in a relatively short timeframe (e.g., the next five years). One common name for this assessment is a **Capital Improvement Program** (CIP). Additional plans may be carried out for longer time periods, perhaps 10 years or even 50 years.

The CIP for Chicago, Illinois, for the planning period between 2008 and 2012 is 374 pages long and is typical of the infrastructure plans developed by large cities. It includes projects worth nearly $2 billion for 2010 alone. The distribution of costs for the City of Chicago's 2010 CIP projects is depicted in Figure 11.12. Note that these capital projects are projects to be completed *in addition* to carrying out the routine maintenance that the city must perform and the operational and administrative expenses of its water treatment facilities, wastewater facilities, and street department.

Although at a much diminished scale, small cities and villages also typically complete similar planning studies and publish a list of projects to be completed in the next 5 years. Shown in Table 11.2 is a listing of projects for a small city (population of 10,000) that are planned to be completed for 2010. This list illustrates the variety of projects that must be completed. Note that the scale of investment is proportional to population (when comparing Figure 11.12 and Table 11.2) and ranges from approximately $100 to $700 per person annually for the two municipalities.

Transportation Planning

Infrastructure planning deals with all aspects of the infrastructure, including transportation projects. However, since transportation planning is a dominant aspect of the infrastructure, we will present it separately.

Transportation issues are often ranked as the number one problem by public opinion polls in metropolitan areas. Specific issues related to transportation problems include traffic congestion, availability of efficient and effective mass transportation, and air pollution associated with the high traffic volumes.

The purpose of transportation planning is to create a plan that will improve transportation in the future. These plans are typically quite far-ranging,

Table 11.2

A Plan for Infrastructure Projects for a Single Year for a Small City

Project Name	Estimated Cost
Transportation	
Greenwood Avenue reconstruction	$ 950,000
Staley Avenue reconstruction	$ 480,000
Elm Street reconstruction	$ 480,000
Madison Street reconstruction	$ 130,000
Furnace Street reconstruction	$ 355,000
Sidewalk repair	$ 55,000
Senior center parking lot	$ 10,000
Water Systems	
Well #5 design	$ 200,000
Well #3 transmission line	$ 25,000
Water meter replacement program	$ 50,000
Sewers	
Sunset Drive easement sewer replacement	$ 50,000
Rehabilitate Bio-Filter Towers (WWTF)	$ 100,000
Northeast Interceptor extension	$ 880,000
Total	$1,370,000

City of Chicago 2010–2014 Capital Improvement Program

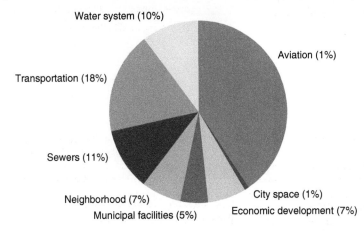

Figure 11.12 Major Components of the 2008 Chicago Capital Improvement Program.

Source: Data from City of Chicago, 2010.

Demographics are the various characteristics that describe a population, such as age distribution, family sizes, and level of household income.

If something is **context-sensitive**, its behavior or characteristics differ in response to its context or setting. Similarly, context sensitive infrastructure is infrastructure that is designed in a way that it "fits in" with its surroundings.

Context-Sensitive Design Example, Interstate I-94

with time frames of 25 to 50 years being common. They are often oriented toward entire regions rather than a specific municipality.

A transportation plan lays out a roadmap (pun intended) of the projects that will best meet the region's needs. Plans may specify a variety of methods, including adding lanes to roadways, adding new roads, adding bus lanes to roads, providing for mass transit needs, and enhancing opportunities for pedestrians and bicyclists. Creating this plan must take into account population projections (discussed at the end of this chapter), availability of funding, safety, security, cost effectiveness, and financing capabilities. The plan must also consider changing **demographics**.

Transportation planners address the triple bottom line through considering **context-sensitive** solutions. Unlike traditional transportation planning, which often had the primary (and only) objective of moving traffic from point A to point B, context-sensitive solutions are solutions that take into account the context in which the project is to be created. Contexts to be considered include aesthetic, archeological, community, cultural, environmental, historic, recreational, and scenic.

The following example illustrates the concept of context-sensitive solutions. In Kentucky, U.S. Highway 27/68, known as the Paris-Lexington Road or the "Paris Pike," was unable to accommodate the increasing traffic and had a high fatality rate. The original plan for reconstruction, formulated in the 1960s, was to replace the two-lane road with a four-lane road and would have required the demolition of historic structures. Local groups protested this plan, and in the 1970s, the project was halted by a Federal District Court injunction. Eventually, a context-sensitive solution was accepted and completed in 2003. This project preserved historic structures and scenic views, and "blended in" with the existing countryside much better than the originally proposed roadway. This solution included conserving the native topsoil, avoiding historic properties, dismantling and reassembling dry stone (i.e., without mortar) walls (Figure 11.13), using timber guardrails to better blend into the surroundings (Figure 11.14), and landscaping with native plants.

Figure 11.13 Reassembled Dry-Stone Walls Along the Paris Pike.

Source: © 2013 Sherman Cahal Works.

Figure 11.14 Timber Guard Rails Used on the Paris Pike.

Source: © 2013 Sherman Cahal Works.

Context-Sensitive Design in Dayton, Ohio

Dayton, Ohio, has is home to the Wright brothers and Wright-Patterson Air Force Base, and bills itself as the birthplace of aviation. A recent reconstruction of Interstate 70 paid tribute to the city's aviation heritage. Highway elements such as retaining walls and bridge piers contain impressions and images of the Wright Brothers' first flight, military fighter jets flying information (Figure 11.15), Apollo 9, the earth and moon, and the American flag.

Figure 11.15 Overpass in Dayton, Ohio.

Source: P. Parker.

Case in Point: Downtown Boston After the Big Dig

When the Central Artery was routed beneath Boston, what was once an interstate (Figure 11.16) was transformed into local roads and a greenway (Figure 11.17). The result is a downtown area that is both pedestrian-friendly and local vehicle-friendly.

Figure 11.17 After the Big Dig: Local Roads and the Greenway that Replaced the Central Artery. The location and angle of this photograph are approximately the same as in Figure 11.16. Note that traffic emerges from the Central Artery Tunnel in the background. The cable-stayed Zakim Bridge can be seen in the background.

Source: Courtesy of the Massachusetts Department of Transportation.

Figure 11.16 Before the Big Dig: Central Artery Traffic in downtown Boston.

Source: Courtesy of the Massachusetts Department of Transportation.

Land Use Planning

Land use planning is the process by which decisions are made on future land uses that are deemed to best secure the well-being of communities.

Urban sprawl is the *unplanned* (and somewhat haphazard) and low density growth of areas surrounding cities.

"Nowhere to Grow"—PBS Video on Urban Sprawl

Land use planning is directly linked to transportation planning. Effective land use planning can make transportation planning much easier by guiding land development such that transportation choices are much more sustainable. The link between transportation planning and land use planning can be illustrated by considering traffic congestion. Traffic congestion is often caused by **urban sprawl**.

Congestion has many negative effects, including increased air pollution, decreased fuel economy, increased non-productive time for idled motorists, and increased anxiety and frustration of drivers. A traditional approach to this problem is to widen the road and add lanes, thus increasing capacity. The downside to this approach is the tremendous cost involved, as well as the fact that such widening projects are often short-sighted with respect to long-term goals.

On the other hand, if urban sprawl were lessened, the problem may be alleviated, and future expansions might be unnecessary. Decreasing urban sprawl through "smart growth" strategies is one of the primary tasks for land use planners.

Smart Growth Plans

Land use plans specify what types of development should occur on certain areas of land. A land use plan is the key portion of a **comprehensive plan**, also known as a **smart growth plan**. The purpose of this plan is to provide communities with information and policies that will guide future community decisions. These plans have a relatively long time frame (20 years is typical) and should reflect the uniqueness of the community. The plans are described in reports that typically include sections on housing; transportation; agricultural, natural and cultural resources; economic development; and land use. At their best, they are a powerful means of ensuring that development is as sustainable as possible. At the same time however, such plans are accused of being a "socialist" tool that provides the government with too much power at the expense of private landowner rights.

A **greenfield** is a site on which no previous infrastructure has been constructed and is usually a farmfield or wooded area on the edge of a municipality.

Urban sprawl is typified by low density housing, perhaps on the order of two or three homes per acre in so-called **greenfield** development. An example of a greenfield site is shown in Figure 11.18. This low density is common among conventionally designed subdivisions (Figure 11.19). These housing developments are very popular as they provide the homeowners with relatively

Figure 11.18 A Typical Greenfield Site–Agricultural Land That is to be Developed as a Residential Subdivision.

Source: M. Penn.

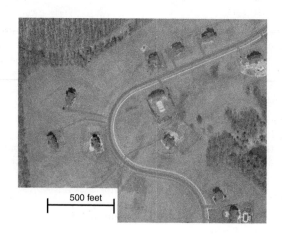

500 feet

Figure 11.19 Low-Density Housing.

Source: U.S. Geological Survey.

private and spacious home sites. However, they are also one of the key drivers to urban sprawl. In order to supply housing at this relatively low density, the developments must be built outside of the city limits. This often necessitates that the homeowners are dependent on automobiles for transportation. The additional use of automobiles is the primary cause of traffic congestion. The problem is compounded by the fact that the cost of adding mass transit, which is expensive at any density, becomes cost-prohibitive when serving low-density housing.

Congestion is not the only negative impact of urban sprawl. Among many other drawbacks, urban sprawl is also very "expensive" for municipalities. Municipalities must provide infrastructure to the outlying developments; given the distance over which utilities must be constructed and the low density of users, providing these services is very inefficient. For example, consider conveying wastewater from the homes. The wastewater will have to travel many miles to the centralized wastewater treatment facility, and the probability of needing a lift station will increase as the **sewer service area** increases.

Additionally, installing the infrastructure is expensive when considered on a per lot basis. For example, every 100 feet of sewer pipe will only serve two or three homes, perhaps between 6 and 12 people. In a higher density area, which serves condominiums, townhouses, or apartments, that same length of sewer pipe could serve many more people, perhaps by an order of magnitude.

 Music Lyrics Reflecting an Increasingly Transient Society, "I'm Not From Here" by James McMurtry

The **sewer service area** is the geographic area from which wastewater sewage flows to a centralized wastewater treatment facility.

Ten Principles of Smart Growth

1. Mix Land Uses. Clustered development works best if it includes a mix of stores, jobs, and homes. Single-use districts make life less convenient and require more driving.

2. Take Advantage of Existing Community Assets. From local parks to neighborhood schools to transit systems, public investments should focus on getting the most out of what is already built.

3. Create a Range of Housing Opportunities and Choices. Communities should offer a range of options: houses, condominiums, affordable homes for low-income families, and apartments.

4. Foster "Walkable," Close-Knit Neighborhoods. These places offer not just the opportunity to walk—sidewalks are a necessity—but something to walk to, whether it's the corner store, the transit stop or a school. A compact, walkable neighborhood contributes to people's sense of community.

5. Promote Distinctive, Attractive Communities with a Strong Sense of Place, Including the Rehabilitation and Use of Historic Buildings. In every community, there are things that make each place special, from train stations to local businesses.

6. Preserve Open Space, Farmland, Natural Beauty, and Critical Environmental Areas. Many people want to stay connected to nature and are willing to take action to protect farms, waterways, ecosystems, and wildlife.

7. Strengthen and Encourage Growth in Existing Communities. Rather than turning forests and farms into developed areas, opportunities to grow in already built-up areas should be investigated.

8. Provide a Variety of Transportation Choices. People can't get out of their cars unless they have another way to get where they're going. More communities need safe and reliable public transportation, sidewalks, and bike paths.

9. Make Development Decisions Predictable, Fair, and Cost-Effective. Builders wishing to implement smart growth should face no more obstacles than those contributing to sprawl. In fact, communities may choose to provide incentives for smarter development.

10. Encourage Citizen and Stakeholder Participation in Development Decisions. Plans developed without strong citizen involvement don't have staying power. When people feel left out of important decisions, they won't be there to help out when tough choices have to be made.

Source: Adapted from smartgrowthamerica.org

Many of the principles of Smart Growth can be achieved by increasing housing densities. Higher densities can be achieved through smaller building lots and multi-family housing. Higher densities can often be achieved

> **Infill** is defined as the development or redevelopment of vacant or underutilized sites in areas that are economically or physically static or declining.

efficiently by redeveloping previously developed areas. This type of development is known as **infill**.

Infill has many benefits, such as often being in close proximity to jobs, amenities, convenience services, and mass transit hubs. However, one of the greatest benefits, as contrasted to urban sprawl, is that infill projects can often utilize existing infrastructure. The extent of these benefits is case-specific, and barriers to infill exist, which include the following list of items, adapted from the Southern Nevada Regional Planning Coalition's Infill Study Plan.

- High land cost (suburban and undeveloped land is much less expensive than urban land)

- Negative public perception (e.g., abandoned areas, deteriorating infrastructure, crime, vagrancy, and graffiti)

- Development risk (developers are more familiar with policies and practices of suburban development compared to infill, and infill areas are more likely to have contamination)

- Aging or inadequate infrastructure (e.g., additional costs to replace sidewalks, improve roads, and improve water and sewer)

- Zoning codes (infill development often requires variances from codes for street setbacks, building height, density, setbacks, and parking—adding to the project timeline and the risk of denial of the development plan)

- Lack of support services (successful infill development depends upon the existence of commercial "support" services in close proximity—grocery stores, medical centers, and business services—which are often lacking)

- Financing limitations (commercial lenders often lack policy and experience to assist with infill development)

Zoning is the key *enforcement tool* of a comprehensive plan. Zoning is the way that communities control the physical development of land by specifying how that land can be used and (re)developed. The primary purpose of zoning is to prevent a landowner from using property in a manner that will negatively impact neighboring landowners. Zoning laws typically designate specific areas in the community for residential, industrial, agricultural, religious, institutional, green space, or commercial development. Goals of these local regulations include:

- Lessening traffic congestion

- Providing safety from fire or other dangers

- Promoting community health standards

- Preventing pollution of streams, lakes, and air

- Facilitating adequate provision for transportation, water, sewage, schools, parks, and other public services

- Promoting economic development

By zoning, the local community can dictate the types of buildings and activities that are allowed in specific regions of the community. Zoning laws may restrict the height, size, and location of buildings on lots; percentage of a lot that can be occupied by a building; size of yards, courts, and other open

Table 11.3

Examples of Zoning Categories	Zoning Category	Examples of Allowed Uses
	Agricultural	Crops, livestock, pasture, associated farm buildings, and residences
	Commercial	Retail and wholesale businesses, restaurants, medical facilities, offices, hotels
	Industrial	Manufacturing, warehousing, packaging, shipping
	Residential	Single family, multi-family (apartments, condos) housing
	Green space	Parks and recreation areas
	Institutional	Churches, schools, universities, police stations, firehouses, utilities, government facilities

spaces; density of population in relation to lot size (e.g., single family vs. multi-family residences), permitted noise levels, lot ingress/egress, and landscaping requirements. Each community defines acceptable activities within a land use zoning category. Typical examples are provided in Table 11.3.

Another tool at the disposal of planners is the power of **eminent domain**. The seizure of property can occur without the property owner's consent, but the property owner is to be fairly compensated for the loss of use and ownership of their property. Eminent domain may be used for a number of infrastructure projects including roadway widening, construction of new roads, and placement of utilities.

Eminent domain refers to the ability of a unit of government (federal, state, or local) to seize private property for public use.

Eminent Domain Approved for Columbia University Expansion

Kelo v. New London

At issue in this 2006 U.S. Supreme Court case was the right of a municipality (New London, Connecticut) to take land, using its eminent domain powers, for the benefit of a *private* entity. Specifically, New London wished to forcibly purchase individual residences; the purchased homes were to be demolished such that a pharmaceutical company (Pfizer, Inc.) would be allowed to build a corporate facility on the purchased land. The rationale was that the increased economic development would help the greater good of the area. This was a contentious issue, since eminent domain is typically applied to taking land by a state, city, or township for *public* use. However, in this case, the land was to be used for the private development for which there would be public benefit. The argument was that this "taking," by helping the overall public good, was a valid cause, and by a 5–4 decision, was upheld by the Supreme Court. Justice Sandra Day O'Connor wrote in the dissenting opinion, "Nothing is to prevent the state from replacing any Motel 6 with a Ritz-Carlton, any home with a shopping mall, or any farm with a factory." Soon thereafter, widespread opposition arose and many states pursued legislative measures to control eminent domain, in some cases restricting its use for government-owned projects.

Regional Planning

The nature of growth is typically outward from city center to suburbs. This inherently means that distinctly separate governments exist for the city and the suburban areas. However, if these individual governments act without considering the impact on neighboring communities, you can easily imagine inefficient and problematic infrastructure conditions. Often, the scenario is more complicated than merely a single city and its surrounding suburbs. In many areas, cities that were once distinct municipalities are melding into large metropolitan areas connected by suburbs. Examples include Tampa–St. Petersburg–Clearwater in Florida and Dallas–Fort

Worth–Arlington in Texas. In these areas, satellite images suggest one single "mass of development" rather than several neighboring communities.

The Mission Statement of the Southern Nevada Regional Planning Coalition

Regional planning commissions must balance the widespread responsibilities of regional planners and the integration of infrastructure management and planning. For example, consider the mission of the Southern Nevada Regional Planning Coalition.

The Southern Nevada Regional Planning Coalition's mission is to bring together all public jurisdictions to coordinate regional planning in a seamless fashion while respecting each member's autonomy. This requires promoting intergovernmental cooperation and trust built on careful planning and accountability, thus enhancing the quality of life in Southern Nevada. The following seven mandated priorities of the Coalition are:

- Conservation, open space, and natural resource protection
- Population forecasts
- Land use
- Transportation
- Public facilities
- Air quality
- Infill development

For example, in Figure 11.20, the Tampa–St. Petersburg–Clearwater area is shown. This area contains more than 100 cities and towns, each with their own governing unit. Each of these cities is regularly making decisions about infrastructure, whether it be mass transit, wastewater treatment, or stormwater management. Moreover, their decisions on these infrastructure projects will likely affect each other. Consequently, in such areas, it is imperative to coordinate regional planning efforts between city, suburb,

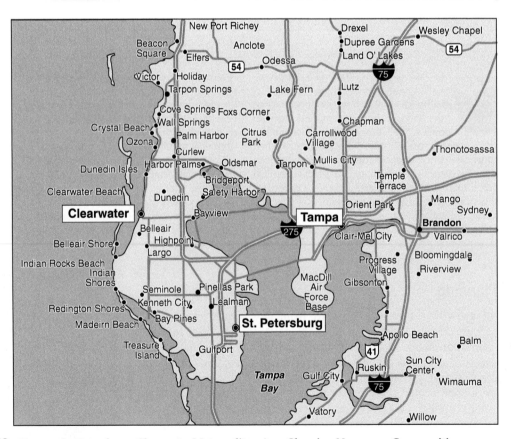

Figure 11.20   Tampa–St. Petersburg–Clearwater Metropolitan Area Showing Numerous Communities.

county, and state entities. An additional complexity arises when metropolitan areas involve multiple states (e.g., Philadelphia, Pennsylvania–Wilmington, Delaware–Camden, New Jersey).

Emergency Planning

In Chapter 18, Security Considerations, we will discuss the various types of emergencies that infrastructure must be able to withstand such as a terrorist attack or a natural disaster. In this chapter, we will briefly discuss the planning that must occur in order to effectively respond to an emergency.

To be effective, an emergency response plan must explicitly state responsible parties, as there is no time to "squabble" during an emergency. Designated responsibility is also important for legal reasons in potential lawsuits. Command and control centers need to be established in areas that are not impacted by the emergency; for flooding, such centers must be located on high ground, and for terrorist targets, the center must be located far away from the target. Hospitals and public safety personnel (police, fire, and ambulance) also have detailed emergency plans that must be compatible with municipal plans.

Horry County Comprehensive Emergency Management Plan

One very important component of disaster management is public education before, during, and after the event. In the case of floods and other weather-related emergencies, this public education must include clear communication of risks and methods to minimize flood damage to persons and property.

Most communities and regions have developed plans for various types of emergencies. For example, Horry County, South Carolina (including Myrtle Beach, a popular tourist area) has plans for hurricanes, terrorism, earthquakes, tsunamis, and mass casualties (weapons of mass destruction or disease epidemics).

There has been much criticism of the failure to adequately evacuate the city of New Orleans prior to landfall of Hurricane Katrina, leaving tens of thousands of stranded residents. The evacuation of Houston, Texas, only weeks later due to Hurricane Rita, left some residents in traffic jams for nearly 24 hours. In response to these widely publicized events, many communities are revising their evacuation and re-entry plans.

Civil and environmental engineers have the training and experience to be integral members not only in the committees that draft these plans, but as responders in emergency situations as well. We encourage you to consider becoming involved in your community's emergency planning; as an engineer, your perspective will be very valuable.

Case in Point: Fargo, North Dakota, Flood Emergency Plan

Fargo, North Dakota, has experienced serious flooding in 1997, 2006, 2009, 2010, and 2011. The 2009 flood (Figure 11.21) captured national media attention as it was compounded by a blizzard and subsequent snowstorms.

The city's flood emergency plan details the steps to be taken when a flood is imminent. The plan has two phases: road closures and dike placement. The road closure plan identifies which roads should be closed based on the flood stage (water level). Emergency dike (earthen or sandbag) placement can be seen in Figure 11.22.

Fargo has an effective public education campaign that is integral to the success of their emergency planning. For example, consider the river stage diagram for Fargo (Figure 11.23). This diagram clearly illustrates to residents the response plan for various flood stages. Without such a plan, many citizens would not be aware of the meaning of reported river stages as reported on television, radio, or the Internet. The city's website also includes interactive flood mapping.

Figure 11.21 Red River Flooding in 2009. Debris in photo has accumulated on a submerged bridge.

Source: Federal Emergency Management Agency/M. Moore.

Figure 11.22 Emergency Dike Constructed During the 2009 Flood in Fargo, North Dakota.

Source: U.S. Geological Survey/J. LaVista.

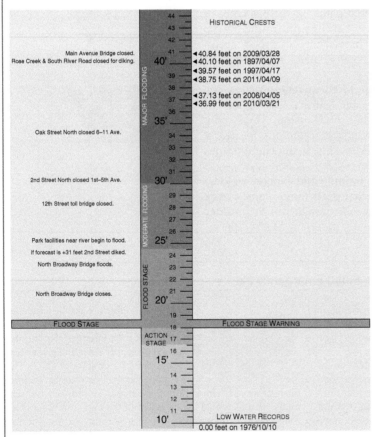

Figure 11.23 Red River Stage Levels. Note the frequency of major flooding in recent years.

Source: Data from City of Fargo, North Dakota.

Population Projections

Understanding the dynamics of human **population growth** is extremely important for engineers and planners. Nearly every type of plan, whether a land use plan, a water supply plan, or a transportation plan, must take into account the population to be served in the future.

One very simple and commonly used method to analyze human population growth is the exponential growth model (Equation 11.1). In exponential growth, the rate of growth (that is, the number added to the population over a fixed time interval) continuously increases. For example, consider the children's puzzle in which they are told that the number of lily pads on a pond double every day. That is, on the first day, the pond has one lily pad; on the second day, the pond has two lily pads; on the third day four lily pads; and so on. Such growth is exponential. The puzzle is phrased as such: "If it takes 30 days for the pond to become half-covered, when will it become fully covered?" The answer is that the pond will be completely covered with lily pads in *only one* additional day. The rate of increase of many things can be approximated by exponential growth, including human population (in certain cases), the consumption of natural resources, and the rate at which computer chip computing capability increases.

$$N = N_o e^{kt} \qquad\qquad (11.1)$$

where

$N =$ population at any time t
$N_o =$ initial population
$k =$ exponential growth rate constant
$t =$ elapsed time

Currently, the population growth rate constant (k) for the world is approximately 1.2 percent (0.012) *per year*, whereas the growth rate constant for the United States is 0.8 percent (0.008) *per year*.

The growth rate constants for individual countries and for the world collectively vary with time for a variety of socioeconomic reasons. Over the last several decades, the growth rate constant for the United States and for the world has been decreasing as is illustrated in Figure 11.24. However, note that even though the population *growth rate* is decreasing, the overall *population*

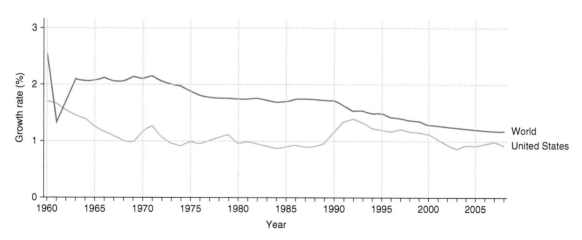

Figure 11.24 **Population Growth Rate Trends for the World and for the United States.**

Source: World Bank.

continues to increase, as any positive value of k, however small, will result in an increase in population. Consequently, the demand for resources will also increase.

The Importance of the Growth Rate Constant

You may have noticed that the magnitudes of the population growth rates do not differ by much in Figure 11.24. For example, in the last 30 years, the world's population growth rate has "only" dropped from approximately 1.7 to 1.2 percent. Yet these "small" changes have a huge effect.

For example, consider that the current population of the earth is approximately 7 billion. Using Equation 11.1, we can forecast what the population will be in 50 years, and doing so with two different values of k will demonstrate the power in k. For example, if k remains at 1.2 percent, the world population in 50 years may be estimated to be:

$$N = N_o e^{kt} = 7 \cdot 10^9 \times e^{0.012 \times 50} = 13 \text{ billion}$$

Alternatively, given a growth rate constant of 1.7 percent (that of 30 years ago), the population in 50 years would be:

$$N = 7 \cdot 10^9 \times e^{0.017 \times 50} = 16 \text{ billion}$$

Thus, a "small" difference of 0.5 percent can have a dramatic effect (3 billion people!) on the population growth.

Also, note that for the case of $k = 1.2$ percent, in the 50th year, the population increases by approximately 150 million—more than 400,000 people per day. (You can confirm this number by calculating the number of people when $t = 49$ years, and subtracting this value from the population calculated for $t = 50$ years.) This huge addition of people in *a single year* illustrates why some people refer to exponential population growth as a population "bomb."

The prediction of future population is of paramount interest to planners. Rather than modeling world population however, most planners are interested in predicting the future population for a village, a city, a county, or a region. Such predictions are fraught with uncertainty and risk, as growth rates are constantly changing, and future growth rates will most likely not be identical to historical growth rates.

At one extreme, imagine being a planner for Las Vegas and needing to predict the city's population in 2020. Historical data is provided in Table 11.4. Note that between 1990 and 2000, the city's population increased by approximately 220,000, nearly *doubling* in the process. Consider the difficulties facing Las Vegas-area planners, who need population predictions for 2020 or perhaps even 2050. The planners must consider whether the population will continue to grow at the same rate as the previous decade, at a lower rate, or perhaps at an even higher rate. And consider that compared to the population increase of the 1990s (85 percent for the decade), a relatively "low" population increase of 10 percent for the 2010–2020 decade will still result in an increase of approximately 50,000 people. Indeed, the "unexpected" economic downturn of 2008–2009 resulted in a significant decrease in the population growth of Las Vegas. As noted in the Design Life section of Chapter 10 (Design Fundamentals), overpredicting population (and thus demand) can lead to "overbuilding" and unnecessary expense.

Perhaps equally difficult is predicting populations for municipalities with *decreasing* populations. Additionally, planners are challenged when predicting future populations for small towns with recently steady populations. A factory closing, a change in the school system, designation of a high-speed rail train stop, or a hospital closure, are just a few of the occurrences that can significantly alter the population of a small town. Thus, for any situation, planners and engineers must use a variety of tools to predict future populations, including using mathematical models judiciously, communicating with civic leaders, and above all, using common sense. In all instances, planners are wise to test multiple scenarios and consider a *range* of possible values.

Table 11.4

Historical Population of Las Vegas, Nevada

Year	Population
1900	25
1910	945
1920	2,304
1930	5,165
1940	8,422
1950	24,624
1960	64,405
1970	125,787
1980	164,674
1990	258,295
2000	478,434
2010	583,756

Given an exponential growth rate constant, the exponential growth model can be utilized to predict future populations in cases where the model provides a "good fit" to existing historical data. However, it is very important to note that *not all population growth is exponential*. In practice, planners and engineers will often use the trend-line tools available to them in spreadsheets to make future predictions (see Cheyenne, Wyoming population sidebar).

Case in Point: Cheyenne, Wyoming Population Growth

Imagine that the year is 1960, and a planner needs to predict the population of Cheyenne, Wyoming, 10 years in the future (1970). The planner collects historic population data as shown in Table 11.5

Table 11.5

Historical Population of Cheyenne, Wyoming

Year	Population
1870	9,118
1880	20,789
1890	62,555
1900	92,531
1910	145,965
1920	194,402
1930	225,565
1940	250,742
1950	290,529
1960	330,066

The data is plotted in Figure 11.25 along with three predictive models: A, B, and C (explained in the following paragraph).

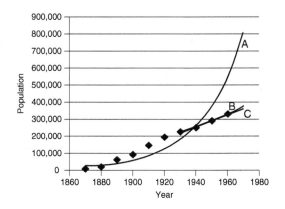

Figure 11.25 **Population Data and Three Predictive Models for Cheyenne, Wyoming.**

One other predictive model (model D, not illustrated in Figure 11.25) will also be analyzed. These four models may be summarized as:

A: exponential trend line fit to the entire historic population data from 1870 to 1960

B: exponential trend line fit to the more recent historic population data from 1930 to 1960

C: linear trend line to the more recent historic population data from 1930 to 1960

D: assumed addition of 20,000 people between 1960 and 1970 (a relatively low value compared to the actual growth of 39, 500 between 1950 and 1960)

Note that these four models are not necessarily "standard" models to use; indeed, there are no standard models used by engineers and planners to predict future populations. But they do represent alternative approaches and provide a range of values. The results are tabulated in Table 11.6, based on extrapolating the trend lines (models A, B, and C) or simple calculation (model D).

These models thus generate a range of 350,000 to 800,000. This is an extremely large range; *however* the model A prediction would be disregarded, given how poorly it models the more recent population data from 1940 to 1960. Thus, the range of values that the planner has some confidence in is 350,000 to 380,000. In reality, the *actual* population of Cheyenne in 1970 was 332,000, *less than even the lowest estimate*. This illustrates the difficulty of predicting future populations.

Table 11.6

Population Predictions of Four Models

Model	Predicted Population, 1970 (rounded to the nearest 1,000)
A	800,000
B	380,000
C	363,000
D	350,000

In the 20th century, the increase in population was accompanied by a trend toward urbanization. This trend is continuing worldwide, and by the year 2050, nearly two-thirds of the world's population is predicted to be living in urban areas. Such urbanization leads to an increased need for adequate infrastructure, and has both positive and negative consequences. One positive effect of urbanization is the associated opportunities for increased efficiency; the concentration of people into small areas makes centralized wastewater treatment and mass transit economically viable, and can be, on a unit basis, more cost effective than non-centralized alternatives. One negative aspect of urbanization is the concentration of pollutants, both air-borne and water-borne, that can have severe environmental consequences if not treated.

In addition to increasing population and increasing urban population, planners and engineers also need to consider changing demographics. For example, consider that the American population is aging. This aging population will affect the workforce, driving patterns, housing types, traffic signage, transit needs, and the way cities are managed.

World Population in 2300

Outro

This chapter illustrates the importance of planning to ensure that infrastructure is effectively and efficiently designed, maintained, and renovated. The information you learned in this chapter about population growth and population modeling will be expanded upon in Chapter 13, Sustainability Considerations, with respect to how the expanding population affects world resources.

chapter **Eleven** **Homework Problems**

Calculation

11.1 Use a spreadsheet to calculate the growth rate percent for each decade of the 20th century for Las Vegas, based on the data in Table 11.4.

11.2 For the following population data for Mobile, Alabama, predict a reasonable range of population estimates for 2030 and for 2050. Clearly explain your rationale.

1950	1960	1970	1980	1990	2000
129,009	202,779	190,026	200,452	196,278	198,915

11.3 For the following population data for Augusta, Georgia, predict a reasonable range of population estimates for 2030 and for 2050. Use at least three scenarios in your analysis. Clearly explain your rationale.

1950	1960	1970	1980	1990	2000
71,508	70,626	59,864	47,532	44,639	195,182

11.4 For the following population data for Kansas City, Missouri, predict a reasonable range of population

estimates for 2030 and for 2050. Clearly explain your rationale.

1950	1960	1970	1980	1990	2000
456,622	475,539	507,087	448,159	435,146	441,545

11.5 For the following population data for Detroit, Michigan, predict a reasonable range of population estimates for 2030 and for 2050. Use at least three scenarios in your analysis. Clearly explain your rationale.

1950	1960	1970	1980	1990	2000
1,849,568	1,670,144	1,514,063	1,203,368	1,027,974	951,270

11.6 For the following population data for Boise, Idaho, predict a reasonable range of population estimates for 2030 and for 2050. Use at least three scenarios in your analysis. Clearly explain your rationale.

1950	1960	1970	1980	1990	2000
34,393	34,481	74,990	102,249	125,738	185,787

11.7 At the time of writing the first draft of this chapter, January 28, 2010, the U.S. population was estimated to be 308,574,880. Use this population clock (available at www.wiley.com/college/penn) to determine the population increase since January 28, 2010. Use the exponential growth equation, to estimate the growth rate constant (hint: convert the time expired into decimal years, for example, 453 days = 1.24 years).

11.8 At the time of writing the first draft of this chapter, January 28, 2010, the world population was estimated to be 6,799,345,235. Use this population clock (available at www.wiley.com/college/penn) to determine the population increase since January 28, 2010. Use the exponential growth equation, to estimate the growth rate constant (hint: convert the time expired into decimal years, e.g. 453 days = 1.24 years).

11.9 At the time of writing the first draft of this chapter, January 28, 2010, the U.S. population was estimated to be 308,574,880. At the time of writing the final draft of this chapter, December 6, 2010, the U.S. population was estimated to be 310,861,399. Over this period of approximately 10 months, what was the increase in U.S. population?

11.10 Perform Internet research to find the populations of some of the largest U.S. cities (use *city* populations, NOT *metropolitan area* populations, the latter of which includes suburbs). Now consider the population added to the United States in the 10-month period referred to in Homework Problem 11.9. This increase in U.S. population is approximately equivalent to the population of which large city?

Short Answer

11.11 List three disadvantages to low-density housing, as compared to high-density housing.

11.12 List one environmental, one social, and one economic impact of the old Central Artery in Boston.

11.13 List the "three Ps" of the triple bottom line.

11.14 Define context-sensitive design.

11.15 List the ten principles of Smart Growth.

11.16 Define zoning.

Discussion/Open-Ended

11.17 If the community in which your university is located were to create a (or update an existing) land use plan, who are the stakeholders?

11.18 For a town of your (or your instructor's) choosing, describe the "livability" of the town. How could planning make the town more "livable?"

11.19 How might the redesign of a roadway on your campus be accomplished in a context-sensitive manner?

11.20 For a community of your (or your instructor's) choosing, identify which of the ten Smart Growth principles is most closely adhered to. Explain your reasoning. Also, identify the principle that is addressed the least in this community.

Research

11.21 Download from the Internet the comprehensive plan for a community of your (or your instructor's) choosing. Answer the following questions:

a. What types of planning, as described in this chapter, are discussed in the comprehensive plan?
b. Who benefits from the plan? Cite specific portions from the plan to support your answer.
c. Who might be adversely affected by the plan? Cite specific portions from the plan to support your answer.
d. If you were a taxpayer in the municipality to which the plan applies, would you be in favor of it? Why or why not?

11.22 Identify five types of "green" infrastructure in addition to those mentioned in this chapter.

11.23 Read the "Emergency Response" section (starting on page 3) of the NTSB report on the I-35 W collapse (available at www.wiley.com/college/penn), and create a timeline that represents the ten most important (in your opinion) emergency response activities.

11.24 Do you think that the Supreme Court made the correct choice in the *Kelo v. New London* case? Explain your reasoning.

Key Terms

- capital improvement program
- context-sensitive
- demographics
- emergency planning
- eminent domain
- exurbs
- greenfield
- heat island effect
- infill
- land use planning
- population growth
- regional planning
- sewer service area
- transportation planning
- urban sprawl
- zoning

References

City of Chicago, 2010. *City of Chicago 2010–2014 Capital Improvement Plan*. http://www.cityofchicago.org/content/dam/city/depts/obm/supp_info/CapitalImprovementProgram2010-2014.pdf, accessed October 1, 2010.

ImagineCalgary. 2006. *imagineCALGARY Plan for Long Range Urban Sustainability*. http://www.imaginecalgary.ca/imagineCALGARY_long_range_plan.pdf, accessed August 8, 2011.

San Antonio Water System. 2009. *2009 Water Management Plan Update*. http://www.saws.org/Our_Water/WaterResources/2009wmp/docs/SAWS2009WaterMgmtPlanUpdate.pdf, accessed July 28, 2010.

Southern Nevada Regional Planning Coalition. *Infill Study Plan*. http://www.snrpc.org/Reports/InfillDevelopmentPlan.pdf, accessed August 8, 2011.

World Bank. Indicators, Population growth (annual %). http://data.worldbank.org/indicator/SP.POP.GROW?page=3, accessed August 8, 2011.

chapter Twelve Energy Considerations

Learning Objectives

After reading this chapter, you should be able to:

1. Identify and manipulate units associated with energy and power.
2. Explain recent trends in U.S. energy consumption.
3. Explain how historical energy and population data can be used to forecast future energy demand.
4. Describe the relationship between energy use and the design and management of infrastructure components.
5. Compare the environmental impacts of various types of energy sources.

Introduction

A country's quality of life and economic growth potential are greatly dependent upon adequate energy that is both safe and secure. In the case of the United States, there are several causes for concern. The American Society of Civil Engineers (ASCE) 2009 *Report Card for America's Infrastructure* assigned a D+ grade to the energy sector. This poor grade was due to concerns about the condition, capacity, safety, and security of our energy infrastructure. Additionally, recent fluctuations in petroleum and natural gas costs have created considerable discussion and debate on energy issues at the national level.

The work of civil and environmental engineers is linked to energy issues in several ways. For example, civil and environmental engineers are involved in the planning, design, construction, and maintenance of power generation and distribution facilities. Moreover, they must be cognizant of the energy demands created by infrastructure (e.g., wastewater treatment facilities and buildings) as well as the demands of existing infrastructure.

Introductory Case Study: The Proposed Power Plant for Surry County, Virginia

As you will learn in this chapter and in Chapter 13, Sustainability Considerations, energy demand is steadily increasing in the United States and worldwide. Moreover, this demand is projected to continue to increase in the next decades. To meet the increased demand, new power-producing facilities are being built across the country. Coal-fired power plants are attractive options for many power utilities in that coal is plentiful and relatively cheap. As of 2010, the National Energy Technology Laboratory lists 31 coal-fired power plants that are either under construction, near construction, or permitted. Once completed, these plants will add 17,000 megawatts (MW) of power to the U.S. grid.

Siting a power plant, and especially a coal-fired power plant, is a very challenging undertaking. Even proposed projects for which the benefits appear to far outweigh the negatives can generate intense opposition and face high hurdles to implementation. For example, consider the proposed Cypress Creek Power Station in Surry Country, Virginia.

Surry County is located in eastern Virginia between the cities of Richmond and Norfolk. A $4 billion, 1,500-MW coal-fired power plant is proposed for construction. The advantages to this proposed location include the following: the proposed site is an abandoned mill site with rail and highway access; coal is currently the most affordable means of producing electricity, and coal is an important part of western Virginia's economy; once operating, the plant would provide 200 permanent jobs to the surrounding rural area (and an additional 2,000 temporary construction jobs to build it); the prospective owner has other facilities in the region with "a good track record." So what's not to like? Those that oppose the plant, including the Chesapeake Bay Foundation, cite concerns over the use of coal and the resulting air and water pollution, increased carbon dioxide emissions and global warming, the "spoiling" of the rural landscape, failure to support new "clean" energy alternatives, etc.

Cypress Creek Power Station homepage

Chesapeake Bay Foundation homepage

Table 12.1

Common Energy and Power Conversions

Energy

$1\ \text{BTU} = 2.928 \times 10^{-4}\ \text{kWh}$
$\qquad = 1055\ \text{joules}$

Power

$1\ \text{hp} = 42.44\ \text{BTU/min} = 745.7\ \text{W}$

Table 12.2

Common SI Prefixes

Multiplication Factor	Prefix	Symbol
10^{12} (trillion)	tera	T
10^{9} (billion)	giga	G
10^{6} (million)	mega	M
10^{3} (thousand)	kilo	k

Background

Energy is the ability to do work. Our bodies extract the chemical energy of the food we eat so that we can go about our daily lives. A coal-fired power plant extracts energy from fuel to create steam to drive turbines and generate electricity.

Energy is the mathematical product of force and distance. Energy is measured in joules (J) in the International System of Units (SI) (i.e., metric) system and British Thermal Units (BTU) in the English system of units. A joule is defined as a newton meter. A newton is the SI system unit of force.

Power is measured using units of watts (W) in SI system or horsepower (hp) in the English system. Power is the *rate* of energy use; for example, 1 watt = 1 joule/second. Energy and power are routinely reported in varying units. Table 12.1 provides some common conversions.

Depending on the scale of consideration (from individual household to global demand), the amount of energy or power required varies by several orders of magnitude. Common SI prefixes are included in Table 12.2. The power requirements of common household appliances (e.g., microwaves and toasters) are often measured in watts. Large electrical power generating facilities in the United States generate on the order of 3,000 MW

numerical example 12.1 Number of Households to be Supported By a Power Plant

Estimate the number of households that can be supported by a 100-MW electrical power generating facility. Assume that each household uses about 1,000 kWh/month.

solution

Like many calculations involving power, an answer can be arrived at by carefully paying attention to units. Given that each house uses about 1,000 kWh/month, we can determine the power requirements of a household as follows:

$$\frac{1,000\,\text{kWh}}{\text{household}\cdot\text{month}}\cdot\frac{10^3\,\text{W}}{\text{kW}}\cdot\frac{1\,\text{month}}{30\,\text{days}}\cdot\frac{1\,\text{day}}{24\,\text{hours}}$$
$$=\frac{1,400\,\text{W}}{\text{household}}$$

The number of households supported by a 100-MW facility can be arrived at by dividing the facility's power generation by this rate of energy use per household:

$$\frac{\left(100\,\text{MW}\cdot\dfrac{10^6\,\text{W}}{\text{MW}}\right)}{1,400\dfrac{\text{W}}{\text{household}}}=71,000\,\text{households}$$

However, this is only the residential energy use, which is approximately 33 percent of the electrical energy that a community needs; the remainder is for commercial and industrial demand. Furthermore, energy production facilities ideally operate below full capacity to extend the life of the facility, reduce maintenance expenses, and provide spare capacity for peaks in demand. Thus, the 100-MW facility can be thought of as an 80-MW facility under typical operation (80 percent of full capacity). If only 33 percent of the power can be used for residential demand, the actual number of households served by the 100-MW facility would be determined as follows:

$$71,0000\,\text{households}\cdot0.80\cdot0.33=19,000\,\text{households}$$

Assuming 2.5 people per household, a population of 48,000 would be supported for residential, commercial, and industrial demand by the facility.

(3×10^9 W) while smaller local facilities generate power on the order of 100 MW.

Electrical power consumption is typically expressed in units of kilowatt-hours (kWh), which is the product of power and the duration for which the power is utilized; thus, kWh is a measure of energy. For example, a 100-watt light bulb in use for 10 hours requires 1,000 watt-hours (1 kWh). The average U.S. household uses approximately 1,000 kWh/month.

Energy Trends and Predicting Demand

Energy is used for transportation, industrial production, and residential and commercial heating, air conditioning, and lighting. Total energy use includes both electricity and fuel. Recent trends for these sectors in the United States are presented in Figure 12.1. Note that unlike the relatively steady increases in

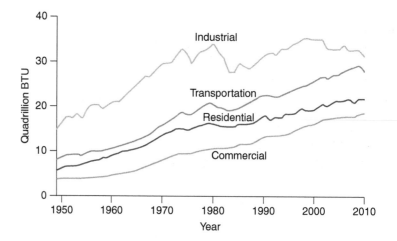

Figure 12.1 Historical Total Energy Consumption by Sector.

Source: Redrawn from EIA, 2008.

transportation, residential, and commercial demand, the industrial demand has actually decreased since the mid-1990s. This can be largely attributed to the shift in the U.S. economy from industry towards services. Notwithstanding recent decreases in industrial use, the total energy consumption in the United States is increasing, and has nearly *tripled* since 1950. Global energy demand will be discussed in detail in the Chapter 13, Sustainability Considerations.

Historical electricity use in the United States is presented in Figure 12.2 for residential, commercial, and industrial sectors. This steady increase demonstrates the need for the construction of new power plants as discussed in the Introductory Case Study. Also note that all three major sectors (residential, commercial, and industrial) are approximately equal in magnitude.

To effectively plan and manage any resource, the future demand for the resource must be known. Moreover, knowing the historical demand can help predict future demand. This principle is applied to energy in this chapter, but applies to many other aspects of the infrastructure, including drinking water use, and traffic volumes. However, graphs such as Figure 12.2 are of limited use for predicting future demand. The ability to predict is improved if the demand is divided by the population to obtain a per capita (or per person) demand, as shown in Figure 12.3. In this figure, the series labeled "Total" is equal to the sum of residential, commercial, and industrial demands.

Demand is the product of resource use per individual and the number of individuals creating demand. Consequently, per capita electrical demands such as those shown in Figure 12.3 can be multiplied by future population

Figure 12.2 Historical U.S. Electricity Use by Sector.

Data source: EIA, 2008.

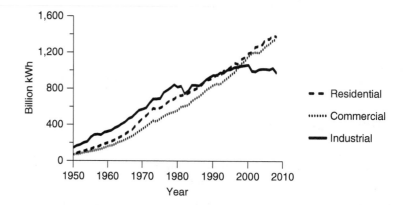

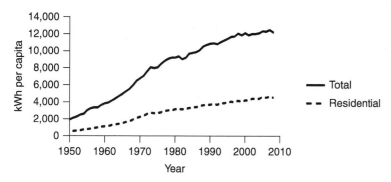

Figure 12.3 Historical U.S. Electricity Use Per Capita.

Data source: EIA, 2008.

projections to obtain estimates of future demand. Many different scenarios are investigated by engineers and planners when forecasting future demands. For example, one scenario might be that population may increase and per capita demand may decrease such that total demand remains constant.

Per capita demands can be used to forecast, as shown in Figure 12.4, and two scenarios of per capita demand projections are provided. The two projections are based on two different interpretations of Figure 12.3. Note that in Figure 12.3, the rate of increase (the slope of the line) for total electricity use per capita has been relatively constant from 1998 to 2008, and that this slope is less than the rate of increase in previous decades. The lower projection in Figure 12.4 is based on the slope of the recent data from 1998 to 2008, while the higher projection is based on the average long-term slope from 1950 to 2008. The latter projection ignores the recent decrease in the slope and likely would result in a gross overestimate; the former projection is a more reasonable projection. In general, recent data are a much more reliable source for making future predictions. However, long-term data sets are valuable for providing a sense of perspective and to help understand whether recent trends are out of the ordinary or in keeping with historical trends.

The projections in Figure 12.4 forecast for a period of approximately 40 years. Predicting the future is always difficult (as Yogi Berra, a professional baseball player and manager said, "It's tough to make predictions, especially about the future"), but the longer the time period associated with the projections, the more uncertain the estimate. Figure 12.4 begs the question, "Will electricity use per capita continue to increase for the next four decades at the

Demand Side Management

When faced with the challenge of meeting projected future demands in energy use, the first solution that is often proposed is to increase supply by building new power facilities. However, the supply and demand relationship can also be balanced out by decreasing demand. At an individual scale, demand can be decreased by increasing efficiency of appliances (e.g., replacing aging and inefficient furnaces and air conditioners with high efficiency units; replacing incandescent light bulbs with compact fluorescent lights) or by conservation efforts (e.g., installing motion-sensing light switches in rooms).

At a larger scale, some power utilities and municipalities practice Demand Side Management (DSM), in which rebates are provided to encourage utility customers to decrease their demand. Austin Energy, which serves Austin, Texas, aims to achieve a reduction in demand of 700 MW by 2020. In essence, the utility will add 700 MW of capacity. Consumers benefit with reduced electricity bills, and the environment benefits due to decreased air emissions.

Austin Energy Residential Power Saver Program Website

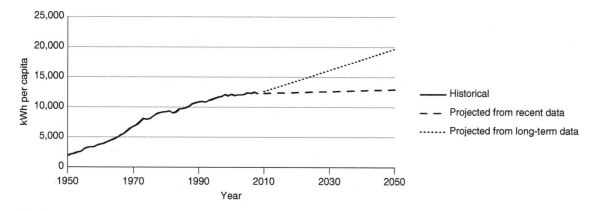

Figure 12.4 Two Projections for Future U.S. Total Electricity Use Per Capita.

rate that it did for the last decade?" You could argue that the per capita use may increase at an even slower rate, or perhaps even decrease.

Entry-level engineers are not responsible for making long-range predictions based on historical data, be it energy consumption, population, or some other quantity. Rather, these forecasts are made by more experienced engineers who have a developed sense of perspective, or even by non-engineers who specialize in planning and forecasting. Also, in many cases, projections are made for several scenarios in order to generate a range of values.

Case in Point: Providing a Range of Population Projections

Future resource demand is a function of per capita demand and the population. Projections for each of these variables have uncertainty, which is compounded when the product is calculated.

Figure 12.5 shows an example for U.S. population projections from the U.S. Census Bureau. The three scenarios illustrate that the prediction of the population in 2050 is highly uncertain; depending on the growth rate, the projections range from 250 million to nearly 500 million.

It is increasingly common to see a range of projections reported for quantities of interest to engineers in the future, including population, electricity demand, number of vehicle-miles traveled, or the quantity of wastewater generated. A range is helpful to planners in that it brackets a range of possibilities. Given a range, it is possible to examine various scenarios in terms of cost, risk, benefit, etc.

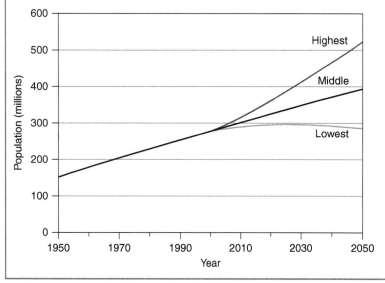

Figure 12.5 Historical and Projected U.S. Population.

Source: Redrawn from the U.S. Census Bureau, 2010a.

"Local" Data

For many civil and environmental engineering projects, data is needed at the scale of the location of the project rather than the national scale. Municipalities often have widely varying characteristics resulting in resource demands that are not "typical"; that is, their characteristics are not accurately described using national averages.

Such local variation is noted in many aspects of the infrastructure, including water demand. For example, residential water use per capita often increases with increasing household income. Water use is also higher in arid regions due to

landscape irrigation. Industrial water use, which is often a significant fraction of a community's water demand, is also highly variable depending on the nature of the industry. Breweries, canneries, and cheese production facilities are examples of high water use industries.

By obtaining recent local data, engineers and managers are able to make better decisions. Large municipalities typically collect and maintain extensive datasets. Small communities are notorious for having sparse data. When data does not exist, engineers must use professional judgment to make estimates.

Sources of Energy

Energy sources are often categorized as renewable or nonrenewable. Non-renewable sources are finite sources such as coal, petroleum, natural gas, and nuclear material. Renewable sources include wind, solar, biomass, and hydropower.

Vehicle transportation is dominated by petroleum, but alternative fuel vehicles are on the market and are expected to comprise a larger fraction of the future fleet. Many mass transit modes use electricity, which may be generated from nonrenewable or renewable sources. As alternative fuel vehicles become more popular, they will require changes to our infrastructure. For example, in order for electric cars to become a mainstream alternative, charging stations (the equivalent of a gasoline station for a traditional car) must be made widely available.

Electricity is generated from a wide variety of fuel sources (Figure 12.6). Coal is by far the most widely used source. Natural gas use has increased recently because many new power plants were built during the 1990s when natural gas prices were low. Nuclear energy is being debated as a potential growth sector. Renewable sources currently account for approximately 10 percent of electrical generation and are expected to increase dramatically in the future. The International Monetary Fund estimates that worldwide, non-hydropower renewable energy (biomass, solar, wind, geothermal, wave, and tide) will increase from 2 percent to 7 percent of electrical generation by 2030. While still a small fraction of the total, this represents a 250 percent increase.

Currently, approximately 20 percent of electricity in the United States comes from nuclear power. Throughout the industrialized world, the fraction of energy that countries obtain from nuclear power varies widely, as shown in Table 12.3. There is variability across the United States as well, presented in Figure 12.7. Nineteen states do not have nuclear power plants, while six states generate more than 50 percent of electricity from nuclear facilities. It is perhaps not surprising that major coal producing states such as Wyoming, West Virginia, and Kentucky do not have nuclear facilities.

Nuclear plant construction worldwide has diminished since its peak in the mid-1980s. But new construction is expected to increase because of increasing natural gas prices and environmental concerns related to fossil fuel use. There have been no new nuclear plants constructed in the United States since 1977; however, in recent years, many license applications for new facilities have been filed.

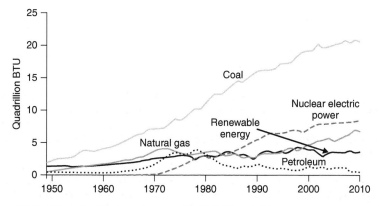

Figure 12.6 **Historical Energy Sources for U.S. Electricity.**

Source: Redrawn from EIA, 2008.

No Nukes in Germany

Following the March 2011 Fukushima nuclear crisis in Japan, Germany announced its plans to phase out all nuclear power plants by 2022. The policy was largely in response to mass anti-nuclear protests across the country. The Environment Minister stated, "It's definite. There will be no clause for revision." These facilities provide more than 20 percent of the nation's energy. The loss of nuclear energy production will be offset by proposed increases in renewable energy sources.

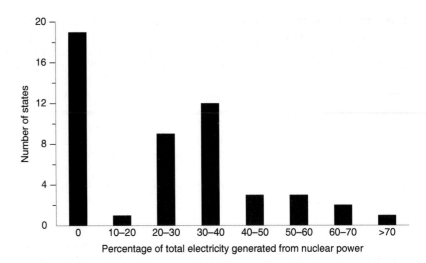

Figure 12.7 **Percentage of Electrical Generation from Nuclear Power in the 50 States.**

Data source: U.S. Census Bureau, 2010b.

Table 12.3

Percent of Electrical Generation from Nuclear Power for Selected Countries

Country	Percent of Electrical Generation from Nuclear Power
France	79
Belgium	55
Sweden	45
Japan	28
Germany	26
United Kingdom	20
Netherlands	4
Norway	0

Data source: International Energy Agency, 2007.

Municipal solid waste (MSW) is another potential source of energy via direct combustion (incineration) of the waste. Waste-to-energy (WTE) facilities combust the MSW to generate electricity, and are more common in densely populated regions (e.g., northeastern United States, Japan, and Europe) where available land for landfills is limited. Indeed, a WTE facility situated on a few acres can replace a landfill of 100 acres. Other benefits of WTE include immediate capture of energy and energy generation from nondegradable organic wastes such as non-recycled plastics. The primary disadvantage of WTE is the potential for air pollution from a wide array of compounds. Air pollution control technology is a rapidly advancing field that is making WTE an increasingly attractive option; however, public resistance to WTE is often fierce.

Environmental Impacts

The most significant environmental impact of energy consumption in the form of fossil fuels is air pollution. Air pollution from the combustion of coal (the primary fuel for producing electricity) includes the following.

- Carbon dioxide (CO_2) is a greenhouse gas implicated in global climate change.

- Nitrogen and sulphur oxides (often abbreviated as NO_x and SO_x, the subscript "x" representing the various chemical forms) contribute to acid rain and respiratory problems. Sulphur content is quite variable in coal deposits, and because of increasingly stringent air pollution control regulations, "low sulphur" coal is in high demand.

- NO_2 also contributes to ground-level ozone formation (i.e., smog), which is a lung irritant. It is important to differentiate between stratospheric (high altitude) ozone, which is beneficial in reducing harmful ultraviolet rays from the sun (and the reason for concern about the "ozone hole"), and tropospheric (low altitude) ozone, which is a health hazard when breathed. The EPA has promoted the slogan "ozone: good up high, bad nearby" to aid in the differentiation.

Security and Energy

More than 30 countries have nuclear power plants, including many developing nations (India, Russia, Pakistan, Armenia, etc.). There are tremendous international security concerns about the possibility of nuclear material falling into the "wrong hands" and being used for weapons. Additional concerns exist regarding the ability of developing nations to properly design, construct, and maintain nuclear facilities so that a disaster such as the 1986 Chernobyl incident, a nuclear reactor explosion releasing radioactive material resulting in thousands of deaths and the relocation of hundreds of thousands of people, will not reoccur. Security will be discussed in more detail in Chapter 18.

- Other elements such as mercury (Hg) are present in trace (i.e., very small) quantities in coal. For example, certain types of coal contain 0.1 mg Hg/kg coal, or 0.1 lb Hg per million pounds (500 tons) of coal. Given that approximately 1 *billion* tons of coal are consumed annually in the United States for power generation, even such trace concentrations result in significant Hg emissions. The EPA estimates that approximately 50 tons *per year* of mercury are released into the environment from coal burning power generation facilities (EPA, 2010). Mercury released into the atmosphere (from power plants, industry, solid waste combustion, and other sources) deposits on land and water and accumulates in fish. More than 3,000 lakes and rivers in the United States (in more than 40 states) have mercury levels in fish that are high enough to merit fish consumption advisories (EPA, 2009); such advisories limit the amount of fish eaten in order to prevent adverse health effects in humans.

- Much like mercury, trace levels of radioactive materials such as uranium and thorium are present in coal. Radioactivity released into the environment from the use of coal for electricity production far exceeds that from nuclear power production (Gabbard, 1993).

Additionally, there are many other potential environmental impacts of fossil fuel use including land, air, and surface water impacts from mining and extraction, refining, transport, storage and disposal of residuals (e.g., ash).

National Listing of Fish Advisories: Technical Fact Sheet (U.S. EPA)

numerical example 12.2 Power Generation by MSW Combustion

Using a "rule of thumb" value of 600 kWh electricity generated from the combustion of 1 ton of MSW, generate an order of magnitude estimate of the power that could be produced if **all** non-recycled waste in the United States was processed through WTE facilities.

solution

Using 2007 U.S. Environmental Protection Agency (EPA) data, 250 million tons of MSW is produced annually, of which 33 percent is recycled. An estimate of the power produced is:

$$\frac{250 \text{ million tons}}{\text{year}} \cdot (1 - 0.33) \cdot \frac{1 \text{ year}}{365 \text{ days}} \cdot \frac{1 \text{ day}}{24 \text{ hr}} \cdot \frac{600 \text{ kWh}}{\text{ton}} \cdot \frac{1 \text{ MW}}{1,000 \text{ kW}}$$
$$= 11,000 \text{ MW}.$$

This is approximately 2 percent of the total electrical power demand in the United States.

Using the typical power usage per home presented earlier in the chapter, the number of households that can be supported by electricity from MSW can be calculated.

$$\frac{\left(11,000 \text{ MW} \cdot \frac{10^6 \text{ W}}{\text{MW}}\right)}{1,400 \frac{\text{W}}{\text{household}}} = 8,200,000 \text{ households}$$

Assuming 2.5 persons per household, the conversion of the nation's non-recycled MSW would support the residential electrical needs of approximately 20 million people, or 6 percent of the population. To put this into perspective, this is more than people who live in New York State.

The proposed Cypress Creek Power Station described in the Introductory Case Study will have significant environmental impacts. According to the Chesapeake Bay Foundation, the power plant will add 1.9 million pounds of nitrogen, 118 pounds of mercury, and 14.6 million tons of carbon dioxide to the atmosphere with potential impacts on the water quality of Chesapeake Bay.

The sources of various air pollutants are presented in Figure 12.8. It can be seen that for some pollutants, transportation is the largest source. Of particular importance are VOCs (volatile organic compounds such as benzene and toluene) and NO_x, which along with sunlight produce ozone. The EPA estimates that in 2007, more than 140 million Americans, or nearly one half of the U.S. population, lived in counties with unsafe levels of ozone, and more than 70 million lived in counties with unsafe levels of particulate matter. In light of this fact, large transportation project proposals are required to evaluate the potential localized air quality impacts.

All types of energy production have disadvantages. Coal is currently the least expensive but the most polluting of the fossil fuels. Limited supplies of, and the reliance by transportation on, petroleum have limited its use in electrical generation. Natural gas is more expensive than coal, but increasingly strict air pollution regulations make it a more attractive option for new electrical power plants. Natural gas is essentially methane (CH_4) with only

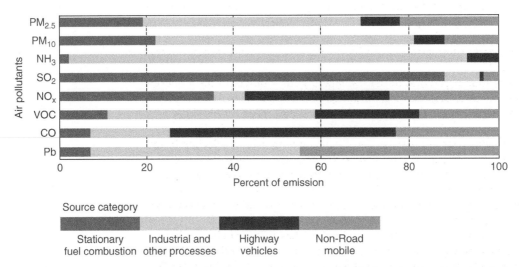

Figure 12.8 Sources of Air Pollutants by Category. Categories include stationary fuel combustion (e.g., electrical power plant), industrial, highway vehicles, and non-road mobile (aircraft, marine vessels, lawnmowers, etc.). PM = particulate matter (the subscripted number designates the size of the particles in micrometers), NH_3 = ammonia, SO_2 = sulfur dioxide, NO_x = nitrogen oxides, VOC = volatile organic compounds (e.g., benzene, toluene), CO = carbon monoxide, Pb = lead.

Source: Redrawn from EPA, 2008.

traces of nitrogen and sulfur; thus, air pollution (with the exception of CO_2) is significantly decreased. Pending carbon dioxide regulations and concerns over global climate change and foreign energy dependence are primary drivers for increasing the use of both renewable energy sources and nuclear energy.

There are minimal direct air emissions from nuclear power plants; the ominous-looking plume emerging from a cooling tower is nothing more than condensed water vapor (Figure 12.9). While there are environmental issues related to the mining and processing of ores, the *disposal* of radioactive waste is of paramount concern.

There are currently no operational long-term storage/disposal facilities for high-level radioactive waste in the world—all storage is temporary. The Yucca Mountain Nuclear Repository in Nevada, after approximately $9 billion in scientific studies and partial construction costs (and numerous project delays), was slated to begin operations in 2017. However, early in 2010, the Obama administration announced the intent to discontinue the project. Many other countries are planning and designing deep geologic storage facilities. These facilities will be required to store highly radioactive waste indefinitely, and quantifying the risks of environmental releases of radioactivity is quite difficult. Long-distance transport of the waste from generation sites (power plants) to ultimate disposal is a potential public health threat if an accident were to occur. For these and other reasons, siting nuclear waste disposal facilities is perhaps the most challenging engineering problem from technical, social, and political perspectives.

Figure 12.9 Cooling Towers at a Nuclear Power Facility. The tower in the foreground is in operation, releasing condensed water vapor.

Source: P. Parker.

Energy as an Infrastructure Consideration

As discussed in Chapter 14, Economic Considerations, and elsewhere in this book, when evaluating project alternatives, the initial construction cost alone should not be used to determine cost-effectiveness. The operation and maintenance (O&M) costs over the design life of the project must also be considered. For an energy-intensive project (e.g., a wastewater treatment facility or skyscraper), projections of energy costs over periods of 20 or more years may be required. Such predictions are particularly difficult with respect to energy costs, given that the cost of energy fluctuates both seasonally (due to varying demand) and annually (due to economic activity, wars, natural disasters, etc.).

Engineers should provide decision-makers with a range of likely energy costs whenever possible, allowing for decisions in light of a level of acceptable risk. For example, a conservative (i.e., low risk) prediction would forecast a high cost of energy, while a more "risky" forecast would predict low energy costs; given both types of forecasts, the client can then make a selection based on their level of acceptable risk. Providing a range of values is necessary, given the severe fluctuations that are common for energy costs. Figure 12.10 presents inflation-adjusted costs for residential natural gas and electricity.

The changes in natural gas and electricity costs also make it necessary for engineers to provide a range of cost predictions to clients. In the last 10 years, electricity costs (averaged for the United States) have risen slightly, whereas natural gas has increased more than 50 percent in inflation-adjusted dollars. For example, note in Figure 12.10 the historical peak prices in the mid-1980s followed by a steady decline. Therefore, an energy-intensive infrastructure project that was built in 1985 would actually have cost less to operate for the next 10 years than might have been expected. Conversely, a facility using natural gas as an energy source built in the late 1990s might have cost considerably more to operate than predicted for its first decade of operation.

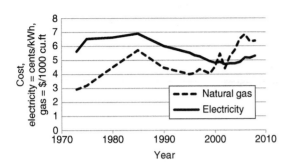

Figure 12.10 Cost of Residential Natural Gas and Electricity in Inflation-Adjusted Dollars.

Data source: EIA, 2010.

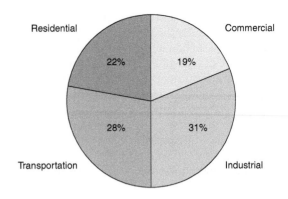

Figure 12.11 Percentage of Total Energy Consumption by Sector, 2008.

Data source: EIA, 2008.

TRANSPORTATION SECTOR ENERGY USE

Energy for transportation accounts for more than 25 percent of total energy use in the United States (Figure 12.11). Historical data for annual total highway miles travelled and per capita miles travelled are presented in Figure 12.12. Total highway miles include passenger cars, motorcycles, freight trucks, and buses. The dramatic increase in miles per capita over the last several decades is apparent in Figure 12.12. Less apparent is the recent leveling of the data, in part due to increased fuel prices. While the U.S. population has increased from 181 million in 1960 to 304 million in 2008 (70 percent increase), the amount of fuel consumed annually has more than tripled (250 percent increase) in the same time period.

Fuel consumption is a function of miles travelled and vehicle fuel efficiency. Data for actual miles traveled divided by fuel use for both passenger cars and for all vehicles combined (including vans, SUVs, and freight trucks) is presented in Figure 12.13. The increase in fuel efficiency is largely the result of the Corporate Average Fuel Economy (CAFE) regulations first enacted in 1975 (also presented in Figure 12.13). Since 1990 the standard has remained 27.5 mpg (miles per gallon); it will increase to 30.2 in 2011, and potentially higher in 2012 if proposed rules are adopted. The actual "achieved" fuel efficiency for

The Future of Freight Rail in the United States

In 2009, the multi-billionaire investor Warren Buffet purchased Burlington Northern Sante Fe (BNSF) Railway Company at an estimated value of $32 billion. Buffet is one of the world's most successful long-term investors. Many consider the BNSF purchase to be a harbinger of greater increases in rail use. In an interview Buffet joked, "This is all happening because my father didn't buy me a train set as a kid" (de la Merced and Sorkin, 2009).

Figure 12.12 Historical Data for Highway Miles Traveled (Total and Per Capita).

Data source: U.S. Department of Transportation, 2010.

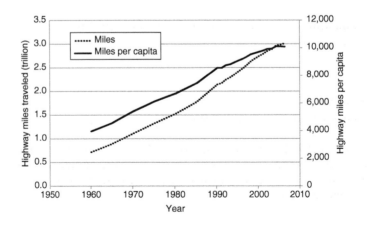

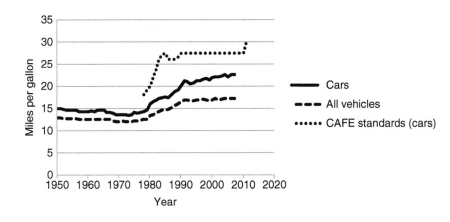

Figure 12.13 Historical Actual Fleet Fuel Efficiencies and CAFE Standards.

Data source: EIA, 2008.

all vehicles on the road is less than the standard because older, less-efficient vehicles are still in use. As these older vehicles go out of service, the achieved fuel efficiency will approach the standard that new cars, collectively as a fleet, are required to achieve.

Transportation energy use can be reduced by the vehicle design, consumer choice of more efficient vehicles, and changing habits (e.g., carpooling or combining errands into a single car trip). However, the planning and design of urban areas can also significantly impact energy use. Urban sprawl (introduced in Chapter 11) often results in increased commuting distances to work. Many new suburban developments include commercial and residential aspects to provide working and shopping opportunities in close proximity to homes. Conversely, many cities are making efforts to get people to move from the suburbs back into the city (where they work). Many communities are also making efforts to promote walking and bicycling.

Energy use for freight transportation depends on the mode. One method for comparing transport modes is the energy required to transport 1 ton of freight a distance of 1 mile (Table 12.4). As mentioned in Chapter 4, Transportation Infrastructure, rail is the fastest growing transport mode in U.S. freight handling, in part due to the fact that it requires 50 percent less energy per ton-mile than trucking. While passenger rail service in the United States lags far behind many countries, the U.S. freight rail system is considered one of the best in the world (The Economist, 2010).

BUILDINGS

The U.S. Green Building Council estimates that buildings (including residential and commercial) in the United States account for:

- 40 percent of total energy use
- 14 percent of the total water consumption
- 72 percent of total electricity consumption
- 39 percent of the carbon dioxide emissions

Consequently, targeting buildings as a source for significant energy savings only makes sense. As a result, many new buildings are being designed to be **green buildings**, which will be discussed in greater detail in Chapter 13, Sustainability Considerations.

Table 12.4

Energy Use Per Freight Ton–Mile for Various Transport Modes

Transport Mode	Energy Use per Freight Ton–Mile (BTU/Ton–Mile)
Barge	1,000
Rail	1,700
Truck	3,400
Air	29,000

Source: CBO, 1982.

Green buildings are designed such that their environmental impacts are much less than traditional buildings.

U.S. Green Building Council: Why Build Green?

Increasing Energy Efficiency in Rental Buildings

Many office and multi-unit residential buildings are operated on a rental basis (e.g., apartments or office space). Inherently, this shifts the cost perspective. A company that intends to build *and* occupy an office building will have a direct cost associated with the energy used to heat, cool, and light the facility and an incentive to minimize that cost. However, a developer that intends to build a structure only to rent the space will be concerned primarily with capital cost, since the operating cost will be passed on to the renter. If the majority of the competing rental structures are equally energy inefficient, this elevated operation cost is uniformly passed on to renters.

However, as modern energy-efficient structures are built to compete for rental demand, the decreased operating cost will impact rental decisions.

A similar situation occurs in many college communities. Older, energy inefficient homes are often rented to college students. The landlord has little economic incentive to increase the energy efficiency of the home as long as utilities are paid by the renters (as is often the case), unless of course, all prospective renters choose other housing options because of expected high energy costs and the house no longer generates rental income.

Figure 12.14 **An Old School Building Converted into Apartments.** Note the lighter colored brick that was used to cover original large window openings.

Source: M. Penn.

Payback period is the time required for the savings from the initial investment to offset the cost of the initial investment. Payback periods are often used as the basis for investments. Longer acceptable payback periods typically make more investment options available for consideration; shorter payback periods often eliminate many options.

Many municipalities are undertaking projects to renovate older buildings (e.g., city offices, police stations, schools) that are not energy efficient. Perhaps you have seen an old institutional building where the original large windows were replaced with smaller windows, such as the example shown in Figure 12.14. The acceptable **payback period** has typically been short for these decisions (e.g., 3 to 5 years). However, some municipalities are now considering longer payback periods in the spirit of being "green" or as a direct effort to minimize carbon footprint.

WASTEWATER TREATMENT

Wastewater treatment facilities are often very energy intensive due to the pumping, aeration, mixing, and disinfection requirements. Energy use accounts for approximately 20 to 30 percent of operational costs and may be 10 to 20 percent of the total life-cycle cost of a wastewater treatment facility. Thus, efforts to reduce energy use through proper equipment specification in the original design, and optimizing energy use in older facilities (e.g., installing more efficient pumps) offers potential for significant cost savings.

However, many treatment facilities have the potential to actually *generate* energy. The organic solids (also referred to as residuals, biosolids, or sludge) that are removed from the wastewater can be digested anaerobically (in the absence of oxygen); such digestion creates methane, which is identical to natural gas. Larger treatment facilities can produce enough methane to cost-effectively use the gas as an energy source (e.g., to run blower engines directly from the gas, to generate electricity via microturbines, or to heat buildings). Often, the gas generation may equate to one-fourth to one-third of the energy required to operate the facility and hundreds of thousands (or millions) of dollars in savings annually. As the processes to capture and utilize the methane become more efficient and as the prices of the equipment decrease due to widespread installation, it is possible for smaller treatment facilities to generate energy cost-effectively.

LANDFILLS

Landfills are another source of "natural" gas. Organic wastes such as paper and food scraps decompose in a covered landfill, and methane is produced. Older landfills were either not designed to capture this gas, or were designed to capture it but were inefficient in doing so. As a result, many projects have been completed recently to retrofit landfills with landfill gas extraction systems. As is the case with wastewater digester gas, landfill gas must be treated to remove impurities and water vapor before combustion to reduce maintenance and extend the life of the equipment. One of many examples is the landfill in Ann Arbor, Michigan. Over a 7.5-year period, 43,600 MWh of electricity, valued at $2.5 million, was generated from landfill gas. This is equal to the electricity requirements of approximately 1,000 homes. (City of Ann Arbor.)

Similar to fossil fuels, the end product of combustion of landfill gas is CO_2. However, the CO_2 produced offsets the CO_2 that would have been generated by the production of electricity from coal or natural gas if the landfill gas was not utilized. Also, it is important to note that the methane generated in a landfill will ultimately escape into the atmosphere if not collected. Methane, per molecule of carbon, is 17 times more potent as a greenhouse gas than CO_2. Thus, energy production from landfill gas minimizes greenhouse gas emissions while extracting otherwise wasted energy. In some cases, landfill gas is merely collected and flared off (combustion without energy capture). While this does not generate energy, it reduces the greenhouse gas impact (CO_2 versus CH_4), nuisance odors associated with landfill gas, as well as the explosive hazard of methane at and around the landfill. Engineers responsible for the design of new landfills now utilize more efficient landfill gas capturing systems. In many cases, new landfills are being designed to recirculate leachate (the liquid that is collected at the base of a landfill) rather than remove it for treatment. This recirculation increases the rate of degradation of the waste and thus the rate of methane generation.

Energy Conservation in the Workplace

Many engineering firms are making efforts to reduce their environmental footprint. This can be accomplished by constructing a new green office building, or measures can be taken to change business practices at their existing facility. Some examples include:

- Providing discounted mass transit fares for employees

- Improving amenities in lunch/break rooms to promote employees bringing their meals rather than "going out" to lunch (i.e. driving to a restaurant)

- Encouraging, or in some cases, requiring employees to turn computers off at the end of the workday

- Installing water-efficient plumbing fixtures

- Allowing employees to work from their home offices

- Installing motion-sensing lights

- Moderating office temperatures (slightly cooler in winter and warmer in summer)

Some firms have volunteer committees established to brainstorm and investigate opportunities for saving energy and water.

Energy Self-Sufficiency

In rare cases such as the Point Loma WWTF (Figure 12.15) in San Diego, California, a facility produces more power from methane recovered from digesters than is required for operation—it is energy self-sufficient and even sells excess energy to the local electricity grid. Few, if any, other WWTFs are energy self-sufficient. Point Loma is one of a few remaining large WWTFs that do not remove BOD (biochemical oxygen demand). As such, the facility's energy needs are much lower because it does not have to continuously aerate large BOD-removal tanks.

In addition to the energy from its digesters, the facility is located on a bluff, 90 feet above the ocean, and utilizes the potential energy of the 175 million gallons/day of wastewater in a 1.4-MW hydroelectric generating facility. This electrical generation helps offset energy costs needed to pump the wastewater from the community to the WWTF.

Figure 12.15 The Point Loma WWTF in San Diego, California, Includes a Hydroelectric Generation Facility to Utilize the Potential Energy of the Effluent.

Source: GeoEye Satellite Image.

Outro

Meeting an increasing demand for energy requires new infrastructure and the modernization of existing infrastructure. Reducing demand through energy conservation and increased efficiency will provide more "net" energy per unit of supply. Increasingly stringent environmental regulations and technological advances, as well as ever-changing political and social considerations will likely shift the mix of future fuel sources utilized to meet demand. Civil and environmental engineers will be critical drivers for change. We can reduce environmental consequences, develop new (and harness underutilized) sources of energy, develop more efficient infrastructure components and systems (and increase the efficiency of existing systems), and promote energy awareness.

chapter Twelve **Homework Problems**

Calculation

12.1 If the per capita municipal solid waste generation rate is 4.1 pounds daily and 25 percent of the waste is recycled (bottles, cans, newspapers, etc.), estimate the potential power production (expressed as MW) from incineration for a city with a population of 1.5 million residents.

12.2 Estimate the energy (expressed as kWh) that the incinerator in Homework Problem 12.1 would generate in one year. Show all work.

12.3 Calculate the number of households that could be supplied by the power in Homework Problem 12.1. What percentage of the city's population does this represent?

12.4 Estimate the number of homes that can be powered by the Point Loma WWTF's hydroelectric facility.

12.5 If all coal combusted in the United States contained mercury at a concentration of 0.1 mg/kg, how many tons of mercury would be released into the atmosphere annually if no removal occurred due to air pollution control? How does your answer compare to the 50 tons of mercury estimated in this chapter? Assuming that the difference in the two values is due to removal of mercury by air pollution control devices, what is the overall national removal efficiency?

Removal Efficiency

$$= \frac{Hg_{input\ to\ power\ plant} - Hg_{output\ from\ power\ plant}}{Hg_{input\ to\ power\ plant}}$$

12.6 Using per capita use and population projections, calculate a range of estimates for electricity use in the United States in 2050. List your assumptions and justify your answer.

12.7 Calculate the cost to operate a 50-hp pump continuously for 1 year if electricity costs $0.08 per kWhr.

12.8 Consider Figure 12.10. Assume you are an engineer in 1985; based on the information available to you in 1985 (i.e., the data from Figure 12.10 prior to 1985), estimate the price of natural gas in 2000. Show all work. Compare this to the actual price in 2000.

Short Answer

12.9 Why is providing a client with a range of possible operations costs for a project a good idea?

12.10 Why is the siting of a nuclear waste disposal site so challenging?

Discussion/Open-Ended

12.11 Why do you think some states do not have nuclear power?

12.12 How has reading this chapter changed your perspective of civil and environmental engineering?

12.13 If household appliances such as refrigerators and washing machines are becoming increasingly efficient, why is residential per capita electricity use still increasing?

Research

12.14 Research one or more of the following (as assigned by your instructor) and discuss their impact compared to conventional approaches. Discuss the pros and cons of the listed item and its conventional counterpart.
a. LED street lighting
b. Geothermal heating
c. Electric vehicles for a municipal vehicle fleet

12.15 Research a LEED certified green building. What level certification (e.g., gold, silver) did it achieve? What are some design features that helped the building achieve this certification?

Key Terms

- energy
- green buildings
- payback period

References

City of Ann Arbor. *Landfill Gas.* http://www.a2gov.org/government/publicservices/systems_planning/energy/Pages/LandfillGas.aspx, accessed August 28, 2011.

Congressional Budget Office (CBO). 1982. *Energy Use in Freight Transportation.* http://www.cbo.gov/ftpdocs/53xx/doc5330/doc02b-Entire.pdf, accessed August 28, 2011.

de la Merced, M. J. and A. R. Sorkin. 2009. "Buffet bets big on railroads' future." *New York Times.* November 3, 2009.

Energy Information Administration (EIA). 2008. *Annual Energy Review 2008.* DOE/EIA-0384.

Energy Information Administration (EIA). 2010. *Monthly Energy Review,* July 2010. http://www.eia.doe.gov/emeu/mer/pdf/pages/sec1_15.pdf, accessed August 28, 2011.

Environmental Protection Agency (EPA). 2008. *National Air Quality Status and Trends through 2007.* EPA-454/R-08-006.

Environmental Protection Agency (EPA). 2009. *National Listing of Fish Advisories: Technical Fact Sheet.* 2008 Biennial National Listing. EPA-823-F-09-007.

Environmental Protection Agency (EPA). 2010. *Control of Mercury Emissions from Coal Fired Electric Utility Boilers: An Update.*

EPA/600/R-10/006. http://www.epa.gov/nrmrl/pubs/600r10006/600r10006.pdf, accessed August 28, 2011.

Gabbard, A. 1993. "Coal Combustion: Nuclear Resource or Danger?" *Oak Ridge National Laboratory Review*: 26(3&4).

International Energy Agency. 2007. *World Energy Outlook, 2006.* http://www.iea.org/weo/2006.asp, accessed August 28, 2011.

The Economist. 2010. *American railways: high-speed railroading.* http://www.economist.com/node/16636101, accessed August 28, 2011.

U.S. Census Bureau. 2010a. *Population Profile of the United States.* http://www.census.gov/population/www/pop-profile/natproj.html, accessed August 28, 2011.

U.S. Census Bureau. 2010b. *The 2010 Statistical Abstract. Energy & Utilities: Nuclear.* http://www.census.gov/compendia/statab/cats/energy_utilities/nuclear.html, accessed August 28, 2011.

U.S. Department of Transportation. 2010. *National Transportation Statistics.* http://www.bts.gov/publications/national_transportation_statistics, accessed August 28, 2011.

U.S. Green Building Council. *Why Build Green?* http://www.usgbc.org/DisplayPage.aspx?CMSPageID=1720, accessed August 28, 2011.

chapter Thirteen Sustainability Considerations

Learning Objectives

After reading this chapter, you should be able to:

1. Define sustainability.
2. Describe how public health is related to economic and social development.
3. Discuss the relationship between growing resource demands and sustainability.
4. Summarize the concept of ecological footprint.
5. Explain why considering sustainability is essential for future engineers.
6. Apply sustainable design principles to an infrastructure improvement project.

Our Common Future ("The Brundtland Report")

Introduction

Prior to the 1990s, sustainability was arguably only a "niche" concern. Today, it is a topic of everyday conversation. In 1987, the Brundtland Commission of the United Nations issued the Report *Our Common Future*. This report included what is widely considered to be the first definition of sustainable development:

> *"Sustainable development is development that meets the needs of the present without compromising the ability of future generations to meet their own needs."*

The "needs" mentioned in this definition refer to the needs of people throughout the world, with highest priority given to the needs of the world's poor.

Mihelcic et al. (2003) define sustainable engineering as "the design of human and industrial systems to ensure that humankind's use of natural resources and cycles do not lead to diminished quality of life due either to losses in future economic opportunities or to adverse impacts on social conditions, human health, and the environment."

The concept of sustainability is reflected in a tradition of the Native American Iroquois to consider the impact of decisions on the next seven generations. Given that the time span for a generation is 20 to 30 years, to consider the seventh generation requires anticipating needs 150 to 200 years in the future. Anticipating effects this far into the future is very challenging, especially when considering that our nation has existed for a time frame that is not much longer than seven generations. Certainly, the American founders could have never foreseen the changes in economic opportunities, technology, social culture, human health, and the environment that occurred in the 150 to 200 years following the birth of the American nation. In fact, the same probably applies to your grandparents with respect to changes that have occurred over the last 50 years. Yet, it is the goal of sustainability, and the objective of sustainable engineering, to ensure that our actions today take into account future generations, despite the challenges involved.

Acting in a sustainable manner is a challenging task for engineers. Yet in one respect, engineers do not have a choice between seeking a sustainable design and ignoring the principles of sustainability. This claim is based on the fact that the first Fundamental Canon of the American Society of Civil Engineers (ASCE) Code of Ethics (see Chapter 17 for an introduction to the Code of Ethics) states that, "Engineers shall hold paramount the safety, health and welfare of the public and shall strive to comply with the principles of sustainable development in the performance of their professional duties." This infers that to be ethical, the engineer must consider sustainability. The NSPE (National Society of Professional Engineers) Code of Ethics is a bit less restrictive, as it states that, "Engineers are *encouraged to adhere* to the principles of sustainable development in order to protect the environment for future generations" (emphasis added).

This chapter will discuss various aspects of sustainability and how they relate to infrastructure engineering. First, we will discuss the various needs of the world. Following this, we will describe the relationship between sustainability and changes in the world's population and energy requirements. The remainder of the chapter will discuss how engineers can incorporate the concepts of sustainability in their work, with a focus on three areas: green buildings, sustainable land development, and water reuse.

Introductory Case Study: The Five Largest Nations

Most of us know that China and India are the two most populous nations. Can you name the next three largest countries? The answer is found in Figure 13.1; note that these five countries account for nearly one half of the world's population.

Resource demand is linked to the wealth of nations, typically measured by **Gross Domestic Product** (GDP), which is the sum of goods and services produced. When

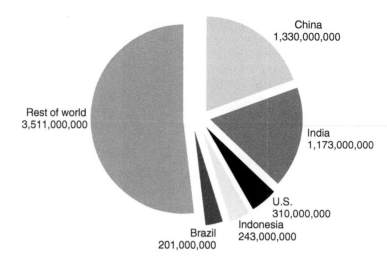

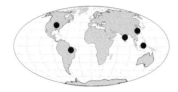

Figure 13.1 A Pie Chart of the World's Population Showing the Five Most Populous Nations, July 2010 Estimates.

Data source: *CIA World Factbook.*

measured in constant (i.e., inflation adjusted) international dollars, these values can be readily compared for different nations. Dividing the GDP by the population provides a per capita (per person) GDP, which is often used as a measure of "average" wealth. The per capita GDP values for the five most populous nations are presented in Figure 13.2. Note that the United States has a much higher value and a relatively steady increase over time. Also note that China's per capita GDP is growing dramatically; its per capita GDP doubled in 9 years from 1998 to 2006, whereas in the United States it doubled in 35 years from 1972 to 2006.

The GDP of each of the five nations, as a percentage of the world's GDP, is presented in Figure 13.3. In the future, the world's collective wealth is expected to be more evenly distributed. Consequently, as globalization continues, the United States and other major industrialized economies such as Germany and Japan, will comprise a decreasing percentage of the world's GDP, and developing nations will comprise an increasing percentage.

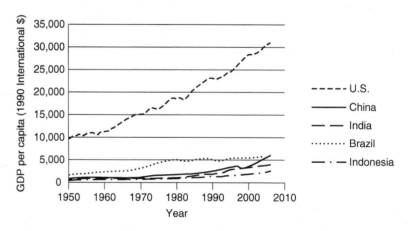

Figure 13.2 Historical GDP Per Capita (Inflation-Adjusted Dollars).

The 1990 International dollar is a common benchmark for historical world economic data.

Data source: Angus Maddison, Groningen Growth and Development Centre.

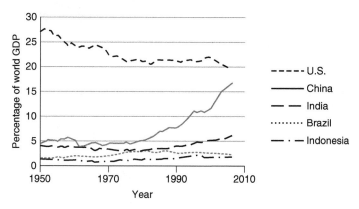

Figure 13.3 **Historical Percentage of World GDP of the Five Most Populous Nations.**

Data source: Angus Maddison, Groningen Growth and Development Centre.

The increase in per capita GDP in developing nations is creating a tremendous need for infrastructure improvements in those countries. It is also leading to extremely high rates of resource consumption that have not been witnessed before in history.

The Needs of the World

The case for sustainability can be made by recognizing two very important characteristics highlighted in the Introductory Case Study: world population is increasing and resource demand is increasing. These two factors make it essential that engineers consider sustainability in their analyses and designs. Population growth was discussed in Chapter 11, Planning Considerations, and one aspect of resource demand, energy use, was discussed in Chapter 12, Energy Considerations.

As mentioned in Chapter 11, world population growth for the past several centuries can be approximated by an exponential growth relationship. In reality, exponential growth does not continue forever, but growth must stabilize as the earth's **carrying capacity** is reached. There are many estimates for the value of earth's carrying capacity. Some experts claim that at its current population of 6.8 billion people, the earth has already surpassed its carrying capacity. Other estimates of carrying capacity range from 7 billion to 50 billion. These estimates arise from predictions concerning the ability of the earth to support people based on needs for energy, food, safe drinking water, land for inhabiting, available places to safely store wastes, etc.

Those experts that propose that earth has already reached its carrying capacity are considering that currently, the earth is not supporting all of its people adequately. For example, the great disparity in life expectancy for 169 nations is shown in Figure 13.4. Life expectancies vary by nearly 50 years; as seen in Figure 13.4, a strong relationship exists between life expectancy and GDP per capita. Another notable feature of this graph is the fact that since 1975, the curve has shifted upward approximately 5 years, signifying availability of better sanitation, greater availability of safe drinking water, and better health care.

The shape of the curve in Figure 13.4 is very instructive. Note that for moderate to wealthy nations, the slope of the curve is nearly level; that is, life expectancy varies minimally for nations with per capita GDPs between

The term **carrying capacity** refers to the maximum number of people that the earth can support.

Figure 13.4 Life Expectancy as a Function of Per Capita GDP. Circles represent individual nations (outliers omitted) for 2005. Solid line represents 2005 trend. Dashed line represents 1975 trend (data not shown).

Source: Adapted from the World Health Organization, 2008.

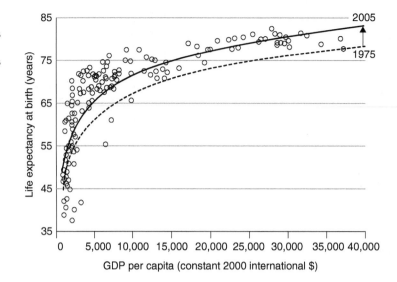

$20,000 and $40,000. However, a significant increase in life expectancy occurs as per capita GDP increases for underdeveloped nations (less than $5,000); indeed, the slope of the line is nearly vertical. Put simply, economies that can support clean water, sanitation, and very basic health care are able to dramatically improve public health.

Another common indicator of public health is infant mortality rate. Infant mortality rate is often expressed as the number of deaths (within one year of birth) per 1,000 births. As with life expectancy, a strong relationship between infant mortality and GDP per capita exists for the countries of the world, as illustrated in Figure 13.5.

Studying how these two indicators of public health (life expectancy and infant mortality) have varied historically in the United States will help us predict how these indicators may change in the future for the entire world. Life expectancy and infant mortality for the United States can be seen in Figure 13.6. Life expectancy in the United States has steadily increased from 50 years in 1900 to 77 years in 2006. The infant mortality rate has dropped from more than 120 per 1,000 (12 percent, or 1 in 8) in 1900, to less than 10 per 1,000 (1 percent, or 1 in 100) since 2000.

Gapminder Website—Visualization of Global Statistics Related to Sustainability

Figure 13.5 Infant Mortality Rates.

Data source: *CIA World Factbook.*

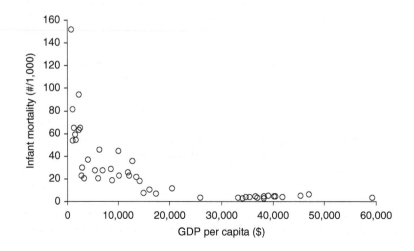

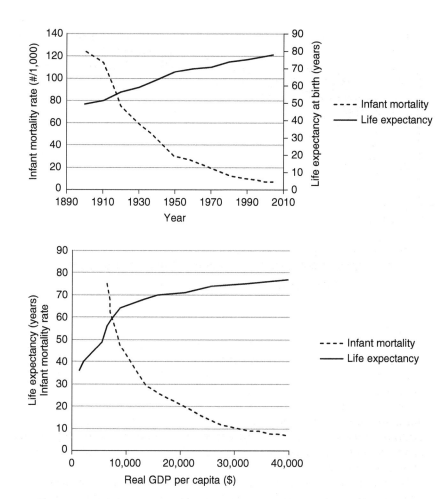

Figure 13.6 Historical Life Expectancy and Infant Mortality Rates in the United States.

Data source: Center for Disease Control and Prevention, Division of Vital Statistics.

Figure 13.7 Life Expectancy and Infant Mortality Rate as a Function of GDP in the United States.

Data source: Department of Commerce, Bureau of Economic Analysis.

These health indicators are also plotted as a function of per capita GDP in inflation-adjusted dollars in Figure 13.7, which helps us predict how these indicators might improve for the rest of the world in the future. The U.S. per capita GDP (inflation-adjusted dollars) has increased from approximately $5,000 in 1900 to approximately $40,000 in 2000. The life expectancy curve can be thought of as a trajectory over time, starting in the late 1800s (low GDP) and moving forward through time to the highest GDP of today. Note that the life expectancy versus GDP trajectory of the United States is very similar to the distribution of nations of the world today shown in Figure 13.4. Thus, many developing countries today are below the level of relative wealth and public health that was present in the United States a century ago.

A large fraction of the world does not have access to clean water or appropriate sanitation, as reported in *Progress on Drinking Water and Sanitation* (WHO and UNICEF, 2008). This report seeks to address two of the United Nations' Millennium Development Goals. The Millennium Development Goals are eight goals to be achieved by 2015 that respond to the world's main development challenges, one of which is to ensure environmental sustainability. Two "targets" are set by the UN for the drinking water and sanitation goals: to increase the proportion of the population with access to safe drinking water sources and to increase the proportion of the population with basic sanitation.

In 2006, only 62 percent of the world's population had access to wastewater treatment (an increase from 54 percent in 1990). Thus 38 percent (more than 1 out of 3), or 2.4 billion people, do *not* have access to sanitation measures

Millennium Development Goals

that ensure minimal contact with human excreta, including nearly 1 out of 5 people that rely on open defecation without pit latrines. This is not only a rural problem. Twenty-one percent of the world's urban population has no wastewater treatment, which is especially problematic due to high population densities and increased risk of spread of waterborne diseases. The Millennium Development Goal is to increase the percentage with basic sanitation to 77 percent by 2015.

With regard to drinking water, the percentage of the world's population that has access to treated drinking water has increased from 77 percent in 1990 to 87 percent in 2006. This is a tremendous achievement and nearly meets the Millennium Development Goal of 89 percent by 2015. Still, nearly 1 in 8 people in the world rely on unimproved water supply such as an unprotected dug well, unprotected spring, or untreated surface water. Only 54 percent of the world's population has drinking water piped to their dwelling, plot, or yard. Thirty-three percent have access to other forms of improved drinking water sources such as public taps or standpipes, protected dug wells, protected springs, and rainwater collection.

The Link Between Sustainability and National Security

Given the strong relationship between a country's economic well-being and the health of its people, it only makes sense to promote economic growth and social welfare in developing countries. Such improvements are not only humanitarian, but are also considered an issue of national security. The United States Africa Command (AFRICOM), established in 2007 by the Department of Defense, is responsible for military operations and relations with over 50 African nations. The following excerpt is taken from the AFRICOM website:

"Africa is growing in military, strategic and economic importance in global affairs. However, many nations on the African continent continue to rely on the international community for assistance with security concerns. From the U.S. perspective, it makes strategic sense to help build the capability for African partners, and organizations such as the Africa Standby Force, to take the lead in establishing a secure environment. This security will, in turn, set the groundwork for increased political stability and economic growth. We are seeking more effective ways to help African nations and regional organizations bolster security on the continent, to prevent and respond to humanitarian crises, to improve cooperative efforts with African nations to stem transnational terrorism, and to sustain enduring efforts that contribute to African unity."

AFRICOM Website

Failed States

Lester Brown, in his book *Plan B*, makes a compelling case for the relationship between government stability and sustainability. He specifically treats the topic of "failing states." A failing state (see Table 13.1) is a country that cannot govern all or part of its territory. Such a government cannot provide health care, food security, education, and may even lose the ability to collect tax revenue. These failing states degenerate into chaos. In terms of sustainability, failing states often have very high population growth rates, minimal economic growth, and decreased interest in environmental protection. And, in terms of national security, such states, with their low education, low employment rates (e.g., an astounding 95-percent unemployment in Zimbabwe), and lack of stability, are feared to be breeding grounds for insurgency and/or terrorism.

Failed States (Continued)

Table 13.1

Top 10 Failed States, 2010 with Comparison to United States

	Failed State Ranking	Infant Mortality (Deaths/1,000 Live Births)	GDP/Capita ($)	Life Expectancy	Unemployment Rate (%)	Education Expenditures (% of GDP)	Population Growth Rate (%)
Somalia	1	107	600	50	NA	NA	2.8
Chad	2	97	1,600	48	NA	1.9	2.0
Sudan	3	78	2,300	53	19	6.0	2.1
Zimbabwe	4	30	NA	48	95	4.6	3.0
Democratic Republic of Congo	5	79	4,100	55	NA	1.9	3.2
Iraq	6	43	3,600	70	15	NA	2.4
Central African Republic	7	101	700	50	8	1.4	2.1
Guinea	8	98	600	58	30	0.6	2.7
Pakistan	9	65	2,600	66	14	2.6	1.5
Haiti	10	58	1,300	61	NA	1.4	1.8
United States		**6**	**46,400**	**78**	**9**	**5.3**	**0.97**

Source: Foreign Policy, 2010 with data from the *CIA World Factbook*.

High population growth rates are common in developing countries, and are considered by some experts to be *the* major challenge for ensuring global sustainability. The distribution of global population is shown in Figure 13.8a. Of particular importance is the percentage of a population that is under 15 years of age, as the number of young people is directly related to future population growth. Note that according to Figure 13.8b, from 1990 to 2000, approximately

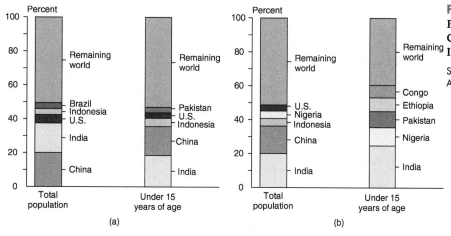

Figure 13.8 Distribution of Global Population, 2000 (a), and Contribution to Global Population Increase, 1990–2000 (b).

Source: Department of the Interior, National Atlas.

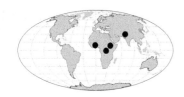

one third of the growth in the world's under-15 population came from only four countries: Nigeria, Pakistan, Ethiopia, and Congo. Consequently, the future populations of these nations are expected to increase dramatically.

Resource Demand

Through economic development, the world is getting wealthier, with more consumption and thus more demand for resources. Historical and projected world average GDP per capita in inflation-adjusted dollars is presented in Figure 13.9. Many experts project the world per capita GDP to grow at approximately 3 percent per year in inflation-adjusted dollars (dashed line in Figure 13.9) while the population is expected to grow at a lower rate of 1 percent. Thus, there will be more wealth per person.

As the GDP per capita continues to increase, we can expect an increase in the consumption of resources such as petroleum and coal. The increasing demand for these finite sources of energy is one of the primary rationales for promoting sustainable development.

Historical energy use for the world, the United States, and China is presented in Figure 13.10. The United States and China are the two largest consumers of energy. The steady increase in world demand is apparent, as is an increase in demand over the last decade. Note that the U.S. demand has remained constant for the last decade, while Chinese demand has more than doubled. Note also that, in the near future, China is likely to overtake the United States as the largest consumer of energy. The historical changes in the U.S. and China energy demand, as a percentage of world energy use, is presented in Figure 13.11.

The U.S. population has increased by 10 percent in the last decade, whereas the population of China has increased 5 percent. In order for U.S. energy

Plan B

One of the most illuminating texts describing the threats facing our world and possible solutions to these threats is *Plan B*, a book written by Lester Brown, president of the Earth Policy Institute. This book has been highly influential among decision-makers at all levels, from local to international. It is highly readable and, perhaps most importantly, the entire text can be downloaded for free from Earth Policy Institute's homepage.

Earth Policy Institute Website

Figure 13.9 Historical and Projected World GDP Per Capita.

Data source: Angus Maddison, Groningen Growth and Development Centre.

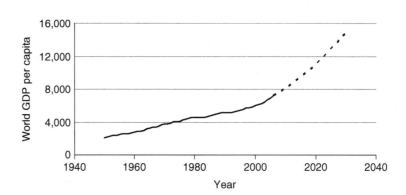

Figure 13.10 Historical Energy Use in the World, U.S., and China.

Data source: British Petroleum, 2010.

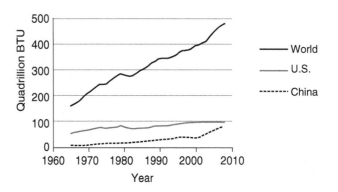

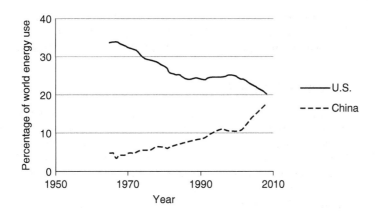

Figure 13.11 Historical U.S. and China Energy Use as a Percentage of World Energy Use.

Data source: British Petroleum, 2010.

demand to remain constant while population has increased, the demand per capita must decrease. Likewise, a doubling of energy demand in China coupled with a 5 percent increase in population results in a dramatic increase in per capita consumption. Per capita energy consumption for several countries and the world average consumption are presented in Figure 13.12.

Highly developed nations clearly have greater per capita demand—a commonly cited statistic is that the United States accounts for only 5 percent (1/20th) of the world's population but is responsible for 20 percent (1/5th) of the world's energy consumption. If the entire world's population consumed energy at the rate of Americans, the world energy demand would be 4.5 times greater.

The dramatic increase in Chinese energy demand is shown in Figure 13.13. Also apparent is that collective energy use per capita in countries other than China has remained relatively constant from 1990 to today; increases in other developing countries are offset by *decreases* in the United States and other industrialized nations as seen in Figure 13.12.

The historical and projected relationship between energy consumption and GDP on a per capita basis is shown in Figure 13.14. In the future, as per capita GDP increases in developing nations, we can expect more energy consumption (dashed line in Figure 13.14) and pollution resulting from increased fossil fuel consumption, industrial byproducts generation, and solid waste generation.

Renewable energy sources were discussed in Chapter 12, Energy Considerations, with regard to pollution. The CO_2 emissions per kWh of electricity

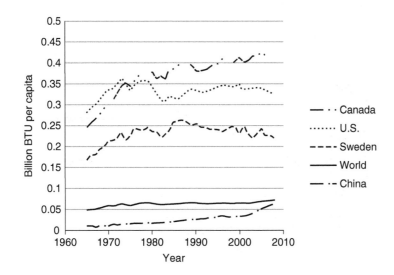

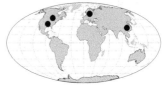

Figure 13.12 Annual Energy Use Per Capita for the World and Several Nations.

Energy data source: British Petroleum Statistical Review of World Energy.

Population data source: Angus Maddison, Groningen Growth and Development Centre.

Figure 13.13 Historical Per Capita
Energy Use for the World, China,
and the World Excluding China.

Energy data source: British Petroleum, 2010.

Population data source: Angus Maddison,
Groningen Growth and Development Centre.

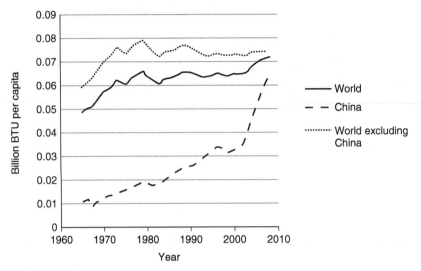

Figure 13.14 Historical and
Projected Relationship Between
Energy Consumption and GDP Per
Capita.

Source of projected energy consumption:
International Energy Agency, 2006.

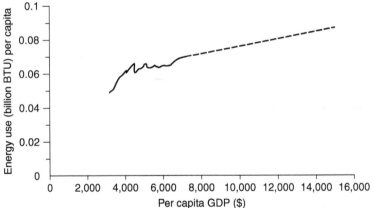

produced for several countries is presented in Figure 13.15. Note that Norway has virtually no CO_2 emissions because its electricity is generated from hydropower. Renewable energy sources, by definition, are far more sustainable than fossil fuels.

The focus thus far in this section of the chapter has been energy, which is perhaps the best single indicator of overall resource consumption, as energy is used to make industrial products, to power vehicles, and to heat, cool, and light buildings. Increasingly, freshwater is becoming a limited resource. In the United States, household water use tends to increase with household income. U.S. per capita water use since 1950 is presented in Figure 13.16; note that use has steadily increased until 1990 (as GDP per capita increased), after which time, water conservation efforts became more common.

The fact that water use increases as income increases is demonstrated in Figure 13.17. Note, however, that the European countries, shown with open symbols, have relatively low water demand due to extensive water conservation efforts. As the income in developing nations increases, the demand for water can be expected to increase, and in many areas freshwater is in limited supply. Many experts claim that in the future, water, rather than oil, will be the major cause of global conflict.

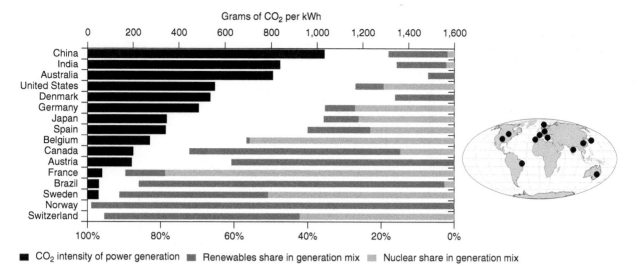

Figure 13.15 Carbon Dioxide Emissions Per kWh of Electricity Generated and Percentage of Electricity from Renewable and Nuclear Energy for Several Nations.

Source: International Energy Agency, 2006.

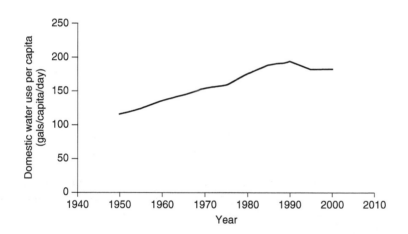

Figure 13.16 Historical Domestic Water Use Per Capita in the United States.

Data source: U.S. Census Bureau.

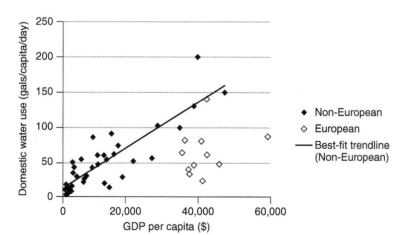

Figure 13.17 Domestic Per Capita Water Use as a Function of Per Capita GDP For Several Nations.

Data source: *CIA World Factbook*.

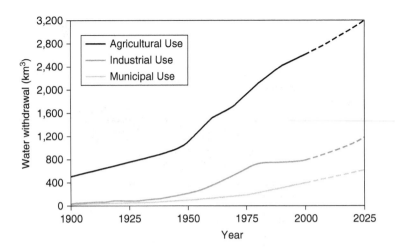

Figure 13.18 Worldwide Water Extraction for Agricultural, Industrial, and Municipal Use.

Source: Data from Shiklomanov, 1999.

One reason that water is so critical is its strong relationship to food production. Agricultural water use accounts for the largest use of water worldwide (Figure 13.18). Countries that have scarce water resources have great difficulty in growing enough food to meet domestic needs, and thus must import food from water-rich countries such as the United States and Canada. In essence, when importing food from a water-rich country, the water-poor countries are also importing water. The water embodied in that food has been termed **virtual water**. For example, it takes about 1,000 L of water to produce 1 kg of wheat and on the order of 16,000 L to produce 1 kg of beef. Virtual water flows for the world are shown in Figure 13.19.

Measures of Sustainability

Quantifying sustainability is a challenging proposition. An attempt to quantify sustainability has been made using measurements such as ecological footprint, energy footprint, and CO_2 footprint. These metrics attempt to quantify the extent of the world's land area that is required to support various types of lifestyle. To calculate your footprint, you can use any number of Internet tools, which vary in degree of technical rigor. The footprint calculators often express your footprint in terms of the amount of land area needed to support your

Figure 13.19 Virtual Water Exports from North America, South America, and Oceania to Other Regions.
Width of arrows are proportional to volume of water exported.

Source: Adapted with permission from Yang et al., 2006.

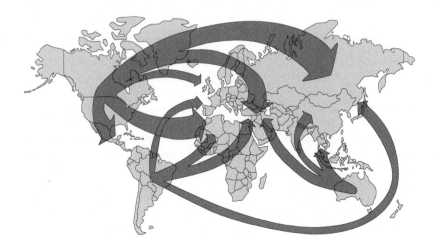

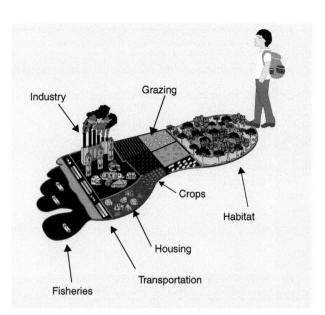

Figure 13.20 The Ecological Footprint.

Source: Adapted from berkleyside.com.

lifestyle; by multiplying that land area by the world's population, you can estimate the "number of worlds" necessary if all human beings lived your lifestyle.

Although these calculations have some shortcomings such as hidden assumptions and being overly simplistic, they offer insight into how our daily decisions relate to sustainability. And although a common system for calculating ecological footprint has not been agreed upon, many of the calculators attempt to measure the quantities shown in Figure 13.20. For example, your ecological footprint depends on how much and what you eat; your transportation requirements, the products you purchase, etc. What should stand out to you is how the infrastructure affects nearly every aspect of our ecological footprint, whether it is the energy infrastructure, the road system, the development of land for residential and commercial use, or the transport of food.

The Role of Civil and Environmental Engineers in Promoting Sustainability

"Paradigm shifts," with regard to engineers' response to environmental problems, are presented in Figure 13.21. Note that we are currently in the era of sustainability and systems engineering (discussed in Chapter 7). You might be surprised to note that only 60 years ago, there was little concern by elected officials and citizens over environmental contamination. The first response was "end-of-pipe" treatment, which as the name implies, focused on treating waste *after* it was created. Such an approach was effective to a certain extent, but we realized that money, energy, and other resources could be saved by *preventing* the pollution from occurring in the first place. In response, **pollution prevention** measures were implemented to decrease the amount of pollution created and thus the need for treatment. This was accomplished by changing industrial processes, performing more regular maintenance of equipment, or changing chemical feed stocks for industrial processes. While advances in pollution prevention are still occurring and there is still room for improvement, we have now progressed to the point where more engineers are seeing their tasks as part of a larger system; engineers must be aware of how the engineered component affects the system (our infrastructure and our world) as a whole, and how that system affects the component.

Figure 13.21 Ways of Dealing with Environmental Problems, 1950 to Present.

Source: Reprinted (adapted) with permission from Davidson et al., 2007 American Chemical Society.

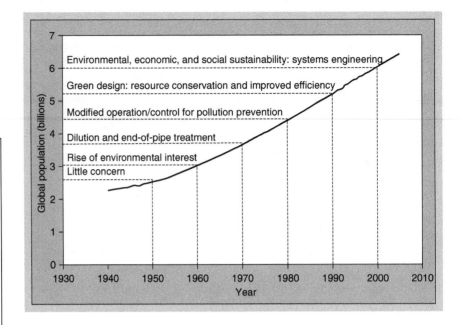

Implementing Sustainability

In 2007, ASCE adopted Policy Statement #418 (below). The policy statement goes beyond defining sustainable design, and beyond including sustainable design in its Code of Ethics. This policy statement identifies implementation *strategies* to meet the goal of sustainable design.

"Promote broad understanding of political, economic, social and technical issues and processes as related to sustainable development. Advance the skills, knowledge and information to facilitate a sustainable future; including habitats, natural systems, system flows, and the effects of all phases of the life cycle of projects on the ecosystem. Advocate economic approaches that recognize natural resources and our environment as capital assets. Promote multidisciplinary, whole system, integrated and multi-objective goals in all phases of project planning, design, construction, operations, and decommissioning. Promote reduction of vulnerability to natural, accidental, and willful hazards to be part of sustainable development. Promote performance based standards and guidelines as bases for voluntary actions and for regulations, in sustainable development for new and existing infrastructure."

The above statement highlights a concept that is stressed throughout this book—*engineers have responsibilities that reach far beyond design.*

It is important to place today's perspectives in a historical context. It is also important to understand that paradigm shifts are gradual. We are currently at the forefront of the sustainable engineering movement. Cliff Davidson of Syracuse University, considered one of the pioneers of this movement, states that, "we have a lot to learn in order to reach our goal of sustainability." Henry Hatch, a past Commander of the U.S. Army Corps of Engineers, stated that sustainable development serves as an "...opportunity for greater public service (and economic gain) and challenge to our traditional education, our methods, our technologies, and even our ethics." He further suggested that the relevancy and reputation of engineering will depend largely on the willingness and demonstrated contribution of the profession to achieving sustainability. The implication here is that engineering *itself* is not sustainable unless the profession takes a lead role in achieving sustainability; in other words, if we do not help solve the problem, society will look elsewhere for solutions.

SUSTAINABLE DESIGN

It is very challenging to grasp what sustainable design actually means. Many students, community leaders, and corporations boast of sustainable designs simply based on the fact that the environmental impact has been "minimized." However, minimal environmental impact does *not* guarantee sustainability.

Sustainable designs should embrace the concept of the **triple bottom line**, which was introduced in Chapter 11, Planning Considerations. Traditionally, the "bottom line" has referred to the difference between revenues and expenses. The triple bottom line emphasizes that the environment and social well-being must also be considered in the equation for calculating the "bottom line." Some refer to this as the "three Ps" (People, Planet, and Profit) as illustrated in Figure 13.22. Note that sustainability is defined as the region where people, planet, and profit overlap. The three Ps also align with what is commonly known as the three pillars of sustainability: environmental, economic, and social demands.

Sustainable solutions may involve technical and/or non-technical approaches. For example, consider a metropolitan area that is faced with a problem of extensive traffic congestion during rush hour in the center core of the city. Solutions can be technical and physical (e.g., expansion of roadway lanes; adding tolling stations; addition of mass transit) or non-technical and non-physical (e.g., providing ride share programs; charging more for use of toll roads during rush hour; enhanced public education).

GREEN BUILDINGS

Green buildings have many characteristics that make them "green" in addition to energy efficiency. These include:

- utilizing renewable energy sources
- reducing stormwater runoff
- reducing the amount of land utilized
- building near existing infrastructure rather than contributing to urban sprawl
- using local materials
- using materials with high recycling content
- providing amenities (e.g., showers, locker rooms) for bicycle commuters
- implementing water conservation techniques
- creating an enjoyable work environment with plenty of natural lighting and individual climate controls
- providing "premium" parking spaces for drivers of electric and hybrid automobiles (Figure 13.23)
- improving indoor air quality through careful selection of materials (e.g., carpets, wall coverings)
- using sustainable materials, such as those that rapidly regenerate (e.g., bamboo)

This brings a wide range of new and exciting design projects to engineers, including green roofs (see "Green Roofs" sidebar), porous pavements, and alternative energy sources.

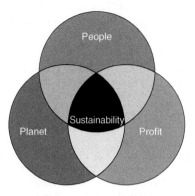

Figure 13.22 **Sustainability and the 3Ps: People, Planet, and Profit.**

 Remember that full-sized versions of textbook photos can be viewed in color at www.wiley.com/college/penn.

Figure 13.23 Reserved Parking Spaces for Fuel-Efficient Vehicles. These parking spaces serve a LEED (Leadership in Energy and Environmental Design) gold-certified building. These spaces are the closest parking spaces to the building, thereby providing an incentive to drivers of such vehicles.

Source: P. Parker.

Green Roofs

Green roofs have long been popular in northern Europe (Figure 13.24) and are gaining popularity in commercial buildings in the United States, particularly in urban centers such as Chicago, Denver (Figure 13.25), and Washington, D.C. It is estimated that over 2 million square feet of green roofs were installed in the United States in 2008, a market that is growing by more than 25 percent per year.

Green roofs may be as simple as small plants and flowers in planters, or relatively complex; the green roof system in Figure 13.26 consists of eight layers and many different materials integrated together as a system. The additional load of the plants, soil, water, and supporting components (up to 150 pounds per square foot) must be evaluated from a structural perspective, especially if being added to an existing building which was not designed for the loading.

Green roofs are "green," both literally and figuratively. Characteristics of green roofs that make them green in terms of environmental impacts include their ability to: insulate buildings, thereby reducing heating and cooling costs; reduce the urban heat island effect (defined in Chapter 11, Planning Considerations); reduce stormwater runoff; provide natural habitats; and reduce noise levels within buildings.

Figure 13.25 **Green Roof on EPA Region 8 Office in Denver, Colorado.**

Source: U.S. EPA.

Figure 13.24 **Green Roof on a Newly Constructed Residence in Norway.**

Source: M. Penn.

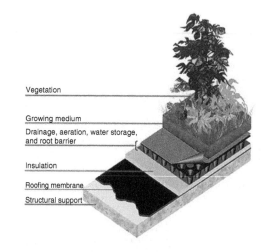

Vegetation

Growing medium

Drainage, aeration, water storage, and root barrier

Insulation

Roofing membrane

Structural support

Figure 13.26 **A Cross-Section of Typical Green Roof Components.**

Source: Courtesy of American Wick Drain.

The potential benefits of green buildings are many, and include:

- protecting water resources by decreasing demand and increasing recharge to groundwater through stormwater infiltration
- improving air and water quality
- conserving natural resources
- reducing heating and cooling costs
- increasing worker productivity
- reducing electricity costs

In order to objectively define what it takes to make a building "green," the U.S. Green Building Council's Leadership in Energy and Environmental Design (LEED) program has created a set of standards. Buildings may be certified at different levels (platinum, gold, silver, or simply "certified"); the level to be achieved depends on the number of points applicants earn in various categories. Representative categories used by LEED are provided below, along with sample methods of obtaining points in those categories. Note that many of the benefits of green buildings listed previously are captured in these rating categories.

- *Sustainable sites*—building near mass transit access, maximizing open space, minimizing the heat island effect

- *Water efficiency*—designing water-efficient landscaping, reducing water use in the building

- *Energy and atmosphere*—improving energy efficiency, using on-site renewable energy

- *Materials and resources*—reusing portions of existing buildings, recycling high percentages (75 percent) of construction waste, including high fraction of recycled materials in the building construction

- *Indoor environmental quality*—using low VOC-emitting adhesives and floor coverings, and enhancing ventilation and natural lighting indoors.

> **Green Retrofitting**
>
> Green buildings are not limited to new structures. The Empire State Building in New York City, which opened in 1931, is planned to undergo a $20-million-dollar retrofit to reduce energy use by nearly 40 percent. Expected *annual* savings in energy costs are greater than $4 million, resulting in a payback period of less than 5 years.

Nationals Park

Although LEED certification most often applies to buildings, it may also be used to certify stadiums, as is the case for Nationals Park (Figure 13.27), the ballpark for the Washington Nationals Major League Baseball team. The park opened at the start of the 2008 season, and is the first LEED-certified major professional sports stadium in the United States. Some methods to obtain its silver certification include: the ballpark uses 15 percent less energy and 35 percent less water than comparable conventional ballparks; 83 percent of the construction waste was diverted from landfills; the site is a former brownfield (contaminated site); alternative transportation is encouraged by the park's proximity to subway and bus services, and a bicycle valet service is provided; a 6,300 square-foot green roof covers a concession and restroom area.

Figure 13.27 **Nationals Stadium.**

Source: Courtesy of Graham Knight at BaseballPilgrimages.com.

SUSTAINABLE LAND DEVELOPMENT

The development of land for residential or commercial use can be a very unsustainable process. Indeed, some believe that the term "sustainable land development" is an oxymoron (a contradiction in terms); that is, it is not possible for the development of land to be sustainable. The rationale to this assertion is that the "supply" of land is finite and the use of it is practically non-renewable.

Indeed, the conversion of native land to some developed use typically has many aspects that affect sustainability. For example, *conventional* land development:

- alters native land irreversibly

- is located away from city centers, therefore requiring the addition of expensive infrastructure (including water, wastewater, communications, and stormwater)

- fragments wildlife habitat

- increases stormwater runoff, thereby leading to increased downstream flooding and decreased replenishment of groundwater aquifers

- requires automobile use by users of the developed land

- does not meet the needs of pedestrians and bicyclists (e.g., see Figure 13.28)

- in an attempt to meet real or perceived market demands, creates large lots resulting in an overall low population density

- exacerbates the heat island effect

Many engineers and their clients are realizing that the conventional methods of land development are unsustainable. As a result, many steps are being taken to make land development less unsustainable. For example, residential land developers are increasing the density of housing by offering smaller lots while at the same time providing enhanced public green spaces. Also, stormwater management is integrated into the design of the developed land through the use of ponds and wetlands. The extent and creativity of the solutions are only limited by the imagination of the designers (and the willingness of the client to pay for these solutions).

Figure 13.28 Pedestrian-Unfriendly Design. This "walking path" along a major arterial has been worn down thanks to a lack of sidewalk. It is unsafe, unsightly, and unlikely to encourage out-of-doors activity.

Source: P. Parker.

WATER REUSE

Fresh water is a critical resource with finite supply. One sustainable management option is to directly *reuse* water. For example, consider a municipality that relies on groundwater as a drinking water source as illustrated in Figure 13.29. In this figure, groundwater is pumped from an aquifer at a rate of 100 mgd (million gallons per day). A fraction of this water is "lost," for example, by leaks in water mains, or by watering lawns, or export as industrial products (e.g., beverages from a brewery). The remaining 80 mgd is used by the community, and after this use it is considered wastewater. This wastewater is treated and returned to the environment. However, the treated wastewater is not returned to the aquifer from which it was pumped, but is discharged to surface water. As a result, the aquifer is depleted over time since the rate of withdrawal typically exceeds the rate of recharge from precipitation.

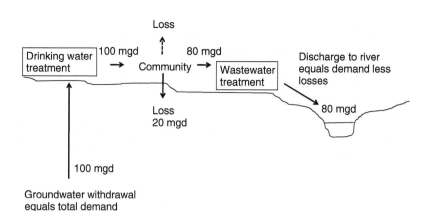

Figure 13.29 A Common Water Use and Discharge Scenario.

This scenario is clearly unsustainable, and many communities are facing the challenge of finding alternative water supplies as their aquifers dwindle.

LEED ND

Just as LEED has a certification system for new construction of *individual* structures, LEED has also developed a certification system for land development called LEED ND (Neighborhood Development). An example of a LEED ND-certified development is "The Gulch," a $400 million, 60-acre development in downtown Nashville, Tennessee that had once been a railroad-corridor, but by the late 1990s was considered a blighted area. The redeveloped area includes retail, dining, and residential options. These options target a wide range of socio-economic classes, with the hope of creating a vibrant and diverse neighborhood.

The Gulch earned several points toward its LEED certification due to its proximity to the central business district and existing infrastructure and public transportation. Nearly 60 percent of businesses and residences are within a quarter mile of bike and walking paths. Also due to its proximity to the central business area, more than 6,000 jobs are within a quarter-mile walk. Additionally, many of the existing buildings have been reused rather than demolished. Phil Ryan, director of the Nashville Metropolitan Housing and Development Agency said, "It has become part of the downtown, where before it was just kind of a trench, a moat between downtown and the commercial areas to the west."

This development is all the more exciting considering the alternative "conventional" style of development. In such a case, the new housing and retail centers would have been built far from the city center. This would result in needing water, sewer, and communications to be extended out to the development. Based on the conventional location, many of the users of the site would need to use automobiles, thus making it difficult for those who do not own cars to use the site, adding to traffic congestion, and increasing tailpipe emissions. Also, in this conventional scenario, the original 60-acre parcel would remain abandoned, decreasing the value of neighboring properties.

A more sustainable alternative to this conventional scenario is **water reuse**. Such a process has been dubbed "toilet to tap." Water reuse is, in a sense, a closed loop in which water is reused repeatedly as shown in Figure 13.30. In this figure, the 80 mgd that was treated and then discharged in Figure 13.29 is instead returned to the drinking water treatment plant. Thus, the only water that needs to be extracted from the aquifer is water needed to compensate for losses (20 mgd in this case).

The sustainability of this scenario can be assessed using the water budget technique, which was presented in Chapter 2. To be truly sustainable, the extraction rate of water from the aquifer cannot exceed the rate at which the aquifer is recharged. If despite the water reuse, the extraction rate still exceeds the aquifer's recharge rate, the volume of losses must be decreased; another source must be utilized; or demand needs to be reduced through water conservation techniques.

Water reuse is the practice of treating wastewater to achieve standards required for drinking water. Drinking water standards are much more stringent than the requirements for wastewater that is to be discharged to a receiving stream.

Figure 13.30 A Water Reuse Scenario. Water demand and losses are the same as the scenario in Figure 13.29.

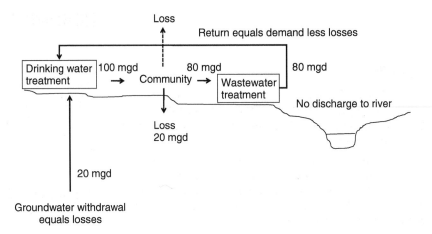

Water reuse is a highly controversial topic, as much of the public finds the idea of drinking treated human wastewater abhorrent. However, reuse may be the only practical alternative for the future. Indeed, it is practiced to a varying degree already. Space Shuttle astronauts drink their wastewater after it is highly treated (this is done to reduce cargo weight). Many communities use treated wastewater for irrigation and some industrial uses. In Chapter 5, Environmental Infrastructure, we highlighted Orange County, California for its wastewater reuse, and presented this in the context of introducing you to the different types of drinking water sources.

Additionally, many other communities already indirectly practice this; for example, Milwaukee extracts water from Lake Michigan, uses it, treats it, and discharges it back into Lake Michigan. The difference between this scenario and the water reuse directly from the wastewater treatment plant is simply a matter of dilution (and of location—the water intake is located far from the wastewater effluent discharge).

Outro

It is notably easier to discuss sustainability with students today than with those from even only a decade prior. We are at a point where conventional (unsustainable) approaches to engineering are being examined with increasing scrutiny. As future engineers and leaders, you will be afforded opportunities to make engineering more sustainable. You should expect to witness changes in both professional engineering approaches (i.e., best practices) and in social behavior. As discussed, sustainability is difficult to quantify. What we think is sustainable today might not be considered sustainable in the future. However, we must strive for sustainability and adopt appropriate adaptive management strategies. The next three chapters of this book are dedicated to the three Ps: Profit (Economic Considerations, Chapter 14), Planet (Environmental Considerations, Chapter 15), and People (Social Considerations, Chapter 16).

chapter Thirteen   Homework Problems

Calculation

13.1   Revisit the Geotechnical Engineering Design Application (Footings) Numerical Example 10.4 in Chapter 10. Calculate the footing width required for the home if it were to include a green roof.

13.2   Assuming a cost of $0.0020/gallon to pump and distribute water in a municipal supply system that loses 17 percent of its water because of leaks, how much money is wasted annually for a city of 2.1 million people with a per capita demand of 150 gallons/day?

13.3   A community's water use of 200 mgd is divided equally between residential, commercial, and industrial users, and 10 percent of residential use is for landscape irrigation. If a water reuse program was implemented to meet half of the industrial demand and all of the residential irrigation, how much water would be reused daily? Express your answer in gallons and in the equivalent number of Olympic-sized swimming pools.

13.4   The five most populous nations comprise approximately 48 percent of the world's population. Perform research to find the populations of the sixth through tenth most populous nations. List these, and calculate the combined percentage of the world's population.

13.5   Using Figure 13.1 and Figure 13.2, calculate the GDP of China and the United States. Do your estimates compare favorably with the information in Figure 13.3?

13.6   Using Figure 13.6, estimate the life expectancy in the United States in 2040.

Short Answer

13.7   List the three Ps of sustainability.

13.8   Define sustainability in less than twenty (of your own) words.

13.9   Why does life expectancy reach a "plateau" for nations with per capita GDP greater than approximately $20,000?

13.10   Why does the infant mortality rate reach an "asymptote" for nations with per capita GDP greater than approximately $20,000?

Discussion/Open-Ended

13.11   How is national security linked to sustainability?

13.12   What do you think are two of the most unsustainable aspects of our society?

13.13   Why are "failing states" potential breeding grounds for terrorism?

Research

13.14   Find information on a large green roof installation and write a one-page report. Your report should provide background on the installation, summarize the design, state the cost, and list the benefits.

13.15   Find information on communities that have considered implementing a "toilet to tap" program and write a one-page report summarizing your findings.

13.16   Research estimates of the carrying capacity of Earth. Write a one-page summary of an estimate that you believe to be reasonable and explain your rationale.

13.17   Research estimates of the carrying capacity of Earth. Write a one-page summary of an estimate that you believe to be *unreasonable* and explain your rationale.

Key Terms

- carrying capacity
- ecological footprint
- green building
- green roof
- gross domestic product

- pollution prevention
- resource demand
- sustainability
- sustainable design

- sustainable land development
- triple bottom line
- virtual water
- water reuse

References

British Petroleum. 2010. *Statistical Review of World Energy 2010.* http://www.bp.com/productlanding.do?categoryId=6929&contentId=7044622, accessed August 2, 2010.

Brown, L. 2009. *Plan B 4.0: Mobilizing to Save Civilization.* New York: W.W. Norton & Company, New York.

Central Intelligence Agency (CIA). *The World Factbook.* https://www.cia.gov/library/publications/the-world-factbook/rankorder/2119rank.html, accessed December 14, 2010.

Davidson, C. I., H. S. Matthews, C. T. Hendrickson, M. W. Bridges, B. R. Allenby, J. C. Crittenden, Y. Chen, E. Williams, D. T. Allen, C. F. Murphy, and S. Austin. 2007. "Adding Sustainability to the Engineer's Toolbox: A Challenge for Engineering Educators." *Environmental Science and Technology* 41(14): 4847–4849.

Foreign Policy. 2010. *The Failed States Index 2010.* http://www.foreignpolicy.com/articles/2010/06/21/2010_failed_states_index_interactive_map_and_rankings, accessed August 29, 2011.

International Energy Agency. 2006. *World Energy Outlook 2006.* http://www.iea.org/textbase/nppdf/free/2006/weo2006.pdf, accessed August 29, 2011.

Maddison, A. (2003). The World Economy, A Millennial Perspective, Paris: OECD; Website: http://www.ggdc.net/MADDISON/oriindex.htm or http://www.ggdc.net/maddison/maddison-project/home.htm.

Mihelcic, J. R., J. C. Crittenden, M. J. Small, D. R. Shonnard, D. R. Hokanson, Q. Zhang, H. Chen, S. A. Sorby, V. U. James, J. W. Sutherland, and J. L. Schnoor. 2003. "Sustainability Science and Engineering: Emergence of a New Metadiscipline." *Environmental Science & Technology* 37(23): 5314–5324.

Shiklomanov, I. A. 1999. *World Water Resources and Their Use.* http://webworld.unesco.org/water/ihp/db/shiklomanov/, accessed August 29, 2011.

World Health Organization. 2008. *The World Health Report 2008.* http://www.who.int/whr/2008/whr08_en.pdf, accessed August 29, 2011.

World Health Organization (WHO) and United Nations Children's Fund (UNICEF). 2008. *Progress on Drinking Water and Sanitation: Special Focus on Sanitation.* http://www.who.int/water_sanitation_health/monitoring/jmp2008.pdf, accessed August 29, 2011.

Yang, H., L. Wang, K. C. Abbaspour, and A. J. B. Zehnder. 2006. "Virtual Water Trade: An Assessment of Water Use Efficiency in the International Food Trade." *Hydrology and Earth System Sciences* 10: 443–454.

chapter Fourteen Economic Considerations

Learning Objectives

After reading this chapter, you should be able to:

1. Explain the relationship between infrastructure and the economy.
2. List common funding sources for public infrastructure projects.
3. Explain economic equivalence.
4. Determine the net present value of simplified infrastructure projects.

Introduction

Consider the fact that you can buy fresh produce during the winter in the northeastern United States at prices that are often not noticeably different from what you would pay in the summer for local produce. One of the reasons that these low prices occur is that very low transportation costs are possible because of the investment that federal, state, and local governments have made in infrastructure. If the condition of the infrastructure that helps move the produce (e.g., the interstate highway system and rail system) was less effective than it currently is, produce would take much longer to reach its destination and the price per pound would escalate. Even worse, if the infrastructure was in very poor condition or nonexistent, you wouldn't even be able to buy fresh produce during the winter.

Alternatively, imagine a hypothetical city with an aging wastewater treatment facility that does not have any spare capacity to accept additional wastewater. A company desires to build a factory in the city, but given the city's aging infrastructure, the factory will have to add its own wastewater pretreatment facility, thus increasing the construction cost to the owners of the factory. The extra cost of adding pretreatment may be large enough that the company decides to build in another city, one which has the infrastructure in place to accept the factory's waste without requiring the factory to pretreat its waste. This scenario also illustrates the relationship between economic activity and the quality of infrastructure.

Introductory Case Study: China National Trunk Highway System

The rapid growth of China's economy has been accompanied by tremendous investment in their infrastructure, and future economic growth can only be possible with continued investment. Since the 1990s, rapid expansion of railways, inland waterways, and roads has occurred. Indeed, China is spending a reported 5 percent (or 9 percent, depending on the source) of their gross domestic product (GDP) on transportation, as compared to approximately 2 percent for most other industrialized countries such as the United States.

The centerpiece of this transportation expansion in China is their expressway system, known as the National Trunk Highway System (Figure 14.1). The purpose of the system is to link "major cities" with each other and with ports. Major cities include provincial capitals and cities with populations greater than 200,000. This highway system is rapidly expanding (Figure 14.2) and is expected to be 110,000 miles long by 2050. By comparison, the U.S. interstate highway system is 47,000 miles long and as seen from Figure 14.2, was not built as quickly as its Chinese counterpart. At its peak construction period (1955–1970), the United States was constructing 2,000 miles per year; between 2003 and 2008, China constructed 6,000 miles per year.

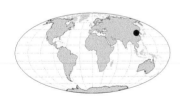

Figure 14.1   Portions of the National Trunk Highway System in Shanghai, China.
Copyright © Prill Mediendesign & Fotografie/iStockphoto.

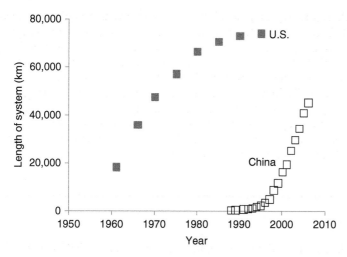

Figure 14.2 **Comparison Between the Growth of China's National Trunk Highway System and the U.S. Interstate System.**

Data source: Chinese National Bureau of Statistics; Federal Highway Administration.

The Relationship Between Infrastructure and the Economy

The relationship between infrastructure and economic well-being is evident at the local, regional, national, and international scale. In fact, we could argue that without infrastructure, we wouldn't have an economy at all. And without infrastructure, a developing country has limited ability to grow its economy, as its economy is tied directly to its ability to move coal from mine to user, agricultural goods from rural farms to metropolitan markets, and light industrial products from factory to consumer. As a result, more international aid to developing countries is being directed to improvements in transportation infrastructure.

Of all infrastructure sectors, the transportation sector may have the largest impact on economic growth. Wayne Klotz, past president of the American Society of Civil Engineers (ASCE), stated "If we don't have the ability to transport people and goods, we don't have an economy." Consider the economic benefits of a properly functioning transportation system as compared to an undersized or otherwise ineffective system:

- The cost to produce goods decreases, due to the decreased cost in obtaining raw materials.

- A properly functioning transportation system results in less congestion, which in turn results in fewer delays and lower fuel consumption.

- The cost of distribution of goods decreases.

- More households have access to more high-paying jobs. This in turn means that more tax revenue is being generated for the various levels of government. This is just one of the many ways in which infrastructure should be looked at as an investment by government. Some studies suggest a **payback period** to governments of less than 4 years for money spent on improving the transportation infrastructure.

PBS Interview with Pennsylvania Governor Highlighting Economic Importance of Infrastructure

Payback period is the ratio of project costs to annual savings. For example, if you pay $3,000 for a new furnace and it saves you $300 per year in heating cost, your payback period is 10 years.

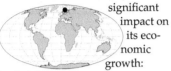
- Potential homeowners have more freedom in housing choices, and greater choice and better access to health services.

- Rural areas can generate wealth by being connected to urban economies, leading to an overall increase in economic activity.

- Safer roads lead to fewer accidents, and accidents are costly to society. The cost to society due to accidents is more than $160 billion, more than twice that of the cost due to traffic congestion (AAA, 2008). Costs associated with accidents are estimated by assessing the costs of medical care, emergency and police services, property damage, and lost productivity and quality of life.

- Building roads creates jobs for construction workers and engineers, as well as other ancillary jobs (e.g., accounting, human resources, planning).

- Tourism is enhanced due to improvements in accessibility and connectivity.

These benefits were part of the rationale for the passing of the nearly $800 billion U.S. Recovery and Reinvestment Act (a.k.a. the "stimulus bill") of 2009. It was apparent to the majority of legislators and the President that in the short term, spending money on infrastructure creates many jobs for transportation and construction workers (who were hit very hard by recession) as well as invests in an infrastructure that will create a more robust economy in the long term. A similar federal program, the Civilian Conservation Corps was instituted in 1933 as part of President Franklin D. Roosevelt's New Deal to provide employment opportunities during the Great Depression. National and state parks, roads, dams, and drainage canals were among the many projects completed by an estimated 3 million enrollees from 1933 to 1941.

Sources of Funds

Infrastructure components (e.g., roads, drinking water distribution systems) are most often owned by the federal, state, or local government rather than private corporations. The government may fund these projects using a variety of sources. At the most basic level, though, the taxpayer eventually pays the bill for any government-funded project.

FUNDING LOCAL PROJECTS

One method of funding municipal projects is through property tax revenues. For municipalities, these revenues are typically generated by property taxes levied on local properties. Property owners pay taxes based on assessed value. The rate of taxation is termed the "mill rate" and is often expressed as tax $/$1,000, or "tax dollars per thousand dollars of assessed value." Therefore, a property with an assessed value of $100,000 in a municipality with a mill rate of $20/$1,000 would pay $2,000 in taxes annually.

Property tax revenues are generated locally and must be divided among various departments in the municipality (including fire, police, water and sewer, streets, libraries, parks, recreation, museums, and a host of other services as well as administration) and pay for regional services such as schools, county services, and community colleges. School districts are typically the largest "consumer" of property tax revenues, accounting for approximately one-half of distributed funds. An example of operating expenses for a

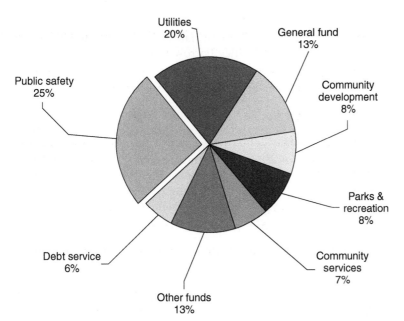

Figure 14.3 **Distribution of Operating Expenses for a City.** "Utilities" include streets, water, wastewater, and stormwater. "Community development" includes engineering, planning, and building.

Source: City of Lake Oswego, Oregon.

municipality can be viewed in Figure 14.3. Operating expenses do not include capital improvements such as building new roads or upgrading wastewater treatment facilities.

Services and projects can also be funded by **user fees** rather than by tax revenues. A toll on a toll road and state park entrance passes are examples of user fees. Many municipalities fund the operation and maintenance of water and wastewater treatment facilities and distribution systems via user fees paid to a water and sewer **utility**. Users pay a fee based on the number of gallons of wastewater generated or water used. This fee is normally set so that the income from users equals the expenses for operations and maintenance.

Increasingly, municipalities are instituting stormwater utilities, whereby residents pay a stormwater user fee. Rather than pay a fee for gallons of stormwater produced (which cannot readily be metered), one method is to have property owners pay based on the amount of imperviousness (e.g., area of rooftop and pavement), since the amount of imperviousness is directly related to the amount of runoff produced. A stormwater utility removes stormwater management expenses from the general tax revenue pool and creates a separate "account." This decreases the likelihood of cuts to stormwater infrastructure projects, as the expenditures are not directly competing with other needs funded by general tax revenue.

A **utility** is an organization that operates and maintains a portion of the infrastructure. The utility is either part of a municipal government, or may be a private business subject to governmental oversight. Examples include water, sewer, electric distribution, natural gas, and telephone.

Special Assessments

Some infrastructure components may be funded through special assessments. For example, if water and sewer services must be extended to a new industrial park, a special assessment (sometimes termed a development impact fee) may be added to the cost of building permits to pay for the infrastructure.

Congestion Pricing

One innovative user fee is congestion pricing, in which automobile drivers must pay a surcharge when using certain roads during times of peak use. The rationale is that the surcharge will discourage some drivers from driving during peak times, thereby reducing congestion. Thus, the solution is to *reduce the demand* for extra roadway lanes rather than *increase the supply*. As such, it is a non-physical (non-structural) solution to the congestion problem. And it has the benefit of making users more aware of the actual costs they are incurring in using the roadway system. Congestion pricing is used in Europe but has not been implemented in North America for several reasons. For example, congestion pricing amounts to a tax, and few politicians are willing to champion such a policy. Nonetheless, several large U.S. cities are exploring this option.

Often the general tax revenues or user fees are not sufficient to fund infrastructure projects, and municipalities must resort to borrowing to fund their projects. Most often, they borrow by selling **bonds**. The advantage to borrowing is that a municipality can tap into a large amount of money to fund infrastructure projects that cannot be funded based on tax revenues in a given year. The drawback is that just like any borrowing (e.g., a home mortgage), interest is charged and must be paid back. For example, a $10 million infrastructure project may be too large for a municipality to fund in a single year. By borrowing that money at say a 5 percent interest rate and paying it back over 20 years, the municipality will only need to pay back $800,000 each year. However, that means that over 20 years, the municipality will be paying a total of $16 million for the $10 million project. The term **debt service** applies to payments made by municipalities on borrowed funds; this often accounts for a significant portion (6 percent in Figure 14.3) of annual budgets.

Bond Ratings

Municipal bonds are sold to investors. Bonds are rated by credit agencies based on the likelihood of default, or inability to repay, as summarized in Table 14.1.

Municipalities and states desire high ratings, as higher ratings provide lower interest rates. When a state or municipality gets into financial trouble, its bond ratings drop and interest rates increase. The increasing rate may correspond to millions of dollars of additional interest payments.

The "grades," or ratings, are deceiving when compared to letter grades that you are accustomed to as a student. The rating of "A" is far from excellent. In March of 2010, only two states, California and Illinois, had A ratings (all others were AA or AAA). Detroit, Michigan, and New Orleans, Louisiana, had ratings as low as BB, which is considered "junk" status. International headlines were made in August 2011 when the U.S. federal credit rating was downgraded from AAA to AA+.

Table 14.1

Bond Ratings as Defined by Standard & Poor's

Likelihood of Default	Rating
Highest quality	AAA
High quality	AA
Upper medium quality	A
Medium grade	B
Somewhat speculative	BB
Low grade, speculative	B
Low grade, default possible	CCC
Low grade, partial recovery possible	CC
Default, recovery unlikely	C

Asset Management

The Federal Highway Administration defines asset management as a "systematic process of maintaining, upgrading, and operating physical assets cost effectively." More and more infrastructure owners (including all levels of government) are undertaking infrastructure asset management projects to better guide their decision-making process concerning how to most effectively invest in the infrastructure. Asset management is carried out by public works

Municipalities also apply for **grants** (funds without obligation to repay) from the state and federal government. For example, the state may help fund a city's cleanup of a brownfield (contaminated) site for redevelopment, or installation of a bike path. Grants often require **matching funds**, whereby the city is required to provide a portion of the funding for a project. In terms of the infrastructure, federal or state governments also provide low interest loans to municipalities for some infrastructure projects.

One funding mechanism that is becoming increasingly popular is **Tax Incremental Financing (TIF)**. TIF is best explained with an example. Consider a piece of property that is not developed. Undeveloped property generates limited revenue for the city, since the taxes generated are based on property value. However, if that land is developed, it will have a much higher property value and will thus generate much more tax revenue. The difference between current (undeveloped) tax and proposed (developed) tax revenues is the "increment" (the "I" in TIF). Of course, to transform the property from undeveloped land to developed land often requires an investment in infrastructure. The purpose of TIF is to borrow money in order to pay for the new infrastructure, and then repay the loan using the incremental tax revenue generated by the improved property.

For example, consider a 20-acre farm field on the edge of a municipality that has a mill rate of $20/$1,000. The farm field is valued at $200,000 and thus generates $4,000 per year in property taxes. A developer proposes to build a "big box" store on this property but cannot afford to pay $500,000 to build the needed infrastructure such as water mains and widening of the road. However, this store will be assessed at $10 million and thus will generate $100,000 in tax revenue annually, or an increment of $96,000 ($100,000 − $4,000) per year. To *encourage* this development, the city agrees to finance the necessary infrastructure through a TIF, rather than requiring the developer to incur the costs. The infrastructure investment can be readily paid off since the incremental amount of $96,000 can be used directly to pay off the loan needed for the $500,000 investment in infrastructure.

In addition to the creation of new jobs and an enhancement to the local economy, an advantage to the municipality is that all of the incremental taxes (new tax revenue generated) can be used by the municipality to repay the loan. Without a TIF, only a portion of the incremental tax revenue would go to the municipality. The remainder would go to support schools and perhaps other regional needs. Once the loan is repaid, the TIF is "closed," and the future taxes will be apportioned to the other agencies. This highlights a drawback to the TIF, in that schools and governmental offices do not get a portion of the incremental tax revenues until after the TIF loan is repaid, yet may need to provide increased services due to the new development.

Thus, local governments have a variety of funding mechanisms to rely upon. This can be seen by re-examining the Capital Improvement Program (CIP) for Chicago that was introduced in Chapter 11, Planning Considerations. The funding sources for projects in the CIP are shown in Figure 14.4. As is typically the case, Chicago is relying very heavily on bonds to fund their infrastructure projects, with approximately 75 percent of the funds generated by bonds. The second greatest contribution is from federal funds.

Given that a community's revenues are based on the value of property in their community, it is tempting for communities to attract growth. However, a community should carefully consider the additional expenses incurred by the growth. For example, a successful industrial park will generate significant property tax revenue. But it may also require upgrades to wastewater treatment facilities, sewer lines, water treatment facilities, water mains, increased road capacity in the vicinity, additional police presence, expanded firefighting capability, etc. Also, it is possible that a new residential subdivision may cost more to support itself than it generates in tax revenue. **Cost of community**

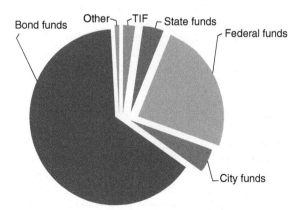

Figure 14.4 Funding Sources for Chicago's Capital Improvement Program (CIP) for 2010–2014.

Data source: City of Chicago, 2010.

service studies seek to quantify the cost that various land uses generate as a means of helping communities plan future growth.

Many consulting engineers spend a significant portion of their time helping clients procure funding for their projects. This may entail helping the client set up a TIF, or it may entail writing grant proposals to federal and/or state agencies. Engineers within each sub-discipline of civil and environmental engineering must be familiar with the current funding mechanisms available to their clients. They also must be proficient in writing clearly and persuasively. The stereotypical engineer seeks out a career in engineering because of strength in mathematics and science and a weakness in written and spoken communication. However, engineers seeking to rise to positions with more responsibility (and higher pay) find that it is essential to be excellent communicators in addition to having excellent technical skills.

FUNDING STATE AND FEDERAL PROJECTS

The discussion of the generation of funds and payment of projects has focused on municipalities to this point, but also applies to state and federal governments. State and federal governments also borrow and debt service comprises 8.5 percent of the 2009 U.S. federal budget. Sales taxes, gasoline taxes, income taxes, vehicle registration fees, "tourist" taxes (e.g., hotels) and alcoholic beverage taxes are among the many types of taxes upon which state and federal governments rely.

Federal and state budgets often contain a "transportation fund," used to fund the various subsectors and components of the transportation infrastructure sector. This transportation fund is funded by gasoline taxes that are factored into the cost of gasoline at the pump. The federal gasoline tax is $0.18/gallon, and states add on their own gasoline tax (ranging from approximately $0.10 to $0.40/gallon). The federal gas tax has not been adjusted for inflation in 20 years; by adjusting for inflation, the $0.18/gallon in 1990 should be $0.30 today.

Unfortunately, transportation funds are sometimes "raided" by legislators during budget shortfalls and used to pay for other services. Thus, tax revenues that were collected for road replacement, safety upgrades, and new road construction is diverted to non-infrastructure projects.

The source of funds for the Wisconsin DOT (Department of Transportation) is depicted in Figure 14.5. Wisconsin's gas tax ($0.33/gallons) is the largest funding source for transportation projects. Transfers of federal money to the state DOT is also a sizeable portion.

Figure 14.5 **State of Wisconsin Funding Sources for Transportation Projects.** The total budget is $6.8 billion for 2009–2011.

Data source: Wisconsin Department of Transportation.

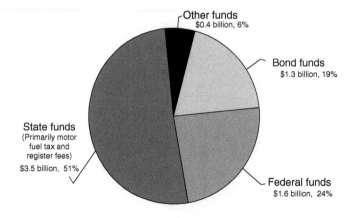

Other funds
$0.4 billion, 6%

Bond funds
$1.3 billion, 19%

State funds
(Primarily motor
fuel tax and
register fees)
$3.5 billion, 51%

Federal funds
$1.6 billion, 24%

Unlike the initial funding of the U.S. Interstate System, China is financing their National Trunk Highway System primarily through bonds to be paid back by tolls. This is seen by some as "risky," as those provinces with less traffic (and therefore less tolls) may not be able to pay back the bonds.

Public works can also be managed and financed by the private sector. Many aspects of the infrastructure are privatized, including water and wastewater utilities and transportation components. The umbrella term for such privatization enterprises is **public–private partnerships (PPP)**. PPP can come in many forms, but two common methods for financing infrastructure projects are **operations and maintenance concessions** and **long-term leases**.

With an operations and maintenance concession, a private company operates a public work. Such a partnership may be as simple as a municipality contracting roadside mowing rather than conducting this work using public employees, or as complex as contracting with a private water company to operate a municipality's drinking water treatment facility.

Long-term leases are becoming increasingly common. Two recent high-profile long-term leases are the Chicago Skyway toll road and the Indiana Tollway. The former is an 8-mile, highly traveled (50,000 vehicles per day) toll road, the lease for which was sold for $1.8 billion, whereas the latter is a 157-mile toll road that traverses the state of Indiana and was sold for $3.8 billion. As part of the lease agreement, the private companies can collect tolls in exchange for maintaining and operating the road. These two leases are long-term (99 and 75 years respectively). These leases were also noteworthy (and controversial) due to the fact that the leasers were foreign-based investment companies. However, these sources of funds are very attractive to states and cities that are struggling with budget shortfalls, as leasing the roadway provides a significant upfront payment that can be used for other infrastructure projects.

There is considerable debate currently over PPPs. One advantage of a PPP is that private firms can operate systems more cost-effectively and often have more technical and managerial expertise than may be available at the local government level. For example, a private company may operate several water treatment facilities, and hire one or two "experts" that can manage several facilities cost-effectively. Such on-staff experts, who would not otherwise be affordable to the municipality, are more able to respond to ever-changing regulations. On the other hand, the private sector is a profit-driven sector, which may lead to conflict with the not-for-profit perspective of a municipal government.

Overview Article on Privatization of Water Utilities

Design-Build as a Form of PPP

The vast majority of local infrastructure projects are delivered by the traditional **design-bid-build** method. In this method, the design is completed by an engineer and then the construction is awarded through a competitive bid process to a contractor, who then builds the project. An alternative project delivery method is **design-build**. Rather than an owner contracting out for design work of an infrastructure component separately from the construction (build) phase, these two are combined in design-build. Thus, the project owner (e.g., a county or city) can partner with a private firm, the latter which designs *and* builds the infrastructure component. This method can provide cost savings and other efficiencies in that the design-build provider can better design the component, knowing that they will also be in charge of building it.

Design-build has been expanded to **design-build-operate**, in which the private sector will also operate and maintain the component for a contracted period of time, perhaps 20 to 30 years after construction. Design-build and design-build-operate methods account for approximately 10 percent of infrastructure project delivery in the United States (ACEC, 2008).

Consideration of Project Economics

Engineering design nearly always requires the engineer and client to choose between several alternative designs. The scale of the decision may be between "simple" options such as choosing between light-emitting diode (LED) and mercury vapor street lights, or may be between complex treatment processes for a wastewater treatment facility. One criterion that is used to select the final design option in nearly every situation is cost. Given that sources of funding are typically limited and that competition for these funds can be quite intense, cost is often the most important criterion.

However, as we discussed in Chapter 10, Design Fundamentals, and will repeat in future chapters, economics is not the *only* criterion to consider; in other words, the lowest cost alternative is not necessarily the most appropriate choice. A client might select an option that is significantly more costly than the lowest cost option based on past experience, public acceptance (real or perceived), aesthetics, or sustainability.

When comparing multiple options, engineers have to make sure they are comparing the entire costs *over the life* of each project, not only the upfront costs. Upfront costs include the costs involved in initial purchasing of equipment and buildings as well as the cost to design, construct, and/or install the option. Other costs that must be considered are costs associated with operations, maintenance, rehabilitation, and replacement. In many cases, these costs can equal many times the upfront costs. Consequently, to compare among alternatives, a thorough accounting of *all* of the costs over the project's life span must be conducted, termed **life-cycle cost**. In many cases, the life span of the competing options will vary.

Consider choosing between a concrete road and an asphalt road. The concrete road will be much more expensive initially, perhaps twice the upfront cost of asphalt. However, the perspective changes when operations and maintenance costs are considered. For a time frame of 40 years for example, the asphalt road may need to be resurfaced twice while the concrete road requires minimal maintenance.

Environmental Life-Cycle Analysis (LCA)

The life-cycle cost analysis described in this chapter evaluates the quantifiable monetary costs associated with a project. An environmental life-cycle analysis (LCA) seeks to evaluate all of the environmental impacts associated with a product or process. This analysis is applied to the product or process from its "birth" (e.g., raw material extraction) to its ultimate disposal. Thus, it has a much longer viewpoint as compared to life-cycle *cost* assessment, which only includes capital costs, operations and maintenance costs, and abandonment.

Consider comparing a cloth diaper to a disposable diaper. For the cloth diaper, many environmental impacts exist. A few of these are fertilizer and pesticide use during growth of the cotton; wastewater discharges and air emissions at the facilities involved in the process of turning raw cotton into a diaper; greenhouse gas emissions in transportation; detergent use in laundering. For the disposable diaper, an LCA would inventory similar impacts as for the cloth diaper, with a major difference being that disposable diapers are made of plastics derived from petroleum or natural gas and do not degrade in landfills.

A similar analysis can and should be performed on many aspects of the infrastructure including asphalt versus concrete pavements; LED versus conventional street lights; solid waste incineration versus landfilling; mass transit versus automobile use. LCA is becoming a more common approach to evaluating design options, and in some cases is required by regulation. A senior vice president from a leading multinational engineering consulting firm recently stated, "Perhaps only 10% of our employees *use* LCA routinely, but *all* of our employees must *understand* LCA."

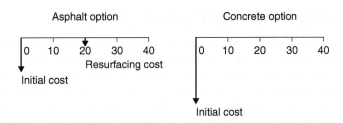

Figure 14.6 Cash Flow Diagrams for Asphalt and Concrete Pavement Options.

The upfront costs and operating costs can be presented in a **cash flow diagram**. In a cash flow diagram, revenues and expenses are represented as arrows. Expenses include initial investments and maintenance costs and are represented by arrows pointing down. Revenues include tolls, sales, and the salvage value at the end of use and are represented by arrows pointing up. The base of the arrow is placed at the end of the year in which the expense or revenue occurred, and the length of the arrow is proportional to the magnitude of the cost or revenue.

Consider Figure 14.6, in which cash flow diagrams comparing asphalt and concrete paving options are depicted. In year 0, both projects incur a large expense, the design and installation of the pavement. For the concrete project, this arrow is twice as long as the asphalt option. For this simplified example, no other costs are assumed for the concrete option, while the asphalt option incurs an additional cost in year 20, when resurfacing occurs.

Note that in Figure 14.6, all the arrows are pointing down. That is because these options only have expenses associated with them. If this road were a toll road, upward-pointing arrows would be used at each year to represent the annual revenue due to tolls.

As another example, consider a cash flow diagram, from a developer's perspective, for developing a parcel of land into residential housing lots. The developer purchases the land for $500,000 in year 0. The cost of adding the infrastructure (roads, sewer, etc.) is $2.5 million. The project will be divided into three construction phases: $1.5 million will be spent on construction in year 0, $0.5 million in year 3, and $0.5 million in year 8. Lots are projected to sell at a steady rate of 15 lots per year for 10 years, at a cost of $50,000 per lot (equal to $750,000 each year). The cash flow diagram for this development scenario is provided in Figure 14.7.

But, how do we compare between the alternatives shown in Figure 14.6? How do we know which pavement type is most cost-effective? Or, how does the developer know how well the scenario shown in Figure 14.7 compares to other development options? It might be tempting to simply add up the arrows in Figure 14.6, but this would ignore an extremely important concept known as the **time value of money**.

To understand the time value of money, consider being given two options: you can receive a gift of $1,000 today or a gift of $1,000 in 10 years. Ignoring any effects of inflation, you would most certainly prefer to be given the $1,000 today. Having that money today allows you to use the money for investment or consumption purposes; this is the concept of the time value of money.

Now, consider being offered $1,000 today or $2,000 dollars in 10 years. If you are rational, you would consider how much interest you could earn in 10 years by investing the $1,000 today, and compare that value to the $2,000 option. If the initial $1,000 gift is predicted to grow to say $2,500, perhaps by investing in a bank certificate of deposit (CD), you will most likely accept the $1,000 today. (The term **economic equivalence**, or simply **equivalence**, describes the fact that, for this example, $1,000 today and $2,500 in 10 years are completely the same; in other words, if you were offered $1,000 today

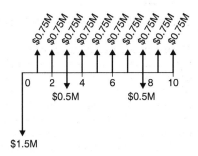

Figure 14.7 Cash Flow Diagram for Residential Land Development Scenario.

Creative (Desperate) Sources of Funding

Illinois was one of the hardest-hit states by the recent recession. Facing a $13 billion budget deficit in 2010 and a low bond rating, alternative sources of funding were sought. One alternative considered was to sell half of the $300 million that the state receives annually as part of a multi-state settlement with the tobacco industry. By doing so, Illinois would be giving up a fixed long-term revenue source for a lump sum to help solve current financial troubles. The State's Budget Director summarized the proposal by saying (to a hypothetical investor), "Will you trade me $1 billion now for the right to get $150 million a year?"

or $2,500 in 10 years, you would have to flip a coin to choose, if your only criteria was economic value.) This analysis, in which the value of two options is compared at some future time, is known as a **future value** analysis. For this example, we would say that the future value of the $1,000 gift is $2,500.

Conversely, you might consider a **present value** analysis. In this case, for a given interest rate, you would calculate how much the $2,000 in 10 years is equivalent to today; that is, you would calculate the *present* value of the future $2,000 gift. This can be accomplished by using Equation 14.1.

$$PV = \frac{F}{(1+i)^n} \tag{14.1}$$

where

PV = present value
F = future value
i = interest rate
n = number of time periods

The units on the number of time periods (n) and the interest rate (i) should correspond. That is, if the interest rate used is an annual interest rate, the time period must be in years.

numerical example 14.1 Present Value Calculation

What is the present value of a $2,000 gift that you will receive in 10 years, using an interest rate of 5.00 percent?

solution

$$PV = \frac{F}{(1+i)^n} = \frac{\$2,000}{(1+0.0500)^{10}} = \$1,230$$

Therefore, $2,000 in 10 years is equivalent to $1,230 today for an interest rate of 5.00 percent.

A present value analysis can be readily applied to a cash flow diagram to compare among life cycle costs. This can be accomplished by computing the **net present value**, which is the difference between the sum of the present value of the revenues and the sum of the present value of the expenses (Equation 14.2).

$$NPV = \sum PV_{\text{revenues}} - \sum PV_{\text{expenses}} \tag{14.2}$$

where

NPV = net present value
PV_{revenues} = present value of the revenues
PV_{expenses} = present value of the expenses

To use a net present value (NPV) analysis, the present values of all the expenses and revenues must first be calculated individually. When comparing alternatives, the NPV for various options can be computed; the option with the highest NPV is the best option in terms of economics.

---

numerical example 14.2 Present Value Analysis

Conduct a present value analysis for the land development scenario presented in Figure 14.7. Use an interest rate of 6.00 percent for the analysis.

solution

The PV of all the expenses and all of the revenues must be individually calculated. A sample calculation is shown here for the present value of the $500,000 infrastructure expense in Year 3.

$$PV = \frac{F}{(1+i)^n} = \frac{\$500,000}{(1+0.0600)^3} = \$420,000$$

The remaining PVs are calculated in a similar manner and are summarized in Table 14.2; you should be able to confirm the values presented using a calculator or spreadsheet.

The sum of the present value of the expenses and the sum of the present value of the revenues is provided in Table 14.2. From this, Equation 14.2 can be used to compute the NPV of this development option.

$$NPV = \sum PV_{\text{revenues}} - \sum PV_{\text{expenses}} = \$5,521,000 - \$2,734,000 = \$2,787,000$$

So, this project has a positive NPV, which is most likely encouraging to a developer. However, it may not be a large enough NPV for a developer to decide to make the investment when considering the risk of declining property values or other economic uncertainties. For example, the recent housing market "bubble burst" had a tremendous impact on the U.S. economy. Housing demand, and thus demand for new development is cyclical (Figure 14.8) and not readily predictable. Many

---

Table 14.2

Present Value of Expenses and Revenues

Present Value of Expenses		Present Value of Revenues	
Land purchase	$ 500,000	Year 1 lot sales	$ 708,000
Initial infrastructure	$1,500,000	Year 2 lot sales	$ 667,000
Year 3 infrastructure	$ 420,000	Year 3 lot sales	$ 630,000
Year 8 infrastructure	$ 314,000	Year 4 lot sales	$ 594,000
		Year 5 lot sales	$ 560,000
		Year 6 lot sales	$ 529,000
		Year 7 lot sales	$ 499,000
		Year 8 lot sales	$ 471,000
		Year 9 lot sales	$ 444,000
		Year 10 lot sales	$ 419,000
SUM	$2,734,000	SUM	$5,521,000

developers (and engineering companies relying on them for projects) went bankrupt or severely reduced the number of employees during the downturn of the housing market. For developers, there is a substantial financial risk when investing in land and infrastructure necessary for development.

Figure 14.8 **New Housing Starts in the United States.**

Data source: U.S. Census Bureau, 2010.

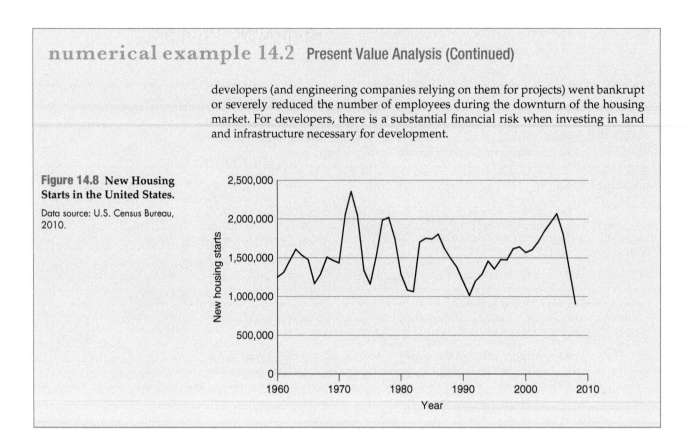

Now, we should reconsider Figure 14.6; neither of the options have any revenue. This is typical of infrastructure projects; projects for the public welfare do not necessarily generate direct revenue. Even though we have shown at the beginning of this chapter that they make a significant impact on the economy, these are difficult to show as "revenues" in a cash flow diagram, even if they could be calculated. For such public works projects, the NPV analysis is still applicable and the analyst would still select the largest NPV; in other words, the project with the *least negative* NPV would be the chosen alternative, ignoring other selection criteria.

Outro

This chapter has introduced you to two very important questions related to economics: "how will we pay for it?" and "which option is the least costly over the project life cycle?" Keep in mind that often, but not always, cost is not the most important criterion by which alternatives are compared.

Also, please reconsider Figure 14.2, which demonstrated the rapid increase in the extent of the Chinese expressway system. Note that according to this figure, expressway length has doubled in 5 years! There are many implications to this doubling that will affect the environment, society, human health, and welfare in potentially beneficial or adverse ways. These are each subjects of following chapters. For example, vehicle miles traveled will likely mirror this trend, as will associated traffic fatalities, fossil fuel consumption, tailpipe emissions, etc. Positive impacts include economic benefits noted in this chapter, most notably increased mobility and opportunity for Chinese citizens.

Calculation

14.1 Given an annual interest rate of 5.00 percent, how much is $500 today equivalent to in 10 years? In 1 year? In 100 years?

14.2 Given an annual interest rate of 7.00 percent, how much is $1,700 today equivalent to in 10 years? In 6 months? In 18 months?

14.3 Given an annual interest rate of 5.00 percent, how much is $5,500 in 10 years equivalent to today? What if the interest rate is only 0.5 percent?

14.4 Given an annual interest rate of 3.00 percent, how much is $45,000 in 6 months equivalent to today? What if the interest rate is only 0.500 percent?

14.5 How much should you place into a 5-year certificate of deposit (CD) bearing 3.00 percent annual interest in order to have the CD worth $6,000 in 5 years?

14.6 If you place $600 from your first paycheck into a Roth IRA earning 4.00 percent annual interest, how much will it be worth when you withdraw it in 40 years?

14.7 Repeat the previous problem, except now take $100 from each of your first 12 monthly paychecks and determine what they will collectively be worth in 40 years.

14.8 A certain wastewater treatment unit process costs $50,000 to purchase and install, and will cost $500 per year to operate. Given an interest rate of 4.00 percent and a design life of 10 years, what is the net present value of this scenario?

14.9 Reconsider Numerical Example 14.2. What would the net present value be if lots took twice as long to sell (i.e., $375,000 in lot sales each year) for 20 years.

14.10 Reconsider Numerical Example 14.2. What would the net present value be if lots sold at the same rate but the sales price was $40,000 per lot instead of $50,000 per lot?

14.11 Reconsider Numerical Example 14.2. What if the designer did not design the subdivision such that it could readily be constructed in phases? Rather, the construction of the entire subdivision infrastructure must occur in a single phase. How does this affect the NPV?

Short Answer

14.12 Why might it *not* be a good idea for a community to grow?

14.13 List five ways that the quality of the transportation infrastructure affects the economy.

14.14 List one benefit and one drawback for a municipality to take out a bond to fund a project.

14.15 Compare an environmental life-cycle analysis (LCA) to a life-cycle cost analysis.

14.16 Explain the concept of *equivalence*.

14.17 What do the upward-pointing arrows on a cash flow diagram represent? The downward-pointing arrows?

Discussion/Open-Ended

14.18 Describe the infrastructure in your community that supports bicycle use. If you were to assign this infrastructure a grade, what grade would you give? List the criteria you used to assign this grade. How would improving this infrastructure help the economy?

14.19 Describe the infrastructure in your community that supports automobile use. If you were to assign it a grade, what grade would you give? List the criteria you used to assign this grade. How would improving this infrastructure help the economy?

14.20 Do you agree with this quote from the chapter? "If we don't have the ability to transport people and goods, we don't have an economy." Give some reasons why someone might agree with this statement and some reasons why someone might disagree.

14.21 Do you think public-private partnerships (PPPs) are a good idea? Give three benefits and three drawbacks to the private acquisition of the Indiana Tollway discussed in this chapter.

Research

14.22 As introduced in Chapter 5, stormwater detention ponds are used to store excess runoff and release it in a controlled manner. What are the life-cycle costs associated with a detention pond?

14.23 Research the use of permeable pavements. Explain how they work and provide a typical cross-section. What are the life-cycle costs associated with permeable pavement? Limit your response to one page.

14.24 How might the environmental impacts of an asphalt pavement compare to the environmental impacts of a concrete pavement? List and describe at least five environmental impacts for each. List your sources.

14.25 Use the Internet to determine the current value of the U.S. debt. List your source(s). Express this debt as a $/person.

14.26 How does your state fund transportation projects? What are the sources of funds? How much is spent? What other interesting information did you find when researching this topic? Limit your response to one page.

14.27 How does your state fund wastewater treatment projects? What are the sources? How much is spent? What other interesting information did you find when researching this topic? Limit your response to one page.

Key Terms

- bonds
- cash flow diagram
- cost of community service
- debt service
- design-bid-build
- design-build
- design-build-operate
- economic equivalence
- equivalence

- future value
- grants
- life-cycle analysis (LCA)
- life-cycle cost
- long-term leases
- matching funds
- net present value
- operations and maintenance concessions

- payback period
- present value
- public–private partnerships (PPP)
- tax incremental financing (TIF)
- time value of money
- user fees
- utility

References

American Automobile Association (AAA). 2008. "Crashes vs. Congestion –What's the Cost to Society?" http://www.aaa newsroom.net/Assets/Files/20083591910.CrashesVsCongestion FullReport2.28.08.pdf, accessed August 29, 2011.

American Council of Engineering Companies (ACEC). 2008. *Design & Construction Industry Trends Survey, 2008-2009.* Washington, DC.

CDM. *Infrastructure Asset Management—A Practical Guide for Utility and Public Works Directors.* http://www.cdm.com/NR/rdonlyres/ 163702E9-1399-45CB-AE71-12574E632247/0/APracticalGuidefor UtilityandPublicWorksDirectors.pdf, accessed September 23, 2010.

City of Chicago. 2010. *City of Chicago 2010–2014 Capital Improvement Program.* http://www.cityofchicago.org/ content/dam/city/depts/obm/supp_info/CapitalImprovement Program2010-2014.pdf, accessed August 29, 2011.

City of Lake Oswego: Finance Web Site. http://www.ci.oswego. or.us/FINANCE/Budget05-07/Message.htm accessed August 29, 2011.

National Atlas of the United States. 2009. Overview of U.S. Freight Railroads. http://www.nationalatlas.gov/articles/transportation/ a_freightrr.html, accessed September 23, 2010.

U.S. Census Bureau. 2010. *The 2010 Statistical Abstract: Construction & Housing: Authorizations, Starts and Completions.* http://www.census.gov/compendia/statab/cats/construction_ housing/authorizations_starts_and_completions.html, accessed September 23, 2010.

chapter Fifteen Environmental Considerations

Learning Objectives

After reading this chapter, you should be able to:

1. Explain the need for environmental protection.
2. Identify potential environmental impacts of infrastructure projects.
3. Compare current and historical engineering considerations of environmental impacts.
4. Describe how environmental regulations "drive" infrastructure engineering.
5. Describe how environmental complexity can lead to unforeseen effects of projects.

Introduction

Components of the natural environment and of the environmental infrastructure were discussed in Chapters 2 and 5 respectively. The need for a systems perspective was emphasized in Chapter 7. Here we will discuss the integration of the built and natural environments into one system, and investigate the potential environmental impacts of developing the civil infrastructure.

Dramatic changes have occurred over the last four decades with respect to scientific understanding, public opinion, and regulation of the environment. Compared to the nearly complete lack of regulations in the 1960s, engineers are now faced with a dizzying array of regulatory requirements at the local, state, and federal levels. These regulations are typically intended to protect the environment at a cost that society can reasonably bear. However, given the vast variability of environmental conditions, situations often arise wherein a required action is expensive but only marginally beneficial; in extreme cases, the action may even be detrimental to the environment.

Introductory Case Study: Wetlands

Wetlands are often perceived as a stumbling block to completing construction projects in a timely and cost-effective manner. Once thought of as wastelands and mosquito breeding grounds, wetlands were extensively filled and/or drained for construction projects. Large-scale filling of wetlands occurred during the building of cities such as Chicago, San Francisco (Figure 15.1), and Boston.

Draining of wetlands is accomplished by lowering the groundwater table using ditches or buried pipes ("drain tiles"), which is common for agricultural purposes, and accounts for the majority of wetland loss. It is estimated that less than one half of the pre-settlement (early 1800s) wetlands exist today nationwide, with some states having lost more than 90 percent of wetlands. Today, wetlands are valued by engineers as vital environmental features (Figure 15.2) and are highly regulated. We will discuss wetlands throughout this chapter, as well as many other environmental components.

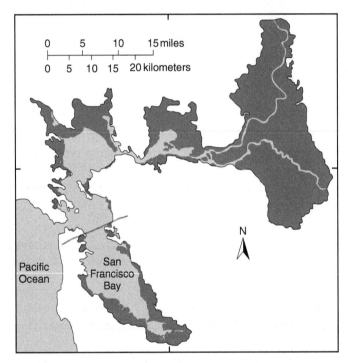

Figure 15.1 Historical Loss or Altering of Wetlands in the San Francisco Bay Region. It is estimated that approximately 95 percent of the original wetland habitat in the region has been lost or altered by filling and levees (dark shaded areas). As much as one-third of the losses were caused by filling for urban development. Much of the remaining wetland areas are protected, and significant restoration efforts are being undertaken.

Source: U.S. Geological Survey, redrawn from Atwater et al., 1977.

Environmental Protection

As highlighted in Chapter 2, Natural Environment, the three major components of the environment are land, air, and water. Within these components lies the supporting habitat for all forms of life.

Figure 15.2 **A Roadside Sign that Informs Drivers of Wetland Benefits.**

Source: M. Penn.

Environmental awareness has increased dramatically in the last several decades, leading to increased public concern over the potential impacts of infrastructure projects. In step with this awareness, the number of environmental regulations has also increased. Most projects cannot begin construction until environmental studies and permit applications have been completed.

Nearly every construction activity or use of the built environment has some impact on the natural environment. Consider the following generic effects on the natural environment due to infrastructure development.

- Loss of habitat—The construction of infrastructure components, unless taking place in a previously developed area (infill), will require land. This land, whether a forest, a pasture, a park, or a riparian (riverbank) area, will previously have been home to a number of plants and animals. The development of the area often irreversibly alters these habitats or may fragment them so that habitats that were once connected are no longer connected (Figure 15.3).

- Decreased **biodiversity**—As land in its natural state is converted to infrastructure components, a loss in biodiversity often accompanies the loss in habitat. A high level of biodiversity is desirable for many reasons. From a biological perspective, a diverse ecosystem is a healthy ecosystem, and is able to withstand and recover from a variety of diseases and extreme weather events. Also, a diverse set of plants and animals is necessary to ensure the productive cycling of nutrients in the environment.

> **Biodiversity** is the diversity of plants and animals within a region. A rainforest or a northern hardwood forest has high biodiversity, while a golf course or cornfield has very low biodiversity.

- Increased contamination of air, water, and soil—Nearly all aspects of the built environment contaminate air, water, or soil. Examples include emissions from coal-fired power plants, the release of contaminants from a leaking landfill liner, emission of CO_2 from automobile tailpipes, or release of zinc from the tires of automobiles.

- Impacts on the natural hydrologic cycle—With many types of development, runoff from rain events increases as a result of the development,

which leads to decreased recharging of groundwater and increased runoff and flooding potential downstream.

Few rational people would argue that these impacts are desirable. However, there is considerable controversy over how widespread the negative impacts are; how harmful to the environment and to humans these impacts actually are; and to what extent the impacts should be regulated by the federal, state, and/or local government.

Case in Point: Benefits of Wetlands

Protection of wetlands leads to many benefits for society. Increased understanding of these benefits has led to rules that regulate filling or draining of wetlands. The following list, modified from a list provided by the Minnesota Department of Natural Resources, provides some benefits of wetlands:

1. **Erosion control.** Wetland vegetation reduces erosion along lakes and stream banks by reducing erosional forces associated with wave action.

2. **Fisheries habitat.** Many species of fish and shellfish utilize wetland habitats for spawning, food sources, or protection.

3. **Flood control.** Wetlands can slow runoff water, minimizing the frequency at which streams and rivers reach catastrophic flood levels. Wetlands also provide storage of flood waters.

4. **Ground water recharge.** Some wetlands serve as a source of groundwater recharge. By detaining surface waters that would otherwise quickly flow to distant lakes or rivers, the water can percolate into the ground and help ensure long-term supplies of ground water.

5. **Natural filter.** By trapping and holding water, wetlands store nutrients and pollutants in the soil and vegetation, allowing cleaner water to flow in to the body of water beyond or below the wetland.

6. **Rare species habitat.** More than 40 percent of threatened or endangered plant and animal species in the United States live in or depend on wetlands.

7. **Recreation/aesthetics.** Wetlands are great places to canoe, hunt, fish, or explore and enjoy nature.

8. **Source of income.** Wetlands provide economic commodities such as cranberries and fish and may add value to residential developments.

9. **Wildlife habitat.** Many animals depend on wetlands for homes and resting spots. Amphibians, reptiles, aquatic insects, and certain mammals need wetlands as a place for their young to be born and grow.

10. **Education.** Wetlands provide ideal locations for classroom ecological studies.

It is important to note that the above listed benefits are *potential* benefits. Individual wetlands may or may not provide these benefits.

Impacts on the environment can be minimized through sustainable[1] design and construction of the infrastructure although few, if any, infrastructure projects can be *completely* free of environmental impacts. In some cases, infrastructure projects can have positive effects on the environment. For example, a wastewater treatment facility may discharge effluent that is cleaner (with respect to some pollutants) than the river to which it releases.

You should note that all impacts on the environment affect, directly and/or indirectly, human welfare. Contaminated air leads to respiratory and other illnesses in humans. Contaminated water ingested by humans can lead to any number of diseases and illnesses. Contaminated surface water may also lead to contaminated food sources such as fish and shellfish. Many possible pathways exist by which contaminants in soil can be ingested by humans: they may be ingested by animals that are eaten by humans, children may ingest the soil directly, contaminants may be taken up by plants that are consumed by people, and contaminants may make their way into the groundwater that serves as a drinking water supply. Polluted runoff leads to reduced recreational opportunities as beaches must be closed to swimming in response to high concentrations of pathogens. Alterations to the hydrologic cycle result in lowered groundwater levels, and thus higher costs to consumers as extracting and treating water becomes more expensive. Loss of habitat leads to loss of recreational opportunities (e.g., hunting, fishing, boating) and reduced biodiversity. There is substantial concern over decreased biodiversity; at the most basic level, it must be understood that humans are supported by the natural environment that is an extremely complex system of chemical, biological, and ecological interactions. A change to any component (e.g., species extinction) can have unforeseen cascading effects on humans.

Debates over the required level of environmental protection are common. As with most considerations surrounding infrastructure projects, the answer is often not black or white, but rather in the gray area between the two extremes. It is likely that you know friends, family members, or neighbors with drastically differing views on the environment. At one extreme are those that believe natural ecosystems are more important than human welfare. At the other extreme are those who believe that the natural environment is to be completely controlled and dominated by humankind and as such, regulations are virtually unnecessary.

The nature of these debates continuously evolves over time. For example, as the population becomes increasingly urbanized (Figure 15.4), there is concern that the majority of Americans are "losing touch" with nature. Also consider

Fish or Jobs?

In the industrialized Fox River valley of Wisconsin, when regulatory efforts to require treatment of industrial wastes before release to the river were first proposed, a common phrase used in opposition to these regulations was "What do you want: fish or jobs?" The implication of this question was that the additional cost of treatment of the industrial wastes would lead to increased production costs and fewer employment opportunities.

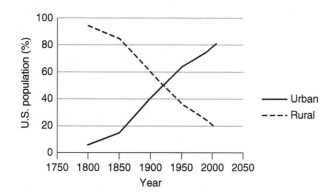

Figure 15.4 **Percentage of U.S. Population Living in Urban and Rural Areas.**

Data source: U.S. Census Bureau.

---

[1] Recall from Chapter 13 (Sustainability Considerations) that sustainable engineering is concerned with minimizing environmental *and* social impacts.

that members of the environmentalist movement (derogatorily referred to as "tree huggers") are increasingly joining forces with conservationists (e.g., hunter associations such as Ducks Unlimited and Pheasants Forever) to protect the natural environment. Additionally, when economic times are good, the environment is of greater concern to more people, while during recessions, environmental concern lessens (Figure 15.5). Additionally, affluent communities are often less tolerant of pollution, and wealthy nations typically have much more stringent environmental standards than developing nations.

When polled about environmental concerns, Americans routinely rank drinking water pollution highest (Figure 15.6). Note that infrastructure projects directly or indirectly impact all of the concerns listed in Figure 15.6.

Figure 15.5 Changing Public Priorities of Two Major Issues. Note the decline in prioritization of environmental protection during the recessions of 2002–2003 and 2008–2010.

Data source: Pew Research Center, 2010.

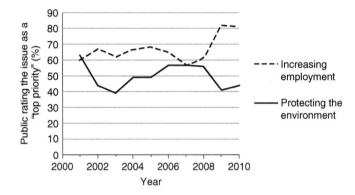

Figure 15.6 Polling Results of Public Environmental Concerns.

Source: Courtesy of Gallup.

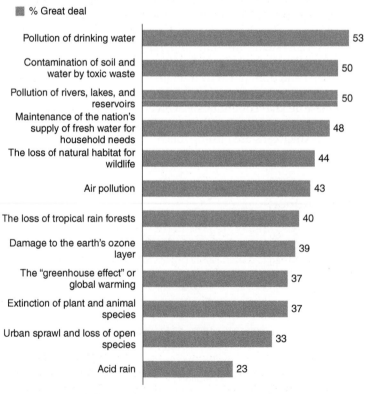

I'm going to read you a list of environmental problems. As I read each one, please tell me if you personally worry about this problem a great deal, a fair amount, only a little, or not at all.

▓ % Great deal

Pollution of drinking water	53
Contamination of soil and water by toxic waste	50
Pollution of rivers, lakes, and reservoirs	50
Maintenance of the nation's supply of fresh water for household needs	48
The loss of natural habitat for wildlife	44
Air pollution	43
The loss of tropical rain forests	40
Damage to the earth's ozone layer	39
The "greenhouse effect" or global warming	37
Extinction of plant and animal species	37
Urban sprawl and loss of open species	33
Acid rain	23

March 6–9, 2008, Gallup Poll

Prior to 1970, the environment was often ignored in engineering decisions. The rapid industrial and population expansion preceding this period led to deplorable environmental conditions. In most cases, discharges of pollutants to the environment were unregulated. Many major rivers did not support healthy fish populations, and in some cases did not support any fish at all. Air quality in many areas significantly impacted human health, and thousands of toxic waste dumps existed across the country. The Clean Water Act and the Clean Air Act (with subsequent amendments) and countless other legislative acts have significantly improved the environment. Today, nearly all discharges to the environment are regulated. However, environmental degradation is rampant in developing countries, much as it was in the United States prior to the 1970s.

Rivers of Fire

Prior to enactment of the Clean Water Act of 1972, many rivers in the United States were "open sewers" containing partially or non-treated municipal wastewater and industrial wastes. River fires were not uncommon. Most often cited is the Cuyahoga River fire of 1969 in Cleveland, Ohio. The event captured international news coverage and became a symbol for poor environmental conditions in the United States. Petroleum slicks and floating debris were the fuel source for the fire. In the 1960s, the river was anaerobic (no oxygen present in the water) and essentially devoid of fish.

Today the Cuyahoga River supports forty species of fish and improved water quality has made the riverfront an attractive place for residential and commercial development. In "The Flats," a popular nightlife district of renovated warehouses, many people are reminded of the past while relaxing after work or on the weekends (Figure 15.7).

Picture of the Cuyahoga River Fire

Figure 15.7 Burning River Pale Ale, a Nod to Local Environmental History.

Source: M. Penn.

Environmental Impacts

It is impossible to detail each and every potential environmental impact, but in this section we will illustrate representative impacts of infrastructure projects on the environment. Rather than classifying air, water, and soil impacts separately, we will highlight potential impacts of common infrastructure development activities.

LAND DEVELOPMENT

Developing land for industrial, commercial, or residential uses directly affects land through the alteration of surface topography, removal of topsoil, increased imperviousness by the addition of rooftops and paved surfaces, and loss of wildlife habitat. Increasingly, suburban development encroaches

Natural Demolition

The extensive damage of hurricanes Katrina and Rita in 2005 resulted in an estimated 50 million cubic yards of waste debris in the state of Louisiana alone, with a 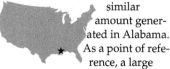 similar amount generated in Alabama. As a point of reference, a large professional sports stadium would hold a volume of approximately 2 million cubic yards. The sheer volume of waste, and its spatial distribution over thousands of square miles, created significant challenges for hauling and disposal, which continued for years after the hurricanes (requiring on the order of 1 million truck hauls). Neighboring states requested the waste (as a source of income) but wood debris could not be exported from the area in an effort to control the spread of a problematic insect, the Formosan termite. An additional challenge came from the fact that conventional landfills are not designed or regulated to accept hazardous materials. As a result, waste debris had to be manually screened for hazardous substances (e.g., medical waste, solvents, pesticides). The costs associated with the disposal of the debris are unknown, but is on the order of billions of dollars.

upon neighboring agricultural land, decreasing land available for production of food. Air quality may be affected by increased vehicle emissions, dust, CO_2 emissions, and industrial air emissions. Water quality may be affected by stormwater runoff carrying particulates, toxins, and nutrients. Nearby surface waters may also be affected by increased flows during storm events, leading to more erosion and flooding. During construction, exposed land can erode and construction vehicles can "track" soil offsite. Additional environmental impacts include light pollution and noise pollution.

BUILDING DEMOLITION

The demolition of buildings and other structures (e.g., sports stadiums) results in the generation of airborne particulates and the potential release of other hazardous compounds such as asbestos and mercury, requiring abatement. If not recycled, large volumes of waste materials are generated and must be landfilled (Figure 15.8).

WASTEWATER TREATMENT

Wastewater treatment facilities (WWTFs) are designed to treat residential, commercial, and industrial wastewater before discharging to the environment. Many of the impacts listed previously for "land development" apply to WWTFs. Additionally, there is a direct impact on the surface water (e.g., a river) to which the treated wastewater is discharged. There are potential thermal impacts, as the temperature of the wastewater is typically higher than that of the receiving body of water. Also, if the wastewater originated as groundwater used for drinking water, the WWTF discharge will increase the flowrate of the river compared to if the river itself was used as the water source.

Today, WWTFs are regulated so that the treated wastewater, when mixed with the receiving water, will cause minimal adverse impacts on humans or wildlife. However, there are many unregulated substances such as pharmaceuticals for which there is increasing concern.

In some cases, WWTFs can affect air quality as well. Some industrial pollutants found in wastewater are volatile (i.e., they have a tendency to migrate from water into the air). If these chemicals are present in high enough

Figure 15.8 Demolition Debris.

Source: M. Penn.

Video of Demolition of Thirty-Story Building

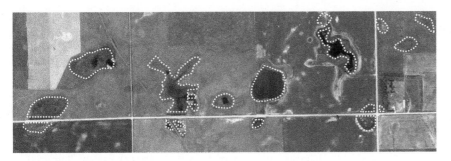

Figure 15.9 **Roadways Crossing Wetlands (Wetlands Outlined).** Prairie potholes (small lakes, ponds and wetlands) are common in eastern South Dakota, North Dakota, and western Minnesota. It is virtually impossible to route roads in this setting without affecting wetlands.

Source: U.S. Geological Survey/National Map.

 Photographs of Highways, Wildlife, and Mitigation Measures

concentrations, they may become a source of air pollution as they volatilize from treatment tanks that are open to the atmosphere.

HIGHWAYS

Highways require land space for travel lanes, shoulders, medians, and interchanges that may take agricultural land out of production or eliminate CO_2 sequestration (uptake) and wildlife habitat from forested land. In order to ensure safe driving conditions in sloped topography, extensive excavation of soil and/or rock is required, altering landforms. The alignment of the road in some cases results in the filling of wetlands (Figure 15.9). As with buildings and parking lots, runoff from the pavement carries pollutants, decreases infiltration of rain into the groundwater, and increases storm runoff. Highways may bisect wildlife habitat and lead to animal (and human) injury and mortality as a result of collisions. Approximately 200 human fatalities and 20,000 human injuries occur annually from wildlife-vehicle collisions. Nearly 40 percent of these collisions occur to people in the 15- to 24-year-old age group.

Air pollution results from vehicle emissions and dust. Noise and light pollution also results from vehicles and roadside lighting. Highways are also corridors for additional residential, commercial, and industrial development leading to a myriad of potential impacts stated above.

DAMS

Dams are typically built for the benefits of hydropower, flood protection, and/or water supply. Dams directly affect the flow of rivers; changing the flow patterns can change river ecology. Dams can act as barriers to fish passage for seasonal migration or spawning. Land inundated by the reservoir formed behind the dam may act as a source of nutrients and other pollutants adversely affecting water quality. Accumulation of sediment, or siltation, occurs upstream of the dam due to lower velocity in the reservoir. Increased siltation also occurs downstream of the dam; with the dam in place, the peak flows are decreased due to the storage capacity of the reservoir, and these less violent flushes do not effectively resuspend and transport sediment. Water temperature increases in the reservoir and downstream of the reservoir during warmer weather, potentially affecting fish and other aquatic species. Increased evaporation occurs due to the increased surface area of the reservoir and increased temperature, leading to the loss of water for downstream users. Nutrients and toxins present in the river will accumulate in the reservoir. Some research suggests that decomposing organic material in the sediments accumulated behind dams is a significant source of methane, a greenhouse gas.

Creative Roadway Lighting

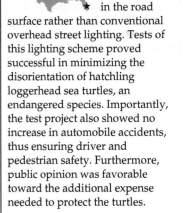

A 1-mile stretch of Highway A1A in Boca Raton, Florida is now lit with LED lights embedded in the road surface rather than conventional overhead street lighting. Tests of this lighting scheme proved successful in minimizing the disorientation of hatchling loggerhead sea turtles, an endangered species. Importantly, the test project also showed no increase in automobile accidents, thus ensuring driver and pedestrian safety. Furthermore, public opinion was favorable toward the additional expense needed to protect the turtles.

In addition to many of the environmental impacts listed for dams, the recently constructed Three Gorges Dam in China (the world's largest concrete structure) is causing massive landslides from eroding banks of the newly inundated land. These landslides are displacing some farmers for a *second* time (those who were moved to higher ground because of the original displacement due to inundation by the reservoir). Some predict that the number of people ultimately displaced because of the dam may double due to landslides.

Additional concerns are based on a phenomenon termed **reservoir induced seismic activity**, whereby dams may cause earthquakes due to the extra water pressure (created by the newly added weight of the water accumulated behind the dam) in cracks in the ground under or near the reservoir. The increased pressure of the water in the cracks may "lubricate" or widen faults which were already under tectonic strain, but were previously prevented from slipping by the friction of the "unlubricated" rock surfaces.

The Good (Bad?) Ole Days

There are countless examples of infrastructure projects from "the good ole days" that have inadvertently had adverse impacts on the environment. A few examples are provided here.

The Everglades in the State of Florida is one of the world's truly unique ecosystems—historically including a massive wetland "river" estimated to be 50 miles wide. More than 50 percent of the original wetlands are no longer in existence, due to conversion to agriculture and urban land. Efforts to protect this resource have strengthened recently, but are often in conflict with increasing development demands for rapidly growing human resident and tourist populations in southern Florida. The population of southern Florida has increased by 25-fold, from 200,000 to over 5 million since 1930, a growth rate approximately ten times that of the United States. In 1947, Everglades National Park was formed and has tripled in size since initial establishment, now encompassing 1.4 million acres.

The Comprehensive Everglades Restoration Plan

In 2000, the U.S. Congress authorized the Comprehensive Everglades Restoration Plan, one of the largest environmental restoration projects in history, with a 30-year time frame and a $10.5 billion budget. According to the Comprehensive Everglades Restoration Plan, "*With no change the region soon will experience frequent water shortages. There will be continued degradation of the Everglades, coastal estuaries, fisheries and other natural resources. Flooding will become more frequent.*"

As part of the restoration plan, 240 miles (400 km) of canals and levees that were installed in the 1940s to control and divert water will be removed. In other words, this project may be described as a complex engineered effort to *undo* previous engineering work that failed to account for environmental impacts. Nearly 2 billion gallons per day of stormwater runoff will be treated to remove nutrients and other contaminants. This runoff will be stored and redirected to more closely resemble pre-settlement (i.e., natural) flow patterns to Florida Bay, a large, shallow and delicate ecosystem between the tip of the mainland peninsula and the Florida Keys (Figure 15.10). More than 200,000 acres of land have been purchased (50 percent of project goal) to control land use and attendant runoff. Measures will be taken to control invasive and exotic species. Habitat improvement through the use of prescribed burns that simulate natural wildfires will be employed.

Another example of undoing previous engineering work is dam removal. Several major dams have been removed and many more removals are under

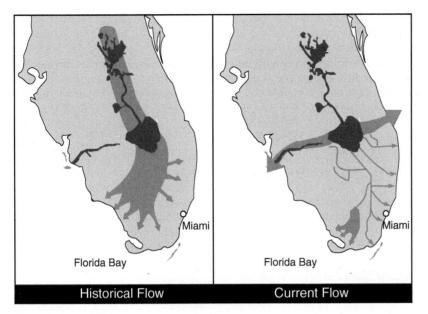

Figure 15.10 Historical and Current Water Flow Through the Everglades.

Source: Redrawn from USACE, 2010.

Historical Flow | Current Flow

Florida Bay | Florida Bay

Miami | Miami

consideration to eliminate negative environmental impacts, especially in the Pacific Northwest where the purpose is to protect native salmon and trout.

In November 2009, an agreement was signed by the U.S. Secretary of the Interior to raze four dams along the Klamath River, and is touted as the largest dam removal project in history. The deal was finalized after a decade of negotiations between Native American tribes, farmers, the fishing industry, the hydroelectric power company, and others. Additional pressure was brought to bear by the governors of Oregon and California and the Bush Administration after a commercial salmon fishery collapse in 2006. The power company that owned the dams was facing costly modifications to the dams to comply with the Endangered Species Act, and estimates suggested that dam removal was the least expensive option. With additional consideration of environmental and social benefits, the decision to remove the dams was made with support of shareholders (stock owners) of the hydroelectric company. A 3-year study will be conducted to evaluate the feasibility of dam removal, and removal is expected to begin in 2020 with an estimated cost range of $200 to $400 million.

The primary benefit of removing these dams will be the restoration of a 300-mile salmon run necessary for spawning and thereby maintaining one of the largest salmon fisheries in the contiguous United States. There is hope that substantial recreational tourism will provide additional economic benefit. Potential adverse impacts of the removal include the increased use of non-renewable electricity generation to replace the renewable power generation lost, higher short-term electricity rates for users, and limited water availability for irrigation.

Environmental Regulations

In 1969, the National Environmental Policy Act (NEPA) was one of the first major environmental policies enacted, largely due to increased public outcry over projects that had historically been undertaken without regard for the environment. NEPA was designed to require the assessment of environmental impacts of federally sponsored projects and, importantly, to

---

The Good Ole Days are Alive and Well

When presenting a design for a roof-top wind-powered electrical generator, a mechanical engineering student was asked "What about the potential impact of this device on birds?" Without hesitation, the student unfortunately responded "Birds? I don't care about birds."

provide opportunities for public input. States now have similar policies for state-funded projects.

An **Environmental Impact Statement (EIS)** is an extensive process that typically requires coordination of several agencies. A common misperception is that an EIS is *only* concerned with the natural environment. In actuality, an EIS also encompasses economic and social considerations, which are discussed in the previous and following chapters, respectively. EIS reports are often massive, typically numbering several hundred pages, and sometimes thousands of pages as in the case of the EIS for "the Big Dig" in Boston. While daunting to students, researching an EIS for a large project can be very enlightening. An **Environmental Assessment (EA)** is a scaled-down version of an EIS, which typically documents an agency's decision to not perform a full-scale EIS because of a "Finding of No Significant Impact" (FONSI). It is very important to note that EIS's do *not* directly lead to the approval or denial of projects; that is, they are not "decision documents." Rather, these reports only *inform* decision-makers and the public of the impacts of alternatives.

The Emergency Wetland Resources Act of 1986 and subsequent regulations have curtailed wetland losses discussed in the introduction of this chapter. Before developing sites, developers must now identify whether or not wetlands exist. Identifying the boundaries of wetland areas is termed "delineating," and is conducted by trained engineers and scientists (Figure 15.11). In many cases, the presence of wetlands is obvious (standing pools of water and cattails), but wetlands include areas with groundwater slightly below the ground surface as well as areas that only periodically flood or retain water. If wetlands are present at a site, several agencies may become involved (e.g., U.S. Army Corps of Engineers, U.S. Fish and Wildlife Service, state environmental protection agency). Regulatory options by these agencies include:

- denial of any filling of the wetland (i.e., the proposed site plan must be altered to avoid any ground disturbance within the delineated area, or in the extreme case, planned development of the site is abandoned);

- approval of filling of the wetland as long as the developer restores a nearby degraded wetland;

- approval of filling of the wetland as long as the developer constructs a new wetland nearby to replace the "lost" wetland; or

- approval of filling of the wetland without either of the two mitigation options listed previously.

Figure 15.11 **A Stake Labeled "Wetland Boundary" Tied with Flagging Tape.** Pink tape is routinely used to delineate the boundary between wetland and non-wetland areas of a construction site.

Source: M. Penn.

Permits and approvals for the second and third bullet items above, termed **mitigation** options, can potentially cause significant delays for construction projects, and must be accounted for in project timelines. In some cases, the presence of wetlands will be a primary factor considered in the decision of whether or not to purchase property for development. There have been many cases of illegal (i.e., unpermitted) development that occurred on wetlands out of ignorance, resulting in costly after-the-fact fines.

Additionally, the *quality* of a wetland is an important factor in regulatory decisions. Some wetlands are of high quality, for example, by virtue of providing important habitat for rare species. Other wetlands may be of minimal environmental importance, such as a degraded wetland that consists of a monoculture of invasive, nuisance plants and that provides minimal flood protection.

Other major legislative actions such as the Clean Water Act, the Clean Air Act, the Water Pollution Control Act, the Comprehensive Environmental Response, Compensation, and Liability Act (Superfund), and the Resource Conservation and Recovery Act have dramatically changed the way that we manage wastes and led to the rapid expansion of the field of environmental engineering over the last several decades. New and pending regulations on carbon dioxide will have a significant impact on the planning, design, and management of infrastructure components and systems.

Environmental Complexity

Ecosystems are complex communities consisting of organisms and the natural environment upon which they depend. As a system, all components must be protected to ensure ecosystem function and health. Historically, environmental managers erroneously looked only for direct impacts on a desired "endpoint" (e.g., a game fish such as walleye). An infrastructure project such as a WWTF, marina, or bridge may not seemingly have a direct impact on game fish. However, if the project adversely affects their food sources (or their food sources' food sources), or their spawning habitat, game fish populations may decrease.

A key element of the previously mentioned Everglades management plan is the Florida panther, a subspecies of mountain lion, which once ranged in eight states over the entire southeast United States, but now has a range reduced to approximately 10 counties in southern Florida. It is considered one of the most endangered large mammal species in the world.

The current population is estimated at approximately 100 animals, and was thought to be as low as 30 in the early 1990s (Figure 15.12). The population decreased dramatically over the last century due to habitat loss, hunting, and bounties. With a small population, limited genetic variability also threatens the panther. Conversion of woodlands to agricultural grazing lands is the primary reason for habitat loss. While loss of habitat is most significant, degradation and fragmentation of remaining habitat also pose major threats to the panther's future. Urban sprawl has also contributed to habitat loss.

In 1958, the State of Florida declared the panther an endangered species. It is now federally listed as a *critically threatened species*. The preferred prey for the panther is whitetail deer, the population of which has also declined, for similar and other reasons. In areas with low deer populations, raccoons are increasingly important to the panther diet. However, raccoons feed on fish and crayfish, and as a result have elevated mercury content through **bioaccumulation** (Figure 15.13). This figure illustrates that the concentration

The Changing Basis for Decisions

The U.S. Army Corps of Engineers was asked to undertake a study of flood diversion for the Waccamaw River, which passes through

Conway, South Carolina, after Hurricane Floyd caused damage in 1999. The proposed project involves constructing a canal to divert excess floodwaters away from Conway. The project had previously been studied in 1930, 1951, and 1965 and all studies concluded the canal to be infeasible. However, Conway hired the Corps of Engineers to look at the feasibility one more time. Again, the canal was found to be cost-prohibitive. According to Corps Spokesperson Glenn Jeffries, "The idea is getting a little worse each time we look at it, because we're looking at more environmental impacts and impacts on other areas as well that weren't necessarily a concern in the 1930s."

Horry County, South Carolina Concept Study of a Flood Reduction Diversion Canal

Bioaccumulation is the process whereby contaminants in the environment are taken in by plants or animals and accumulate in higher members of the food chain.

Figure 15.12 A Roadside Sign in the Everglades Indicating the Population of Florida Panthers in the Late 1990s. Note the vandalism of the sign in which the vandals used spraypaint to change the number remaining from 30 to 29.

Source: M. Penn.

of mercury in walleye is 1 million times greater than the concentration of mercury in phytoplankton.

Increased mercury levels in panthers may be linked to decreased health and reproductive success, but studies have not yet proved or disproved this hypothesis. This scenario linking the terrestrial and aquatic food chains demonstrates ecological complexity.

We also must be concerned with complexities of the *physical and chemical cycling* of pollutants in the environment. A pollutant discharged to the air may not remain there; it can end up in surface waters, in the fish that live there, and in the humans that eat the fish (recall the discussion on mercury in Chapter 12, Energy Considerations). Pollutants disposed on or in the ground may seep into groundwater and potentially contaminate drinking water wells or rivers and lakes. There are countless other examples.

In some cases, chemicals end up in places we wouldn't necessarily expect. For example, PBDEs (polybrominated diphenylethers) are flame retardants used on fabrics and furniture. Prior to recent phase-outs of their use, concentrations of PBDEs were increasing worldwide in human breast milk, seals, fish, and in virtually anything that has been analyzed for PBDEs. You might wonder how this contaminant is getting in the environment. Most likely, it is not in the environment due to direct releases by industry, but rather by emissions from solid waste incinerators, unregulated ("backyard") trash burning, and by "dust" (fragments of fabrics from wear and tear). Like mercury (Figure 15.14), PBDEs accumulate in fish to thousands of times the concentration in the water. Moreover, there is uncertainty as to what the potential health impacts are from eating these fish, or for infants drinking breast milk, or for any other pathway. The government or PBDE industry never conducted studies to determine the health effects because people were not assumed to be eating their furniture and no one foresaw the other pathways by which PDBEs would enter the natural environment. While regulated in the European

Figure 15.13 Bioaccumulation of Mercury in the Onondaga Lake (Syracuse, New York) Aquatic Food Chain. The levels in walleye (a highly desirable fish for consumption) are more than double the federal mercury concentration limit for safe consumption of 1 μg/g. Relative size of plankton are exaggerated as shown.

Source: Modified from Mihelcic, 1999. Reprinted with permission of John Wiley & Sons, Inc.

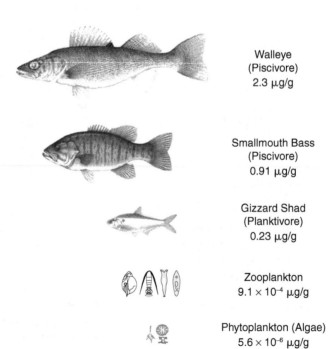

Walleye
(Piscivore)
2.3 μg/g

Smallmouth Bass
(Piscivore)
0.91 μg/g

Gizzard Shad
(Planktivore)
0.23 μg/g

Zooplankton
9.1×10^{-4} μg/g

Phytoplankton (Algae)
5.6×10^{-6} μg/g

Union, most manufacturers in the United States have voluntarily chosen to use alternative substances for reducing fire risk in light of pending regulations at the federal or state level.

Figure 15.14 A Fish Consumption Advisory Sign Warning Anglers of Elevated Mercury Levels in Fish in Rural South Dakota.

Source: Courtesy of J. Stone.

Transportation Impacts on Panthers

Vehicle collisions have been one of the most significant sources of Florida panther mortality. Interstate 75, which bisects the Everglades region, was designed in some stretches with wildlife underpasses and fencing to reduce opportunities for panthers to cross over the highway. However with increased traffic, increased panther population, and only partial fencing (Figure 15.15), collision deaths (Figure 15.16) have increased and represent a loss of *more than 10 percent of the entire population*. Between 2007 and 2009, deaths due to collisions averaged 14 per year as compared to about 1 per year from 1978 to 1998.

Figure 15.15 An Unfenced Portion of I-75 in the Everglades.

Source: M. Penn.

Figure 15.16 A Florida Panther Killed by a Vehicle Collision.

Source: Courtesy of the Florida Fish and Wildlife Conservation Commission/D. Land.

Outro

Development of our infrastructure affects the natural environment in many ways. Efforts to minimize environmental impacts are often driven solely by regulation, but arguably should be driven by sound economic, scientific, social, and ethical considerations.

Environmental aspects of infrastructure projects are quite controversial, scientifically, economically, and politically. However, our understanding of the environmental impacts of infrastructure projects is continually improving due to the establishment of long-term data sets, improved monitoring, and increased scientific knowledge of environmental complexities. These all provide the basis for better decisions. In future classes, you will learn methods to minimize the environmental impacts discussed in this chapter, while providing needed local, regional, national, or international infrastructure improvements.

chapter Fifteen Homework Problems

Calculation

15.1 Using Figure 15.1, approximate the area (in square miles) of wetlands filled in the San Francisco Bay area.

15.2 Using Figure 15.13, how much greater is the mercury concentration in walleye compared to gizzard shad?

Short Answer

15.3 Why have cities such as San Francisco historically filled in their wetlands?

15.4 How does Figure 15.3 illustrate "habitat fragmentation?"

15.5 In what ways does land development affect the environment?

15.6 How will removal of dams on the Klamath River:
a. help the environment;
b. hurt the environment;
c. benefit society;
d. be detrimental to society?

15.7 How does bioaccumulation affect humans?

15.8 As noted in this chapter, many aquatic species have high mercury concentrations. What is the primary source of this mercury?

15.9 The State of Minnesota recently had a public education campaign stating "Eat smaller fish." What is the basis for this advice?

15.10 Summarize the threats to the Florida panther.

Discussion/Open-Ended

15.11 Do you agree with the statement, "that the majority of Americans are 'losing touch' with nature"? If this statement is true, what are some implications for environmental decisions related to the infrastructure?

Research

15.12 Explain how infrastructure projects affect each environmental impact listed in Figure 15.6. Limit your answer to two to three sentences for each impact.

15.13 During the 1960s and 1970s, Lake Erie was known as a "dead lake." Why? What measures were taken such that it is no longer considered "dead?" Limit your response to one page.

15.14 How was the solid waste from the World Trade Center, destroyed by the 9/11 terrorist attacks, disposed of? Describe the process by which it was handled.

15.15 Research a dam removal project not discussed in this chapter. Summarize the reasons for removal and public opinion toward the project. Limit your response to one page.

15.16 Locate a wetland of your choice (or of your instructor's choice) and summarize the environmental benefits, giving specific examples of benefits.

15.17 What other issues besides environmental protection become of lesser concern to the general public during economic recessions? Cite your references.

Key Terms

- bioaccumulation
- biodiversity
- environmental assessment (EA)
- environmental impact statement (EIS)
- mitigation
- reservoir induced seismic activity

References

Atwater, B. F., S. G. Conard, J. N. Dowden, C. W. Hedel, R. L. MacDonald, and W. Savage. 1977. "History, Landforms, and Vegetation of the Estuary's Tidal Marshes." In: *San Francisco Bay: The Urbanized Estuary. Investigations into the Natural History of San Francisco Bay and Delta With Reference to the Influence of Man*, pp. 347–386. Pacific Division of the American Association for the Advancement of Science. http://www.estuaryarchive.org/cgi/viewcontent.cgi?article=1031&context=archive, accessed August 29, 2011.

Mihelcic, J. R. 1999. *Fundamentals of Environmental Engineering.* Hoboken, NJ: John Wiley & Sons.

Minnesota Department of Natural Resources. *Benefits of Wetlands.* http://www.dnr.state.mn.us/wetlands/benefits.html, accessed August 29, 2011.

Pew Research Center. 2010. "Public's Priorities for 2010: Economy, Jobs, Terrorism." The Pew Research Center for the People and the Press. http://people-press.org/report/584/policy-priorities-2010, accessed August 29, 2011.

U.S. Army Corps of Engineers (USACE). 2010. "Water Flow Maps of the Everglades: Past, Present, Future." http://www.evergladesplan.org/education/flowmaps.html, accessed August 29, 2011.

U.S. Geological Survey. *Coastal Wetlands and Sediments of the San Francisco Bay System: USGS Fact Sheet.* http://pubs.usgs.gov/fs/coastal-wetlands/, accessed August 29, 2011.

chapter Sixteen Social Considerations

Learning Objectives

After reading this chapter, you should be able to:

1. Explain the relationship between the infrastructure and human health.
2. Describe how infrastructure projects affect neighborhood cohesion.
3. Define environmental justice.
4. Define social impact assessment.
5. Explain the importance of considering the social impacts of infrastructure projects.

Introduction

Building, maintaining, and modifying infrastructure affects society in a number of ways, both positively and negatively. For example, in Chapter 11, Planning Considerations, we discussed the Big Dig project in Boston; one of the societal impacts was that 20,000 people were displaced as a result of the construction of the Central Artery in the 1950s. The social impacts of this displacement were many, as homes, businesses, neighborhoods, livelihoods, and ways of living were completely disrupted. On the other hand, the Central Artery project helped people by decreasing travel time (although only in its first few years of operation). The fact that the detrimental impacts were more numerous than the benefits is typical of what happens when infrastructure is planned and built without properly considering social implications.

Today, the planning and design of large infrastructure projects typically include many **stakeholders** through public workshops, informational meetings, and advisory groups. This increased incorporation of community input over the last few decades represents a dramatic improvement to the process of planning and constructing infrastructure projects.

Infrastructure affects society not only when a project is completed, but also during construction. As such, society may be adversely affected in many ways, including household relocation, business relocation, reduced neighborhood cohesiveness, deterioration in lifestyle, human health impacts, and reduced access to facilities and services. On the other hand, beneficial social effects may include improved access to facilities and services, including jobs, health care, and education; decreased travel times; improved public health; and associated economic benefits that "trickle down" to individual members of society. Note that some aspects of the social environment may be both positively and negatively affected simultaneously (e.g., human health).

The term **stakeholder** is commonly used for those that have a stake in the outcome of the project. Stakeholders include landowners, businesses, environmental advocacy groups, regulatory personnel, and elected officials.

Introductory Case Study: Three Gorges Dam

The Three Gorges Dam (Figure 16.1), recently constructed in China, is a project that is similar in scale and cost to the Big Dig project, but with greater societal impacts. This dam was constructed on the Yangtze River and is not only the largest dam ever built (approximately 1.5 miles wide and 600 feet tall), it is the largest concrete structure in the world (over 36,000,000 cubic yards of concrete). The dam was completed in 2003 at an approximate cost of $20 billion, and after several years of water accumulation behind the dam, the created reservoir flooded 250 square miles of land, as illustrated in Figure 16.2. The hydroelectric power generation, at approximately 20,000 MW (megawatts), is nearly ten times the production of the Hoover Dam and nearly 10 percent of China's electricity needs.

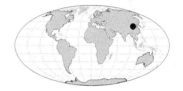

The primary purposes of the Three Gorges Dam are flood control and hydroelectric power generation. The dam provides more than 700 billion cubic feet of flood storage capacity, which can be released slowly from the dam to prevent downstream flooding for a 100-year event and to reduce flooding from larger than 100-year events. This represents a dramatic improvement over the 10-year event protection that existed

Figure 16.1   The Three Gorges Dam.
Source: Prill Mediendesign & Fotographie/iStockphoto.

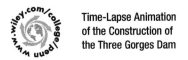
before the dam was constructed. The Yangtze River (the third longest river in the world) is prone to flooding; a flood in 1998 killed more than 2,000 people and left millions of people homeless.

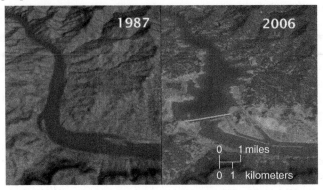

Figure 16.2   Reservoir Formed By the Three Gorges Dam.

Source: NASA/Goddard Space Flight Center Scientific Visualization Studio and U.S. Geological Survey.

The construction of the dam led to many other benefits. River navigation is greatly improved, as a series of locks are also part of the project (Figure 16.3). These locks allow ocean-going vessels to travel a total of 1,500 miles inland to the city of Chongqing (approximately 300 miles upstream of the dam), and thereby open the region to economic growth. Tourism and recreational opportunities have also been created. Given the scale of the project, its completion is a tremendous source of pride to this developing country.

However, many negative social impacts are associated with many dam construction projects, and are magnified on projects of the scale of the Three Gorges Dam. Over 1 million people from 13 cities, 140 towns, and 1,350 villages were displaced from the land inundated by the reservoir; this is 60 times more people than were displaced by Boston's Central Artery project of the 1950s. Additionally, thousands of archaeological and religious sites have been inundated. After completion of the dam and subsequent reservoir formation, landslides from erosion of newly saturated soils began to occur. As a result, the Chinese government has estimated that in the future, *millions* of additional people may have to be relocated (Yardley, 2007).

Figure 16.3   Lower Portion of the Locks of the Three Gorges Dam.   Parallel locks provide separate two-way travel (left = upstream, right = downstream). Each of the five "steps" of the parallel locks (only two of which are shown in the photo) provide approximately 120 feet of elevation change.

Source: M. Penn.

The environmental impacts of dams presented in Chapter 15, Environmental Considerations, are also predicted to occur at the Three Gorges dam. An additional environmental concern for this case is the flooding of pre-existing industrial sites that were highly contaminated. Also, there is concern that wastewater receiving partial or no treatment will enter and accumulate in the reservoir (and potentially impact wildlife) rather than being flushed to sea. An estimated $8 billion is being spent to try to mitigate the water quality impacts. Extinction of several species is also a possibility.

Welcoming the Challenge

Engineers have long been stereotyped as being overly focused on technical issues. The need for engineers that are holistic thinkers has never been greater, and this need will increase in the future. However, many engineers shy away from political and social considerations because of unfamiliarity, associated uncertainties, and the difficulty in quantifying these considerations. Yet, most engineers like to solve problems and relish challenges; thus it is ironic that they would not embrace social and political issues surrounding infrastructure projects. Valero and Vesilind, in *Socially Responsible Engineering*, state:

"We are not naïve in thinking that all engineering students and practicing engineers will simply embrace what some may perceive to be "soft" subject matter. In fact, we believe that the approaches applied to date in raising awareness of societal issues have been woefully inadequate ... In particular, they seem to treat issues like justice as sidebars where the engineer is asked to suspend engineering realities to think about social issues. This approach is unfortunate and ignores the essence of engineering since issues like ethics and social justice are even more mathematically challenging than "typical" engineering problems ... Compared to many design problems, social issues frequently have more variables, exhibit initial and boundary conditions that are extremely difficult to define, require sophisticated mathematical approaches, call for creative optimization schemes, and hardly ever have a singular solution."

They further state:

"The face of engineering is changing. We must remain highly competent in our application of mathematics and the natural sciences, but engineering is more than that. The engineer today and in coming decades must adapt to a changing world ... An essential part of an engineer's social contract is that we be trusted as professionals. Such trust must be based on both sound science and the appropriate societal application of that science."

Social Impact Assessment

Engineering projects should not be judged on their economic, political, or engineering/technical aspects alone, but also must be judged according to their environmental impacts and social impacts. As discussed in Chapter 15, Environmental Considerations, an environmental impact statement/assessment analyzes not only the potential environmental effects a project might have, but also its potential societal and economic impacts. However, the emphasis of these assessments is typically environmental. A social impact assessment (SIA) attempts to analyze the effects a project might have on society. Quantifying environmental impacts can be quite challenging, and being able to assess the relative significance of those impacts is often difficult. But SIAs can be many times more challenging to complete, given the greater difficulty in quantifying social impacts as compared to environmental impacts. Perhaps as a result of this difficulty, procedures for SIAs are not standardized as compared to EIAs, and are not commonly carried out in North America. However, we are discussing it here to help provide a framework to the discussion of social impacts, and believe that SIAs will become more common in the future.

The Interorganizational Committee on Principles and Guidelines for Social Impact Assessment (ICPGSIA, 2003) offers this definition of social impacts:

"By social impacts we mean the consequences to human populations of any public or private actions that alter the ways in which people live, work, play, relate to one another, organize to meet their needs and generally cope as members of society. The term also includes cultural impacts involving changes to the norms, values, and beliefs that guide and rationalize their cognition of themselves and their society."

Displacement

In the process of bringing energy to industries, providing irrigation to farm fields, and widening roads around cities, land development is required. That land is often inhabited by people. Thus, an infrastructure project may require displacing people from their homes and business owners from their businesses. This displacement is completely non-voluntary on the part of the residents and businesses: they are literally forced to move from their homes.

Case in Point: Pruitt-Igoe Public Housing

The Pruitt-Igoe housing development in St. Louis, Missouri, was a modernist monument, emblematic of advances in fair housing and progress in the war on poverty. Regrettably, Pruitt-Igoe has become an icon of failure of imagination, especially imagination that accounts properly for the human condition.

Although we think of such public projects in terms of housing, they also often represent elements of environmental justice. Contemporary understanding of environmental quality is often associated with physical, chemical, and biological contaminants, but in the formative years of the environmental movement, aesthetics and other "quality of life" considerations were essential parts of environmental quality. Most environmental impact statements have addressed cultural and social factors in determining whether a federal project would have a significant effect on the environment. These factors have included historic preservation, economics, psychology (e.g., perception of open space, green areas, and crowding), aesthetics, urban renewal, and the *land ethnic*. In his famous essays, posthumously published as *A Sand County Almanac*, Aldo Leopold argued for a holistic approach: "A thing is right when it tends to preserve the integrity, stability and beauty of the biotic community. It is wrong when it tends otherwise."

The land ethic was widely disseminated about a decade after the Pruitt-Igoe project was built, so the designers did not benefit from the insights of Leopold and his contemporaries. However, the problems that led to the premature demolition of this costly housing experiment may have been anticipated intuitively if the designers had taken the time to understand what people expected. Then we must ask who was to blame. There is plenty of culpability to go around. Some blame the inability of the modern architectural style to create livable environments for people living in poverty, largely because they "are not the nuanced and sophisticated 'readers' of architectural space that the educated architects were." This is a telling observation and an important lesson for engineers. We need to make sure that the use and operation of whatever is designed is sufficiently understood by those living with it.

Other sources of failure have been proposed. Design incompatibility was almost inevitable for high-rise buildings and for families with children. However, most large cities have large populations of families with children living in such environments. In fact, population density was not the problem since St. Louis had successful luxury town homes not too far from Pruitt-Igoe. Another identified culprit was the generalized discrimination and segregation of the era. Actually, when originally inhabited, the Pruit section was for blacks and Igoe was for whites.

Costs always become a factor. The building contractors' bids were increased to a level where the project construction costs in St. Louis exceeded the national average by 60%. The response to the local housing authority's refusal to raise unit cost ceilings to accomodate the elevated bids was to reduce room sizes, eliminate amenities, and raise densities. As originally designed, the buildings were to become "vertical neighborhoods" with nearby playgrounds, open-air hallways, porches, laundries, and storage areas. The compromises eliminated these features; and some of the amenities removal led to dangerous situations. Elevators were undersized and stopped only every third floor and lighting in the stairwells was inadequate. So, another lesson must be to know the difference between desirable and essential design elements. Human elements essential to a vibrant community were eliminated without much accommodation.

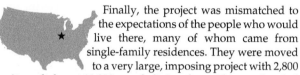 Finally, the project was mismatched to the expectations of the people who would live there, many of whom came from single-family residences. They were moved to a very large, imposing project with 2,800 units and almost 11,000 people living there. This quadrupled the size of the next largest project at the time.

When the failure of the project became overwhelmingly clear, the only reasonable decision was to demolish it, and the spectacular implosion shown in Figure 16.4 became a lesson in failure for planners, architects, and engineers. In the designer

Figure 16.4 **Demolition of Pruitt-Igoe Housing Development Buildings in St. Louis, Missouri.** The development was completed in 1956 and began to be torn down in 1972. (From O. Newman, *Creating Defensible Space*, U.S. Department of Housing and Urban Development, Washington, DC, 1996.

Minoru Yamasaki's own words: "I never thought people were that destructive. As an architect, I doubt if I would think about it now. I suppose we should have quit the job. It's a job I wish I hadn't done."

(Source: *Socially Responsible Engineering* by Vallero and Vesilind. Reprinted with permission of John Wiley & Sons, Inc.)

Sociologists have studied the effect of dislocation on people, and have found that although some people benefit from the move, many of the dislocated people suffer from a variety of ills. Old friends and neighbors are separated. Depending on the distance of the move, a new job might need to be obtained, which was certainly the case for many associated with the Three Gorges Dam. Also, the psychological effects associated with losing land and the disruption to daily lifestyles are significant. Additionally, when displaced people move, they create new demands for housing, jobs, and other resources in the areas to which they relocate. Thus social impacts are not restricted to only those who were displaced.

Neighborhood Cohesion

Sometimes, civil and environmental engineers are called "social" engineers, because so much of their work directly affects how people interact. This can be observed in the design of parks and playgrounds or in new multi-use (residential and commercial) developments, but also in the more mundane decisions regarding the width of a sidewalk, inclusion of pedestrian-friendly crosswalks, selection of plantings and landscaping, etc.

Several aspects of modern residential land development directly affect social interactions. Consider the following examples.

- Homes are typically automobile-centered, such that a garage door is the prominent architectural feature of the portion of the home facing the road.

Figure 16.5 A Typical Modern Suburban Home. Note the contrast from traditional design in Figure 16.8.

Source: M. Penn.

The garage doors have consequently supplanted the front porch as the focal point for the front of house (Figure 16.5), which leads to decreased opportunity for social interactions. Many people living in such homes will admit that they never use the front door.

- Sidewalks are at times placed on only one side of the street as a cost-saving measure.

- The prevalence of cul-de-sacs reduces **connectivity**.

- Roads are often wider than may be necessary; the large widths allow for plenty of on-street parking and navigation by fire trucks and by snow plows in colder climates. However, the wide roads may lead to increased traffic speeds (and decreased pedestrian and bicyclist safety).

- Building lots are large enough that each family can have its own private outdoor space, which precludes the need for shared public open space.

Connectivity is the presence of easily navigated transportation networks within and between neighborhoods; Figure 16.6.

Covenants are restrictions that control some of the activities and the appearance of homes within a subdivision. Examples include that home color must be selected from a predetermined list, laundry cannot be dried on clotheslines, or that pets cannot be kept outside.

The market for homeowners who desire more social interactions in their neighborhoods appears to be increasing. To meet this need, developers have incorporated some best practices. For example, **covenants** can be included to ensure that homeowners include a front porch on their homes and that all garage doors open to the side of the lot; or alleys can be added such that garages are located behind homes. Shared open space can be incorporated into the design that can greatly facilitate social interactions. Road width

Figure 16.6 A Subdivision Near Salt Lake City, Utah, with Limited Pedestrian Connectivity.

Source: U.S. Geological Survey.

Headwater Streets

The headwaters of a river are the most upstream reaches of the river. Headwaters are relatively narrow given the small flowrate of water conveyed. For example, Figure 16.7 shows the headwater of the Mississippi River in Itasca State Park in Minnesota. The river can be safely waded across at this point and is only 20 feet wide. A headwater stream increases in width as the flowrate increases; for example, the Mississippi River is more than 2,000 feet wide at Dubuque, Iowa, and is nearly 3,000 feet wide in Baton Rouge, Louisiana.

Some road designers have made an analogy between headwater streams and "headwater streets." Just as in rivers where the flowrate of water and the width are related, the analogy suggests that the width of roadways should be related to the number of vehicles that the road carries. That is, streets that convey minimal traffic should be designed to be narrower than collectors and arterials. Although the concept of headwater streets seems relatively straightforward, many existing headwater streets such as dead ends are unnecessarily wide, as the code requirements for many cities are relatively inflexible. Additionally, many of the methods by which roads can be made narrower (e.g. providing narrower travel lanes, allowing only one-way traffic, providing parking on only one side of the street, or to not provide parking at all) are not desired by many municipalities which seek extensive on-street parking and are reluctant to allow one-way operation.

Figure 16.7 Headwaters of the Mississippi River.

Source: Courtesy of Visit Bemidji.

can be decreased, which is attractive to developers as narrower roads represent a significant cost savings as compared to the conventional wide roads. However, municipalities are often reluctant to allow narrower roads.

Middleton Hills, near Madison, Wisconsin, has implemented many of these practices. The neighborhood sidewalks are quite popular with residents. Each home (e.g., Figure 16.8) is required to have a front porch, the elevation of which must be at least 2 feet above the sidewalk elevation, thus facilitating conversations between people sitting on the porch and those walking past.

Figure 16.8 Street View of Middleton Hills, Wisconsin. Note that garage doors are not visible from the street.

Source: P. Parker.

Health

One of the greatest societal impacts of basic infrastructure is the positive effect on human health. Nearly 30 years have been added to life expectancy in the United States since 1900 (see Figure 13.6 in Chapter 13). This increase can be attributed to many factors, but William Wulf, past president of the National Academy of Engineering, has proposed that much of the increase can be attributed to the work of civil and environmental engineers through the provision of wastewater treatment and drinking water treatment.

A disparity in life expectancy exists in the world today (see Figure 16.9). Low life expectancies are directly related to many factors, but the primary factors are availability of clean drinking water, proper sanitation, and availability of (and access to) medical facilities. Each of these is directly affected by the presence and quality of basic infrastructure. That is, improved infrastructure will allow for more extensive treatment of wastewater and drinking water, as well as provide for more modern health care in rural areas. Indeed, a shaded map showing the amount of money spent per capita on public infrastructure would most like look very similar to Figure 16.9.

However, the practices undertaken in building infrastructure also may have deleterious effects on public health. For example, consider the following ways in which infrastructure design and planning decisions affect public health.

- Many land-use decisions in the United States have led to urban sprawl, the unplanned, low-density method of developing residential land. This type of land use and the lifestyle it requires is very dependent on the automobile. Yet, on-road vehicles are the leading source of the following air pollutants: carbon monoxide, volatile organic compounds, and nitrogen oxides. The latter two pollutants are significant contributors to the formation of ozone (urban smog). Poor air quality is directly linked to a number of human respiratory illnesses. Therefore, it is not an exaggeration to say that infrastructure planning decisions can directly affect human respiratory health.

- Depending on how the built environment is designed, it can either promote or discourage physical activity. There are many factors that can encourage out-of-doors physical activity in urban areas: the presence and width of sidewalks, the interconnectedness of roads and sidewalks; availability of bicycle lanes; aesthetics and noise levels of outdoor places; extent of

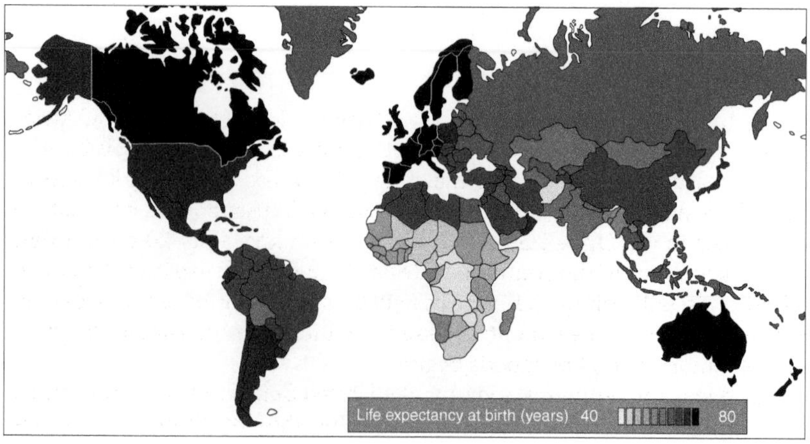

Figure 16.9 **Global Life Expectancy.**

Source: World Bank.

Life expectancy at birth (years) 40 ‖‖‖‖‖‖ 80

Figure 16.10 **Example of Curb Extension.**

Source: Courtesy of Richard Drdul.

parks; real and perceived safety. The decisions surrounding these factors are often made by engineers, or made by non-engineers with significant input from engineers. A pedestrian- and bicycle-friendly infrastructure is very important, as statistics show a significant decline in out-of-doors physical activity over the past few decades and a decrease in number of children walking to school. These findings are correlated with an alarming increase in childhood obesity. Certainly, there are many reasons for these changes, and although a correlation between the amount of physical activity and the "walkability" of neighborhoods exist, engineers and other decision-makers must be cautious to infer causation. However, there is no doubt that engineers can design the built environment to encourage outdoor activity.

- More than 10 percent of all traffic fatalities are pedestrians. Many factors are involved in these accidents, with the most significant being pedestrians and/or drivers with high blood alcohol content. Yet, many design decisions can help reduce automobile/pedestrian accidents. These include setting lower speed limits; implementing traffic calming devices, such as speed bumps, curb extensions (Figure 16.10), pedestrian islands, etc.; providing sidewalks and walking paths; or providing overpasses or tunnels for crossing particularly busy intersections.

Environmental Justice

Based on the previous section and on Chapter 13, Sustainability Considerations, we can conclude that the extent of infrastructure and the quality of human health are related, and that there is not an equal distribution of infrastructure across the world. Such an unequal distribution has historically occurred in the United States. In the late 19th and early 20th centuries, many municipalities in the southern states did not provide African-American neighborhoods with municipal services such as drinking water or access to centralized wastewater treatment; such services literally ended at the "border" where the minority neighborhoods began.

Similar disparities still exist today in the United States today. For example, consider the city of Modesto, California. Within the city limits are a series

The U.S. Environmental Protection Agency (EPA, 2010) states that "**environmental justice** is the fair treatment and meaningful involvement of all people regardless of race, color, national origin, or income with respect to the development, implementation, and enforcement of environmental laws, regulations, and policies. The EPA has this goal for all communities and persons across this Nation. It will be achieved when everyone enjoys the same degree of protection from environmental and health hazards and equal access to the decision-making process to have a healthy environment in which to live, learn, and work."

of "islands"—unincorporated areas that do not have access to city services, including water, sewer, and emergency services. When originally built, these areas were located outside the city limits. However, they are referred to as islands because areas surrounding these islands have been annexed into the city of Modesto. The city has refused to annex the islands because they do not have the required services, and the city will not provide the money to add those services. Unlike newly built subdivisions, the islands do not have centralized sanitary sewer collection and treatment. Other recently developed areas continue to be annexed into the city because they have the needed infrastructure in place—this infrastructure is paid for by the developer who in turn passes the cost on to the homeowners. These new areas are predominantly white and relatively affluent (able to afford new housing), while residents of the islands are more economically impoverished and predominantly Latino.

Thus, within the boundaries of a single city, the more affluent neighborhoods have city services for water and sewer while the minority islands have to rely on wells and septic tanks. Moreover, these islands have substandard (or non-existent) sidewalks, poorer quality roads, increased flooding due to a lack of storm drainage, etc. This same scenario repeats itself across the United States in cases where money is diverted to new development, and therefore diverted away from older, poorer neighborhoods. Clearly, a difference in the quality of services exists, and this was supported by a court case concerning the Modesto example; however, the court did not conclude that the difference was due to discrimination. That is, the lower quality of infrastructure in the islands was not the result of discriminatory actions by the city of Modesto.

The discussion on the islands of Modesto and the sidebar on colonias illustrate an important fact: poorer neighborhoods often have poorer quality infrastructure (possibly resulting in greater health risks) as compared to wealthier neighborhoods. This type of disparity has led to a relatively new field of study termed **environmental justice**.

Colonias

Colonias are unincorporated neighborhoods located in Texas along its border with Mexico. More than 400,000 people live in more than 2,000 colonias, the vast majority of which are Latinos with very low household incomes. These neighborhoods lack many infrastructure services, and are served by outdated septic systems, or in some cases, by no wastewater treatment process at all. Often, potable water infrastructure is not available either; in certain instances where water distribution is available, homes are not allowed to hook up to the system because the homes are so dilapidated they do not meet county building codes (Figure 16.11).

Figure 16.11 An Example of a Colonia Residence.

Source: Courtesy of A. Barud-Zubillaga.

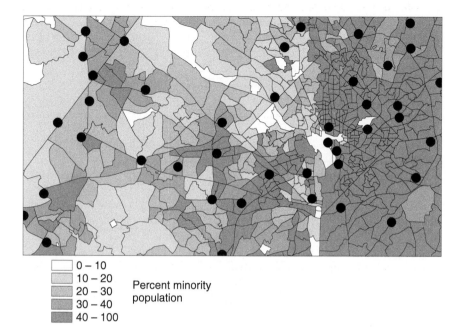

Figure 16.12 Toxic Release Sources (Points) in Arlington County, Virginia.

Source: U.S. EPA EJView.

	0 – 10
	10 – 20
	20 – 30
	30 – 40
	40 – 100

Percent minority population

A variety of environmental justice studies have been performed by researchers investigating links between aviation noise and race, income level, and education level; between air quality and economic status; between the occurrence of toxic releases and race and income level; between the location of facilities for treatment, storage, and disposal of hazardous wastes (TSDFs) and minority population; etc. For example, consider Figure 16.12, which shows the relationship between toxic releases (as categorized by the EPA) in Arlington County and the density of minorities. Note how many of the toxic releases occur in areas with high minority populations.

Some might argue that the preceding examples are discriminatory acts, in which polluters are taking advantage of people with less political influence. Others argue that it is not a matter of discrimination, but a matter of economics; that is, less desirable land uses (e.g., industries, waste storage areas, power plants) are located in areas with lower property values because the cost of development is less. As such, this illustrates the need to consider the triple bottom line when making decisions rather than the single bottom line of economics.

As a result of a 1994 Presidential Executive Order, every federal agency (e.g., Federal Highway Administration, EPA, branches of the military) has made environmental justice part of its mission by identifying and addressing the effects of all programs, policies, and activities on "minority populations and low-income populations." Moreover, environmental justice analyses are now common for large infrastructure projects, and often consider social factors (e.g., property buyouts, noise, traffic) as well. These analyses aim to identify whether or not minority or low-income populations are disproportionately negatively affected by a proposed project. In some cases, additional populations are considered, such as the physically handicapped, those primarily using public transportation, etc. If impacts are identified, these populations are provided opportunities to participate in decision-making to feasibly mitigate the problem. An environmental justice decision flowchart is presented in Figure 16.13 to demonstrate the process of determining adverse impact.

EJView

The U.S. Environmental Protection Agency has a map utility named EJView that allows users to create maps based on the geographic areas and data sets they choose. Data sets include demographics, health information, toxic releases, etc. Figure 16.12 was created using EJView.

EJView

Environmental Justice Transportation Case Studies

Figure 16.13 Environmental Justice Flowchart.

Source: Adapted from the Air Force, 1997.

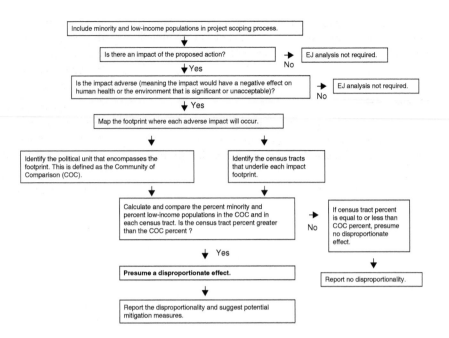

Outro

The following chapter (Chapter 17, Ethical Considerations) considers the relationship of ethics to engineering. We also contend that considering social impacts is an ethical consideration. For example, the assessments described in this chapter (i.e., environmental justice assessment, social impact assessment) are often seen as "additional" steps to be taken in building and managing the infrastructure. To some people, perhaps those of the unaffected majority, these steps may seem like a "waste of time," especially in cases when extensive analysis shows no disproportionate impact. But these analyses are indeed critical to ensuring that social needs are addressed, and that adverse project impacts and the provision of new infrastructure services are distributed equitably. Although these analyses may be required by law, we believe they should be performed voluntarily in light of professional ethics. Moreover, ensuring equity adds to the public trust of the engineering profession.

chapter Sixteen Homework Problems

Short Answer

16.1   Define environmental justice.

16.2   Compare and contrast social impact assessment to environmental impact assessment.

Discussion/Open-Ended

16.3   How does infrastructure positively affect public health? What are some potential negative public health impacts?

16.4   Do you think that environmental justice assessments should be done prior to building *any* infrastructure project? Explain.

16.5   Illegal aliens from Mexico are subjected to high doses of pesticides working in farm fields. Is this an example of environmental injustice? Why or why not?

16.6   Residents living under flight paths from airports are exposed to significant noise pollution. Often, these flight paths pass over minority communities at a higher frequency than they do over non-minority communities. What might be some reasons for this difference?

16.7 Read the paper entitled "Environmental Injustice: Case Studies from the South," available at www.wiley .com/college/penn. Write a one-page reflection on this paper.

Research

16.8 For one of the traffic calming methods listed (as chosen by your instructor), explain how it works; list its benefits; list its drawbacks. Cite all information sources.

a. bump-outs
b. traffic circle
c. speed bumps
d. raised crosswalks
e. neckdowns
f. center islands
g. chicanes

16.9 Read one of the case studies of your (or your instructor's) choice from the FHWA's Environmental Justice website (also available at www.wiley.com/college/penn) and write a one-page reflection.

16.10 Write a one-page paper summarizing the infrastructure (or lack thereof) in the Colonias.

Key Terms

- connectivity
- covenants
- displacement
- environmental justice
- neighborhood cohesion
- social impact assessment
- stakeholders

References

Department of Air Force. 1997. *Guide for Environmental Justice Analysis with the Environmental Impact Analysis Process.* http:// www.afcee.af.mil/shared/media/document/AFD-070830-060.pdf, accessed August 29, 2011.

Environmental Protection Agency. 2010. *Environmental Justice: Compliance and Enforcement.* http://www.epa.gov/environmental justice/, accessed August 29, 2011.

Interorganizational Committee on Principles and Guidelines for Social Impact Assessment (ICPGSIA). 2003. "Principles and Guidelines for Social Impact Assessment in the USA." *Impact Assessment and Project Appraisal*, 21(3): 231–250.

Valero, D. A. and P. A. Vesilind. 2007. *Socially Responsible Engineering: Justice in Risk Management.* Hoboken, NJ: John Wiley & Sons.

World Bank. *Indicators, Life Expectancy at Birth, Total (Years).* http://data.worldbank.org/indicator/SP.DYN.LE00.IN/countries? display=map, accessed August 29, 2011.

Yardley, J. 2007. "Chinese Dam Projects Criticized for Their Human Costs." *The New York Times.* http://www.nytimes.com/ 2007/11/19/world/asia/19dam.html?pagewanted=1&_r=2, accessed August 29, 2011.

chapter Seventeen Ethical Considerations

Learning Objectives

After reading this chapter, you should be able to:

1. Define engineering ethics.
2. Apply the ASCE code of ethics to ethical decisions.
3. Compare and contrast ethical situations that students face to situations faced by practicing engineers.
4. Summarize, from memory, the seven Fundamental Canons of the ASCE Code of Ethics.
5. Define corruption and list various types of corruption.
6. Describe the negative effects that corruption has on society.

Introduction

The study of engineering ethics is the study of "moral values, issues, and decisions involved in engineering practice" (Martin and Schinzinger, 2004). In even more simple terms, engineering ethics is knowing what you ought to do, and doing it.

Engineers are among the most respected and trusted members of society. This claim is based upon numerous public opinion polls that rate engineers as much more trusted than many other professions. This trust and respect has been earned over many years, thanks not only to the profession's contributions to increased quality of life, but also to the ethical behavior of the vast majority of practicing engineers.

For many of the ethical choices with which you are faced, the separation between right and wrong will be quite distinct; that is, the decision will be a matter of choosing between black and white. Of course, this does not infer that the choice will be an easy choice for you. For example, you may be inspecting a job site and be asked to "look the other way" as the contractor uses substandard materials or methods. Or, you may fail to save a spreadsheet of collected data, and to cover up your mistake, you might be tempted to fabricate data. In each of these cases, the ethical decision is relatively easy to identify but not necessarily easy to implement.

On the other hand, for many ethical decisions, it is much more difficult to differentiate between right and wrong and you will be faced with making choices in the "gray" area. For example, perhaps you own some unimproved land and are involved with an infrastructure construction project that will potentially bring utilities (and therefore add value) to your land—what should you do about this potential conflict of interest? The introductory case study also presents a situation in which the right thing to do is not immediately obvious.

Is this chapter a waste of time? In other words, is it possible to "teach" students to be ethical? Instructors cannot make students act ethically; indeed, some people simply do not respect or observe the norms of ethical behavior. Rather, the goal of this chapter is to ensure that you know what ethical behavior "looks" like and where to find guidance to help you make ethical decisions. Then, we hope you will choose to act ethically in your future work.

Introductory Case Study: Political Contributions[1]

Engineers from a variety of engineering firms were invited to participate in a fundraising barbeque held by the deputy secretary of the state Department of Transportation (DOT) for the governor's re-election campaign. Some of the engineers were designated as "sponsors" because they gave $250, $500, or $1,000 at the fundraiser. Among the individuals invited were a number of engineers who work for engineering companies that do business with the state DOT. When questioned, the engineers indicated that their contributions would have no bearing on their firm's selection to do business with the state. At the time of the fundraiser, some of the firms were in the process of negotiating contracts with the state DOT. The deputy secretary of the state DOT has no role in approving engineering contracts. This situation begs the question: was it *ethical* for the engineers to participate in the fundraiser?

Ethics and Engineering

You should note that many of the ethical challenges faced by practicing engineers are very similar to the challenges faced by engineering students. Table 17.1 compares a few of those challenges. Many more examples exist; indeed, most likely every ethical decision with which a student is confronted has a corresponding ethical decision in the "real world."

Nearly every unethical decision can be justified by the person carrying out the decision. And just as the unethical decisions are similar for student and practicing engineer, so are the motivations and justifications for

> ### Ethics in Higher Education
> As a student, you expect your university professors and administrators to act ethically. In October 2009, the Chancellor, the President, and several trustees of the University of Illinois Champaign-Urbana resigned amidst a scandal involving the special admission of students connected to university trustees, lawmakers, and other influential businesspeople.

[1] This case study is from the National Society of Professional Engineers (NSPE) Board of Ethical Review Case No. 06–12. It is reprinted here with permission from NSPE.

the unethical choice. Table 17.2 provides motivations for unethical behavior and illustrates how they might apply to both practicing engineers and students. The reasons (or perhaps excuses?) provided in Table 17.2, can be justified in a number of ways. Common justifications include "no one will find out about it," "everyone else is doing it," or "I'll only do it once."

Table 17.1

Comparison of Unethical Behavior of Students and Practicing Engineers

Unethical Behavior by a Student	Corresponding Unethical Behavior by a Professional
Copying a classmate's homework assignment	Copying and pasting a diagram from the Internet into a report without receiving permission or citing the source
Collaborating with a classmate on a take-home exam even though such collaboration is forbidden	Participating in bid rigging, in which bidders conspire among each other to "take turns" winning bids, thus removing competition from the bidding process and increasing cost to the project owner
Altering data obtained from a lab exercise, thereby ensuring that results appear to be as expected and that the lab work will not have to be repeated	Falsifying data in any number of ways; for example, overbilling a client for the amount of excavation that occurred or not constructing a below-ground utility properly and literally "covering up" the mistake

Table 17.2

Examples of Motivations for Unethical Behavior

Motivation for Unethical Behavior	Example from Engineering Education	Example from Engineering Practice
Competition	In some instances, classes are graded on a curve; this may mean that only a certain number of A's or B's are awarded	Consulting engineering is a very competitive career, and practicing engineers are faced with the ultimate "curve": if a proposal or bid gets accepted, the engineer receives an A; otherwise, he or she in essence receives an F
Pressure to perform	Teammates on a group project may create a large amount of pressure, and seemingly the only way to meet their demands is by cheating	A supervisor may set high criteria for success (and therefore future salary increases), which results in the need (real or perceived) to resort to unethical behavior
Economic dependence	Students are faced with the possibility that if they do not pass a particular class, their graduation will be delayed and therefore they will have to pay for additional tuition. Additionally, if they fail to complete their degree, the economic consequence is not being able to secure an engineering job	In some organizational cultures, the inability to "win" jobs on a competitive basis can lead to termination of employment
Ignorance of what constitutes ethical behavior	A student that copies a photograph from the Internet may not realize that not citing the source is a form of plagiarism	Perhaps because of poor training, a new employee might not realize it is unethical to accept a gift of a certain value from a vendor

Codes of Ethics

In deciding what type of behavior is ethical, engineers can rely on codes of ethics for guidance. Codes of ethics are provided by nearly every professional engineering society. In this chapter we have included the American Society of Civil Engineers (ASCE) Code of Ethics. This code contains four Fundamental Principles and seven Fundamental Canons. You should become very familiar with the Fundamental Canons. The majority of the code is devoted to providing guidance on each of the canons.

Code of Ethics[a]

Fundamental Principles[b]

Engineers uphold and advance the integrity, honor and dignity of the engineering profession by:

1. using their knowledge and skill for the enhancement of human welfare and the environment;
2. being honest and impartial and serving with fidelity the public, their employers, and clients;
3. striving to increase the competence and prestige of the engineering profession; and
4. supporting the professional and technical societies of their disciplines.

Fundamental Canons

1. Engineers shall hold paramount the safety, health and welfare of the public and shall strive to comply with the principles of sustainable development[c] in the performance of their professional duties.
2. Engineers shall perform services only in areas of their competence.
3. Engineers shall issue public statements only in an objective and truthful manner.
4. Engineers shall act in professional matters for each employer or client as faithful agents or trustees, and shall avoid conflicts of interest.
5. Engineers shall build their professional reputation on the merit of their services and shall not compete unfairly with others.
6. Engineers shall act in such a manner as to uphold and enhance the honor, integrity, and dignity of the engineering profession and shall act with zero-tolerance for bribery, fraud, and corruption.
7. Engineers shall continue their professional development throughout their careers, and shall provide opportunities for the professional development of those engineers under their supervision.

Guidelines to Practice Under the Fundamental Canons of Ethics

CANON 1

Engineers shall hold paramount the safety, health, and welfare of the public and shall strive to comply with the principles of sustainable development in the performance of their professional duties.

a. Engineers shall recognize that the lives, safety, health, and welfare of the general public are dependent upon engineering judgments, decisions and practices incorporated into structures, machines, products, processes and devices.

b. Engineers shall approve or seal only those design documents, reviewed or prepared by them, which are determined to be safe for public health and welfare in conformity with accepted engineering standards.

c. Engineers whose professional judgment is overruled under circumstances where the safety, health, and welfare of the public are endangered, or the principles of sustainable development ignored, shall inform their clients or employers of the possible consequences.

d. Engineers who have knowledge or reason to believe that another person or firm may be in violation of any of the provisions of Canon 1 shall present such information to the proper authority in writing and shall cooperate with the proper authority in furnishing such further information or assistance as may be required.

e. Engineers should seek opportunities to be of constructive service in civic affairs and work for the advancement of the safety, health, and well-being of their communities, and the protection of the environment through the practice of sustainable development.

f. Engineers should be committed to improving the environment by adherence to the principles of sustainable development so as to enhance the quality of life of the general public.

CANON 2
Engineers shall perform services only in areas of their competence.

a. Engineers shall undertake to perform engineering assignments only when qualified by education or experience in the technical field of engineering involved.

b. Engineers may accept an assignment requiring education or experience outside of their own fields of competence, provided their services are restricted to those phases of the project in which they are qualified. All other phases of such project shall be performed by qualified associates, consultants, or employees.

c. Engineers shall not affix their signatures or seals to any engineering plan or document dealing with subject matter in which they lack competence by virtue of education or experience or to any such plan or document not reviewed or prepared under their supervisory control.

CANON 3
Engineers shall issue public statements only in an objective and truthful manner.

a. Engineers should endeavor to extend the public knowledge of engineering and sustainable development, and shall not participate in the dissemination of untrue, unfair, or exaggerated statements regarding engineering.

b. Engineers shall be objective and truthful in professional reports, statements, or testimony. They shall include all relevant and pertinent information in such reports, statements, or testimony.

c. Engineers, when serving as expert witnesses, shall express an engineering opinion only when it is founded upon adequate knowledge of the facts, upon a background of technical competence, and upon honest conviction.

d. Engineers shall issue no statements, criticisms, or arguments on engineering matters which are inspired or paid for by interested parties, unless they indicate on whose behalf the statements are made.

e. Engineers shall be dignified and modest in explaining their work and merit, and will avoid any act tending to promote their own interests at the expense of the integrity, honor, and dignity of the profession.

CANON 4
Engineers shall act in professional matters for each employer or client as faithful agents or trustees, and shall avoid conflicts of interest.

a. Engineers shall avoid all known or potential conflicts of interest with their employers or clients and shall promptly inform their employers or clients of any business association, interests, or circumstances which could influence their judgment or the quality of their services.

b. Engineers shall not accept compensation from more than one party for services on the same project, or for services pertaining to the same project, unless the circumstances are fully disclosed to and agreed to, by all interested parties.

c. Engineers shall not solicit or accept gratuities, directly or indirectly, from contractors, their agents, or other parties dealing with their clients or employers in connection with work for which they are responsible.

d. Engineers in public service as members, advisors, or employees of a governmental body or department shall not participate in considerations or actions with respect to services solicited or provided by them or their organization in private or public engineering practice.

e. Engineers shall advise their employers or clients when, as a result of their studies, they believe a project will not be successful.

f. Engineers shall not use confidential information coming to them in the course of their assignments as a means of making personal profit if such action is adverse to the interests of their clients, employers, or the public.

g. Engineers shall not accept professional employment outside of their regular work or interest without the knowledge of their employers.

CANON 5

Engineers shall build their professional reputation on the merit of their services and shall not compete unfairly with others.

a. Engineers shall not give, solicit, or receive either directly or indirectly, any political contribution, gratuity, or unlawful consideration in order to secure work, exclusive of securing salaried positions through employment agencies.

b. Engineers should negotiate contracts for professional services fairly and on the basis of demonstrated competence and qualifications for the type of professional service required.

c. Engineers may request, propose, or accept professional commissions on a contingent basis only under circumstances in which their professional judgments would not be compromised.

d. Engineers shall not falsify or permit misrepresentation of their academic or professional qualifications or experience.

e. Engineers shall give proper credit for engineering work to those to whom credit is due, and shall recognize the proprietary interests of others. Whenever possible, they shall name the person or persons who may be responsible for designs, inventions, writings, or other accomplishments.

f. Engineers may advertise professional services in a way that does not contain misleading language or is in any other manner derogatory to the dignity of the profession. Examples of permissible advertising are as follows:

- Professional cards in recognized, dignified publications, and listings in rosters or directories published by responsible organizations, provided that the cards or listings are consistent in size and content and are in a section of the publication regularly devoted to such professional cards.

- Brochures which factually describe experience, facilities, personnel, and capacity to render service, providing they are not misleading with respect to the engineer's participation in projects described.

- Display advertising in recognized dignified business and professional publications, providing it is factual and is not misleading with respect to the engineer's extent of participation in projects described.

- A statement of the engineers' names or the name of the firm and statement of the type of service posted on projects for which they render services.

- Preparation or authorization of descriptive articles for the lay or technical press, which are factual and dignified. Such articles shall not imply anything more than direct participation in the project described.

- Permission by engineers for their names to be used in commercial advertisements, such as may be published by contractors, material suppliers, etc., only by means of a modest, dignified notation acknowledging the engineers' participation in the project described. Such permission shall not include public endorsement of proprietary products.

g. Engineers shall not maliciously or falsely, directly or indirectly, injure the professional reputation, prospects, practice, or employment of another engineer or indiscriminately criticize another's work.

h. Engineers shall not use equipment, supplies, laboratory, or office facilities of their employers to carry on outside private practice without the consent of their employers.

CANON 6

Engineers shall act in such a manner as to uphold and enhance the honor, integrity, and dignity of the engineering profession and shall act with zero tolerance for bribery, fraud, and corruption.

Code of Ethics (Continued)

a. Engineers shall not knowingly engage in business or professional practices of a fraudulent, dishonest, or unethical nature.

b. Engineers shall be scrupulously honest in their control and spending of monies, and promote effective use of resources through open, honest, and impartial service with fidelity to the public, employers, associates, and clients.

c. Engineers shall act with zero-tolerance for bribery, fraud, and corruption in all engineering or construction activities in which they are engaged.

d. Engineers should be especially vigilant to maintain appropriate ethical behavior where payments of gratuities or bribes are institutionalized practices.

e. Engineers should strive for transparency in the procurement and execution of projects. Transparency includes disclosure of names, addresses, purposes, and fees or commissions paid for all agents facilitating projects.

f. Engineers should encourage the use of certifications specifying zero tolerance for bribery, fraud, and corruption in all contracts.

CANON 7

Engineers shall continue their professional development throughout their careers, and shall provide opportunities for the professional development of those engineers under their supervision.

a. Engineers should keep current in their specialty fields by engaging in professional practice, participating in continuing education courses, reading in the technical literature, and attending professional meetings and seminars.

b. Engineers should encourage their engineering employees to become registered at the earliest possible date.

c. Engineers should encourage engineering employees to attend and present papers at professional and technical society meetings.

d. Engineers shall uphold the principle of mutually satisfying relationships between employers and employees with respect to terms of employment including professional grade descriptions, salary ranges, and fringe benefits.

[a]The Society's Code of Ethics was adopted on September 2, 1914 and was most recently amended on July 23, 2006. Pursuant to the Society's Bylaws, it is the duty of every Society member to report promptly to the Committee on Professional Conduct any observed violation of the Code of Ethics.

[b]In April 1975, the ASCE Board of Direction adopted the fundamental principles of the Code of Ethics of Engineers as accepted by the Accreditation Board for Engineering and Technology, Inc. (ABET).

[c]In November 1996, the ASCE Board of Direction adopted the following definition of Sustainable Development: "Sustainable Development is the challenge of meeting human needs for natural resources, industrial products, energy, food, transportation, shelter, and effective waste management while conserving and protecting environmental quality and the natural resource base essential for future development."

Ethical Decision-Making Framework

When faced with an ethical decision, you should consider the following questions, which are based on the American Society of Civil Engineer's PLUS (Policies, Legal, Universal, Self) decision-making process:

• Does the action hold paramount the safety, health, and welfare of the public?

• Does the action serve the best interests of the client?

- Is the action consistent with the ASCE code of ethics?
- Is the action compliant with the letter *and the spirit* of applicable laws and regulations?
- Does the action satisfy your own personal definition of right, good, and just? Will you be able to sleep at night with your decision?

You should not make ethical decisions in isolation, and should share your concerns with trusted co-workers and friends. Be discrete in these discussions. Professional societies often have an ethics board or a help line to which you can ask questions in anonymity.

If your decision leads to having to report an activity or situation, use the proper chain of command when doing so. For example, if you are certain that a co-worker has falsified data on an environmental permit application, your first place to report this probably is not the state environmental protection agency, the local newspaper, or your Facebook or Twitter account. Rather, you may want to first discuss this unethical behavior with your direct supervisor.

In some (rare) instances, the ethical decision you are faced with may involve keeping your job or not. But walking away and "washing your hands" of the situation is not always the ethical choice either. In some cases, the ethical choice is walking away and at the same time reporting the unethical behavior to appropriate authorities. Few people want to be labeled a "whistle-blower," and such actions are rare in engineering and should be considered a last resort; however, whistle-blowing may be the *only* ethical decision to make in certain situations.

Case Study

Case studies based on real ethical decisions can be very helpful in highlighting the fact that the ethical decisions associated with real situations are, rather than being black or white, often "gray." Ethical decisions are full of competing variables, and ethical case studies help you appreciate this. For example, the engineer has to be an "agent" or trustee to the client, yet must abide by all regulations, while at the same time not forgetting that public welfare must be held paramount. The remainder of this section is the continuation of the Introductory Case Study and illustrates that some interpretation of the code is necessary in real situations. This case study refers to the National Society of Professional Engineers (NSPE) Code of Ethics, which is very similar to the ASCE Code of Ethics.

Ethical and related issues involving political contributions by individual engineers and firms has been an issue that the engineering profession has

Student Cheating

Some research has shown that engineering students are more prone to cheating than students pursuing degrees in other disciplines. This is likely due to many factors, the most significant of which are the demanding workload associated with engineering curricula and the competitive culture in some undergraduate engineering programs. In one study, 96 percent of engineering students reported that they had cheated at least once in one form or another (Carpenter et al., 2006). Forms of cheating included copying from someone else's homework, allowing one's homework to be copied, copying an old term paper, etc. These results are very disturbing, as other research has shown that cheating in school carries over to cheating in the workplace. That is, students that cheat throughout their undergraduate studies are more likely to behave unethically in the workplace. If such carryover occurs in engineering such that 96 percent of engineers behave unethically, the effect on the reputation of the profession will be disastrous.

Whistle-Blowing in Reality

The possibility of whistle-blowing is a very disconcerting proposition for practicing engineers. Consider the following situation, based on the reality in which a young engineer found himself. This engineer was working on a project that was running over budget. Rather than bill his hours to that project, his employer repeatedly required him to bill the hours to a *different* project that was under budget. As a result, work on the overbudget project was billed to a different client. This young engineer was faced with a dilemma; it was clear that this unethical practice was supported by company management, and that questioning the practice might lead to termination of employment. Ultimately, rather than reporting the activity to authorities (the "correct" but difficult choice of action), the engineer chose to find employment with another firm.

addressed and continues to address. The issue has a variety of facets, including the well-acknowledged need for professional engineers to become actively involved in the political process to affect legislative, legal, and regulatory policy. This need is contrasted with the perception of individuals and companies making political contributions with the expectation of, or at least in anticipation of, favorable consideration for future public engineering work. Clearly, this is a complicated issue with no easy solution.

Over the years, NSPE has studied the issue in significant depth and taken a position to encourage all engineers to support political candidates who have demonstrated through their activities a commitment to ethical professional practices. NSPE cautions, however, that consistent with the NSPE Code of Ethics, that it is unprofessional for engineers, either on their own behalf or on behalf of their firms or employers, to make political contributions in the form of either cash or services in a manner intended to influence the award and administration of contracts involving a public authority, or which may have the appearance of influencing the award and administration of contracts involving a public authority. Therefore, consistent with its values, goals, and Code, NSPE:[2]

1. Encourages, endorses, and supports the enactment of public disclosure laws, which identify political contributions to federal, state, and local candidates.

2. Endorses the enactment of laws and rules, administered by state ethics and election commissions and professional and trade licensing boards, which are intended to assist candidates for public office and professionals in avoiding ethical and legal conflicts relating to political contributions.

3. Endorses the establishment of state ethics and election commissions to monitor these laws in those states where such commissions do not currently exist, and advocates that state ethics and election commissions and licensing boards for all professions and trades be empowered to establish rules and limits for contributions.

Furthermore, NSPE endorses that these boards and commissions be empowered to establish penalties and take appropriate enforcement action against parties that fail to follow the requirements of the laws and rules.

It would appear that a $250 contribution or contributions in the range of $250 are well within the definition of nominal contributions to a state gubernatorial campaign and, therefore, clearly fall within what is acceptable under the language of the Code. Notwithstanding the facts and circumstances involved, it is difficult, if not impossible, to believe that a political contribution of this magnitude could create any expectation or other basis to affect the selection or receipt of future state or other engineering work. The board's view might be altered if there was some suggestion in the facts that there was a coordinated effort to have a large number of individuals from one firm make individual contributions to a candidate for public office, resulting in a large contribution from what would be perceived as a special commercial interest. However, those facts do not appear to be present in this case.

In conclusion, it was ethical for the engineers to participate in the fundraiser, given that (1) the contribution amounts were not excessive, (2) there was no linkage between the contributions and the selection of the firms for state work, and (3) the contributions are publicly disclosed.

[2]Excerpted from NSPE Professional Policy No. 146—Political Contributions.

International Corporate Ethics

The World Economic Forum publishes a biannual Global Competiveness Report. As part of the report, a survey is given to over 10,000 corporate executives. The table to the right summarizes the results from the following question, "How would you compare the corporate ethics (ethical behavior in interactions with public officials, politicians, and other enterprises) of firms in your country with those of other countries in the world?" (highest ranking is most ethical):

World Economic Forum Global Competitiveness Report

Rank	Country	Rank	Country
1	New Zealand	14	Germany
2	Sweden	15	United Kingdom
3	Finland	16	Hong Kong
4	Denmark	17	Ireland
5	Singapore	18	Barbados
6	Switzerland	19	United Arab Emirates
7	Netherlands	20	Belgium
8	Norway	21	Chile
9	Iceland	**22**	**United States**
10	Canada	23	Oman
11	Australia	24	Japan
12	Luxembourg	25	France
13	Austria		

Corruption

According to Transparency International (an international society dedicated to stopping global corruption), corruption is the "misuse of entrusted power for private gain." Corruption includes **bribery**, **extortion**, **fraud**, **collusion**, and **money laundering**.

One of the industries worldwide in which corruption is most common is the construction industry, which is intertwined with engineering practice. A report entitled "Corruption Prevention in the Engineering and Construction Industry" provides reasons why the construction and engineering industry is at high risk for corruption (PricewaterhouseCoopers, 2009). The reasons include:

- large projects (multi-year and multi-million dollar)

- long supply chains (many companies involved)

- interaction with government officials

- work through third-party intermediaries

- work in expanding markets (developing nations) that do not have appropriate anti-corruption regulatory frameworks and cultures.

Ethical case studies often focus on the gray areas as mentioned previously. But in reality, a significant amount of corrupt activities are carried out by individuals who are well aware that their actions clearly violate every code of ethics as well as some laws. For example, consider this description of a type of corruption known as "price fixing," from the Anti-Corruption Training Manual (Stansbury and Stansbury, 2008).

Transparency International Webpage

Bribery is making payment to people in authority as a means of influencing their decisions. The payment may be in cash, but also may consist of a vacation or other gifts.

Extortion is a type of blackmail, where threats are made unless demands are met. The threats may include release of embarrassing information, a refusal to grant permits, or physical harm. Demands are often for payments. Yielding to such threats may make you guilty of bribery.

Fraud is simply deception. It includes concealment of defects or fabricating or falsifying evidence.

Collusion occurs when two parties cooperate to deceive another party. Bid rigging is an example of collusion.

Money laundering is the conversion of money from an illegally-obtained source (e.g. from fraud) into a form that appears to be legitimate.

"A group of contractors who routinely compete in the same market secretly agree to share the market between them. They will each apparently compete on all major bids, but will in advance secretly agree which of them should win each bid. The contractor who is chosen by the other contractors to win a bid will then notify the other contractors prior to bid submission as to its (the pre-selected contractor's) bid price. The other contractors will then bid at a higher price so as to ensure that the pre-selected contractor wins the bid. The winning contractor would therefore be able to achieve a higher price than if there had been genuine competition for the project. If sufficient projects are awarded, each contractor would have an opportunity to be awarded a project at a higher price. This arrangement is kept confidential from the project owners on respective projects, who believe that the bids are taking place in genuine open competition, and that they are achieving the best available price. The project owners therefore pay more for their projects than they would have done had there been genuine competition."

Corruption Perception Index

Transparency International publishes an annual Corruption Perception Index (CPI), an index that ranks countries according to the extent to which corruption is perceived to exist among public officials and politicians.

The CPI is based on a scale of 0 to 10; the lower end of the scale indicates a perception of high level corruption, while the high end represents a perception of low level corruption. The CPI results have shown over the years that the incidence of corruption is higher in Africa and in war-ravaged countries. New Zealand is perceived to be the least corrupt while the United States ranks 19th in the 2009 rankings.

Corruption Perception Index

Unfortunately, in some countries, engineers do not uphold an ethical code. A lack of an ethical code combined with corrupt governments results in widespread corruption. This corruption can take many forms, but as Transparency International points out, regardless of which form it takes, it always results in projects that are unnecessary, unreliable, dangerous, and/or over-priced. This can lead to loss of life, enhance poverty, and negatively affect the economy.

Transparency International estimates that 10 percent of all construction activity worldwide is lost to corruption. Given a worldwide construction activity valued at $5 trillion, the losses add up to a staggering $500 billion annually; this money could potentially fund schools, hospitals, roads, water projects, etc. ASCE has taken a strong stand against corruption, and the ASCE code of ethics Fundamental Canon 6 now reads: "Engineers shall act in such a manner as to uphold and enhance the honor, integrity, and dignity of the engineering profession and shall act with *zero-tolerance for bribery, fraud, and corruption*" (emphasis added).

ASCE's Combating Corruption in Engineering and Construction Charter

ASCE has created the following "Engineer's Charter."

We, the undersigned, as leaders in the global engineering community, recognize that corruption of all forms diverts resources from projects intended to raise living standards, threatens sustainable development, impoverishes communities, and tarnishes the reputation of the profession. We hereby join in the battle against bribery, fraud, and corruption in engineering and construction worldwide. We acknowledge, as fundamental principles of professional conduct that engineers as individuals must:

- Ensure that they are not personally involved in any activity that will permit the abuse of power for private gain.

- Recognize that corruption occurs within the public and private sectors, in the procurement and execution of projects, and among employers and employees.

- Refuse to condone or ignore corruption, bribery, or extortion; or payments for favors.

- Urge professional engineering societies and institutions to adopt and publish transparent, enforceable guidelines for ethical professional conduct.

- Enforce anti-corruption guidelines by reporting infractions by any participant in the engineering and construction process.

Further, we pledge to support the formal adoption of these principles by our professional organizations; build professional and public support for zero tolerance for bribery, fraud and corruption; seek transparency in all dealings with public officials and private owners; and coordinate our efforts with the work of Transparency International, the Partnership Against Corruption Initiative of the World Economic Forum, the World Bank, World Federation of Engineering Organizations, FIDIC (International Federation of Consulting Engineers), and other local or global organizations seeking the same goal.

Thankfully, the widespread culture of corruption that exists in many developing countries does not exist in the United States. As Transparency International points out, "People are as corrupt as the system allows them to be. It is where temptation meets permissiveness that corruption takes root on a wide scale." The "system" in the United States and many other developed countries is such that clear codes of ethical conduct are part of the engineering culture and meaningful penalties exist for unethical behavior (e.g., having one's Professional Engineer's [PE] license revoked or being sentenced to jail time).

This is not to say that corruption does not occur in the United States and other developed nations. New York City, Chicago, and New Orleans are but three of many cities notorious for such activity, often centered around organized crime in construction, trucking, and solid waste management. Note, however, that organized crime's success is often dependent on corrupt local officials and law enforcement personnel.

Corruption also occurs at the corporate level. Price-fixing schemes are often uncovered. In 2005, executives from two construction companies in Wisconsin pleaded guilty to bid-rigging over $100 million in state road contracts. This sent shock waves through the industry in a state that prides itself on its "squeaky clean" reputation. In October of 2009, an investigation began into price-fixing in Montreal, Canada, after a media report that construction costs may be as much as 35 percent higher than competitive "fair market" prices. In September 2009, the Office of Fair Trading in London imposed $200 million (U.S. dollar equivalent) in fines to over 100 construction firms in England that colluded on building contracts. In 2008, the FBI indicted over 60 individuals from the Mafia, construction companies, and labor unions in New York City for extortion in several large construction projects (amongst other charges such as money laundering and illegal gambling). These, and similar acts, serve to lessen the public trust of engineers and the construction industry. It is particularly maddening to discover that elevated prices are paid during periods of economic downturn when limited funding is available.

Public Perception of Corruption in New Orleans

New Orleans has a notoriously corrupt local political and legal system. As a former editor of the major New Orleans newspaper put it, "Corruption in New Orleans had long been a spectator sport, a point of roguish civic pride in a decadent town" (Horne, 2010). But post-Katrina frustration has led to less acceptance of the "old way of doing business." In 2008–2009 as part of the development of a community Master Plan, residents were asked to rank their top concerns. You might think that given that the city was recently devastated by Hurricane Katrina, that levees and taxes would top the list; yet results shown in the accompanying table revealed that corruption was much more important than levees or taxes (American Planning Association, 2010).

Concern	Percent of Respondents Ranking as Highest or Second Highest Priority
Crime	54
Education	30
Corruption in city government	23
Jobs and economic growth	21
Affordable housing	13
Neighborhood development	13
Levees	12
Health care	9
Race relations	6
Local road conditions	5
Taxes	2

Corruption and Organized Crime in the United States

The World Economic Forum's biannual Global Competitiveness Report, cited in a previous sidebar, also surveys corporate executives regarding corruption and organized crime. Results suggest that extensive corruption occurs in the United States. For example, for the following question, "In your country, how common is diversion of public funds to companies, individuals, or groups due to corruption?", the United States ranked 28th. New Zealand was ranked 1st (least corrupt), Venezuela was ranked last, 133rd (most corrupt). For the following question, "Does organized crime (mafia-oriented racketeering, extortion) impose costs on businesses in your country?", the United States ranked 72nd. Luxembourg was ranked 1st (least influenced by organized crime), and El Salvador was ranked last.

Corruption Humor

Three contractors were visiting a tourist attraction on the same day. One was from Illinois, another from Wisconsin and the third from Minnesota. At the end of the tour, a security guard asked what they did for a living. When they all replied that they were contractors, the guard said, "Hey, we need one of the rear fences redone. Why don't you guys take a look at it and give me a bid?" So they all went to the back fence.

The Minnesota contractor took out his tape measure and pencil, did some calculations and said, "Well I figure the job will run about $900: $400 for materials, $400 for my crew, and $100 profit for me."

Then the Wisconsin contractor took out his tape measure and pencil, did some quick figuring and said, "Looks like I can do this job for $700: $300 for materials, $300 for my crew, and $100 profit for me."

Without so much as moving, the Illinois contractor said, "$2,700." The guard, incredulous, looked at him and said, "You didn't even measure like the other guys! How did you come up with such a high figure?"

"Easy," he said. "$1,000 for me, $1,000 for you, and we hire the guy from Wisconsin."

(This joke has circulated for years, with different states substituted depending on the joke teller.)

Personal Experience with Corruption in Ghana

Dr. Sam Owusu-Ababio, a professor of Civil and Environmental Engineering at the University of Wisconsin-Platteville, returns often to his native Ghana. The following are his observations on what he has learned of corruption in the transportation sector in Ghana.

Unlike the other sectors of development, the transportation sector always has large capital expenditures involved, and as a result, it is more susceptible to corrupt practices. It is easier to manipulate or hide some transactions given the absence of information and communication technology that could otherwise be used to improve outdated financial and project monitoring or auditing processes. Electronic databases or integrated database systems that can facilitate record keeping and the monitoring process are lacking. This makes it difficult to find the trail of specific fraudulent transactions, especially in agencies where poor practices exist in keeping records.

Whenever there is a change in government, there is always a massive overhaul of the transportation sector. High-level responsibilities and appointments that deal with planning and funding of transportation infrastructure are sometimes allocated to people with no expertise in transportation planning or administration. Consequently, policy decision-making processes lack the depth of any rigorous planning process that encourages community-wide involvement and seeks to address long-term travel demand forecasting needs, environmental, economical, and social impacts. The political appointments are also perceived to come with a hidden agenda in the form of selecting people who can make decisions that provide more opportunities for funds diversion for personal gain. Eventually, substantial amount of public resources are sunk into projects that fail to adequately address the immediate and long-term transportation needs.

Some specific observations of corruption follow.

- Workplace recruitment, transfers, assignments, and promotions are often based on favors and connections with people in power rather than merit. One can get promoted by being cooperative with corruption practices within an agency. This approach leads to more opportunities for corruption to surface and blossom.

Personal Experience with Corruption in Ghana (Continued)

- There is a lack of supervision and adequate record keeping when it comes to usage of public assets. It is not uncommon to have public assets such as construction equipment or vehicles put to use on a private job to benefit the engineer who oversees the equipment. Vehicles assigned to public officials for public work are frequently used outside the prescribed hours of use. Where materials and parts are not well inventoried, some get stolen by agency employees and sold on the market.

- During the feasibility and design phases of a project, consultants can overdesign or purposefully expand project feasibility components if the fee for consultancy services is tied to some percentage of the overall cost.

- When it comes to contracts allocation, a contractor's expertise may not count; rather, it is the connection with the decision-makers or political figures and ability to pay a bribe that can often determine one's success in securing a contract. The contractor with the connections may receive inside information about the bidding process to help tip the decision in his/her favor. In some cases the contractor may be advised to deliberately lower the bid price to win the contract and then after the contract is won, change orders can be submitted to make up for the bid deficiency. In return, the provider of the inside information can get a bribe, which is often negotiated in terms of a percentage of the contract sum. The negotiation process can create a delay in the awards of the contract and the ultimate start of the project.

- Some contractors may choose to lie about their qualification requirements in order to secure a contract. They borrow construction equipment while attempting to conceal the identity of the owner. They can make financial arrangements with corrupt bankers to provide them with financial documentation to demonstrate their financial capacity, all done to create an illusion of their ability to compete in the bidding process. Occasionally, some contractors after collecting mobilization funds can abandon the project.

- Some contractors may conspire with corrupt engineers to falsify quantities in change orders during project construction. In addition, work certification for compliance may also be compromised to aid the contractor in return for a bribe.

- Processing of claims for payment of completed project work is one of the stages where bribing is practically inevitable. The process is cumbersome and slow; it is set up to require multiple approvals. If one wants to accelerate the process, then some negotiations in the form of bribes will take place along the approval paths to avoid significant delay in payment.

- Road infrastructure maintenance requires periodic assessment of conditions to help identify maintenance needs and prepare corresponding budgets. The field conditions data are sometimes not collected due to unavailability of tools for conducting the survey and lack of trained personnel. However, the corrupt engineers are able to prepare the budget either based on a specified percent increase of previous budgets or use fictitious condition data.

As a result, corruption steals resources from the transportation system and directly or indirectly affects the quality of the system components. The main impacts as observed in Ghana are as follows:

- Roads are built to less desirable standards and tend to deteriorate rapidly (Figure 17.1).
- There is overcrowding of public transportation systems since the road system is poor.
- There is increased potential for more accidents because of inadequate resources to manage incidents when they happen; this can further create congestion.
- There is constant flooding and washing away of roads and bridges (Figure 17.2).
- Maintenance of deteriorated systems components is deferred, requiring more money to fix them at a future date.

Figure 17.1 Deteriorated Roadway in Ghana.

Source: Courtesy of S. Owusu-Ababio.

Personal Experience with Corruption in Ghana (Continued)

Figure 17.2 Impassable Roadway in Ghana.

Source: Courtesy of S. Owusu-Ababio.

Outro

It is important to keep in mind that you are always accountable for your decisions, and that your decisions often affect others. Unethical engineering decisions may not only negatively affect you and your career, but also your coworkers, your employer, your client, and the general public.

chapter Seventeen Homework Problems

Short Answer

17.1 Print out a copy of the NSPE Code of Ethics Examination (available at www.wiley.com/college/penn) and take the test without looking at the answers before completing the exam. Self-grade the exam after you have taken it and hand in the graded copy to your instructor.

17.2 For each of the quiz questions from Homework Problem 17.1, state the pertinent section of the NSPE code of ethics that addresses each question.

Discussion/Open-Ended

17.3 Why do some students cheat? List five reasons/excuses.

17.4 How do students justify cheating? List five ways.

17.5 Which portion(s) of the ASCE code of ethics would you violate by looking at the answers prior to taking the online ethics exam in Homework Problem 17.1 and not reporting that you did so?

17.6 The NSPE's Board of Ethical Review (available at www.wiley.com/college/penn) contains hundreds of case studies. For one of these case studies selected by your instructor, answer the following questions.

a. For the ethical question(s) stated in the case study, what are two possible unethical responses. For each response, provide justifications and counter these justifications with specific text from the ASCE Code of Ethics.

b. Do you agree with the conclusion reached by the Board of Ethical Review? Why or why not?

17.7 The ASCE defines sustainability as "the challenge of meeting human needs for natural resources, industrial products, energy, food, transportation, shelter, and effective waste management while conserving and protecting environmental quality and the natural resource base essential for future development." This definition is provided in the ASCE Code of Ethics. Do you think that an engineer that does not practice in a sustainable manner is unethical?

17.8 Cite specific violations of the ASCE Code of Ethics reported in the "Personal Experience with Corruption in Ghana" section of this chapter.

17.9 Are the reported practices in Ghana unethical if no professional code of ethics exists in that country?

17.10 Other than adopting a code of ethics, what can be done to change engineering practice in Ghana to make it more compliant with a code of ethics such as that of ASCE?

Key Terms

- bribery
- code of ethics
- collusion
- corruption
- extortion
- fraud
- money laundering
- whistle-blower

References

American Planning Association. 2010. "By the Numbers." *Planning*: 76(1), 37.

Carpenter, D. D., T. S. Harding, C. J. Finelli, S. M. Montgomery, and H.J. Passow. 2006. "Engineering Students' Perceptions of and Attitudes Towards Cheating." *Journal of Engineering Education*, 95(3): 181–194.

Horne, J. 2010. "Trump Card." *Planning*, 76(1), 35–36.

Martin, M. and R. Schinzinger. 2004. *Ethics in Engineering*. New York: McGraw-Hill.

PricewaterhouseCoopers, 2009. *Corruption Prevention in the Engineering & Construction Industry*. http://www.pwc.com/en_GX/gx/engineering-construction/fraud-economic-crime/pdf/corruption-prevention.pdf, accessed August 29, 2011.

Stansbury, C. and N. Stansbury, 2008. *Anti-Corruption Training Manual*. Global Infrastructure Anti-Corruption Center/Transparency International (UK). http://www.giaccentre.org/documents/GIACC.TRAININGMANUAL.INT.pdf, accessed August 29, 2011.

chapter Eighteen Security Considerations

Learning Objectives

After reading this chapter, you should be able to:

1. Define infrastructure security.
2. Explain how recent catastrophic events have revealed weaknesses in our infrastructure.
3. Describe steps taken to make three infrastructure sectors (water, transportation, and energy) more secure.
4. Explain risk management and risk assessment.

Introduction

We have demonstrated in preceding chapters that infrastructure affects our economy, our health, our environment, and our society. Consequently, it makes sense that we would want our infrastructure to be secure to ensure that the economy continues to grow, that the environment is protected, and that society can function as it "normally" does.

Infrastructure security became a household term following the 9/11 terrorist attacks. Following this tragedy, the news media presented many stories concerning the threat to our infrastructure from terrorist attacks, and the general public began to consider much more seriously the value of our infrastructure in terms of the security of our country. People began to understand that terrorist attacks on water treatment facilities, power grids, ports, roads, information infrastructure, and rail lines could produce even farther-reaching effects than did the 9/11 attacks.

However, infrastructure security refers to security from damages inflicted from both natural *and* human sources, such as:

Natural hazards:

- dam failure
- earthquake
- fire
- flood
- heat
- hurricane
- landslide
- tornado
- tsunami
- volcano
- wildfire
- winter storm

Human-induced hazards:

- terrorism
- vandalism
- accident
- war

Human-induced and natural disasters have several similarities and differences. Both can lead to mass casualties, damage to the infrastructure, may occur with or without warning, and often lead to evacuation or displacement of people. Terrorist attacks, on the other hand, are *intentionally* caused by people. Unlike a tornado, which indiscriminately destroys everything in its path, a terrorist attack will most likely target specific weaknesses, or **vulnerabilities**, in the infrastructure. One factor that complicates recovery from a terrorist attack is that the damaged areas are treated as crime scenes, which greatly complicates emergency services and cleanup activities.

An additional dissimilarity between human-induced and natural catastrophes is that we have a historical record of natural disasters and an increasing science-based understanding of the causes. We cannot predict when they will occur, but we can estimate the probability with increasing confidence. Additionally, we have an ever-increasing ability to monitor potential natural disasters such as earthquakes and hurricanes that can provide advanced warning for emergency response. Terrorist activity, on the other hand, is extremely difficult to predict and monitor. Intelligence information may provide information that suggests an increased likelihood, but the location, probability, and extent are unknowns. Accidents and vandalism are often random in occurrence.

> **Vulnerabilities** are flaws or deficiencies that render an infrastructure component or system susceptible to damage from natural disasters, terrorism, or inadvertent human actions.

Introductory Case Study: Hurricane Katrina

Hurricane Katrina was used in Chapter 7, Infrastructure Systems, to illustrate the implications of the failure to treat the hurricane protection system as a system, and is used in this chapter to illustrate security considerations.

Figure 18.1   A Portion of Flooded New Orleans.

Source: U.S. Coast Guard/K. Niemi.

Hurricane Katrina occurred in August 2005 and caused on the order of $100 billion in damages and nearly 2,000 deaths. Economically, it was the most damaging hurricane in U.S. history. While the damages were widespread along the coast of the Gulf of Mexico, the New Orleans metropolitan area was hardest hit. Multiple levee failures caused widespread flooding, up to 15 feet in depth in some locations, with catastrophic impacts on human welfare and the infrastructure. Once flooded, the water had literally filled a "bowl" (as New Orleans is below sea level) and could only be removed by pumps. Flooded roads were impassible and the Interstate 10 bridge collapsed. Damage to New Orleans in the aftermath of the hurricane is illustrated in Figure 18.1.

Virtually *every* aspect of the infrastructure was severely affected. Electric power was lost and backup generators were flooded; generators that were flood protected ran out of fuel. Root systems of trees uprooted by the high winds broke water mains, resulting in losses of pressure to the water distribution system; combined with pumping systems that also failed, the city did not have adequate pressure for fire protection. In addition, the drinking water treatment plant was inoperable for approximately 1 week.

In order to drain the city, first the pumping stations had to be made operational. Secondly, the failed levee sections had to be temporarily "fixed" via helicopter and sandbags so that the canals could transport water out of the city. Also, the canals had to be cleared of debris (e.g., cars, portions of homes, fallen trees) for conveyance of the pumped water. Within 2 weeks, most of the floodwater had been removed. Then Hurricane Rita struck. While the brunt of the storm hit Galveston, Texas, the storm surge (elevated sea levels) caused additional levee failures and more flooding in New Orleans.

U.S. Coast Guard
Response to Hurricane
Katrina

Analysis of the
Interstate 10 Twin
Bridge's Collapse
During Hurricane
Katrina

Economic Impacts and Infrastructure Security

Threats and hazards, whether they are natural, accidental, or purposeful, can have a significant impact on the economy of a region, country, or even the entire world. Basic economic considerations were presented in Chapter 14; here, some additional concepts relating directly to security are presented. Specifically, we wish to highlight that disasters of equal magnitude will have very different monetary damages based on the extent of infrastructure and the quality of that infrastructure in place at the time of the event. Wildfires in the western United States are a classic example. As more people move into previously remote areas, more homes (and consequently greater property values) are susceptible to fire damage. Thus, the same fire that 50 years ago would have destroyed hundreds of acres of forest will today destroy hundreds of acres of forest, homes and other property, and damage associated infrastructure.

On a larger scale, consider a Category 4 hurricane that hit Miami, Florida, in 1926 and which caused nearly 400 fatalities. Some of the resulting damage from the hurricane is illustrated in Figure 18.2. At the time of the hurricane, the Miami metropolitan area had a population of approximately 100,000 people, whereas today the population exceeds 5 million. As a result, the infrastructure has expanded greatly in the intervening 85 years, as is shown in Figure 18.3. As a result of the increased amount of infrastructure, it is estimated that if the same hurricane hit Miami today, it would result in $100 billion of damage.

The damage to Miami would be *even greater* than the estimated $100 billion if not for the improvement in standards and codes that have made the infrastructure of today much more **resilient** than it was in the 1920s. Using the same logic, you could correctly infer that if Miami had today's building codes for hurricane-proofing in place back in 1926, the damages from the hurricane would have been substantially less.

Unfortunately, in many developing countries, substandard codes and unenforced or lacking building codes are the norm rather than the exception. As a result, developing countries are especially prone to loss of life and casualties from natural disasters. Recent earthquakes and tsunamis have claimed hundreds of thousands of lives in Indonesia, Sri Lanka, China, India, Turkey, Pakistan, Japan, and Haiti.

The September 30, 2009 earthquake in Indonesia left thousands of people trapped in the rubble of structures that were apparently not designed and constructed to meet existing building codes. Dwellings such as those shown in the foreground of Figure 18.4 are obviously susceptible to high winds

2009 California Climate Adaptation Strategy Report

Hurricane Irene

In late August 2011, Hurricane Irene tracked along the eastern U.S. coast causing more than 40 deaths in 13 states ranging from Florida to Vermont. Millions of homes and businesses suffered power outages. Extensive flooding occurred. Preliminary estimates of damages ranged from $7 to $15 billion, making it one of the most expensive natural disasters in U.S. history.

Infrastructure **resilience** is the ability of infrastructure to withstand a threat *and* its ability to recover once the threat has passed.

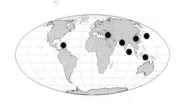

Figure 18.2 Aftermath of the 1926 Miami Hurricane.

Source: Library of Congress.

Remember that full-sized versions of textbook photos can be viewed in color at www.wiley.com/college/penn.

Figure 18.3 Today's Miami Skyline.

Source: Courtesy of M. Wade.

Figure 18.4 A Residential Area in Bangladesh.

Source: Courtesy of J. Jordens.

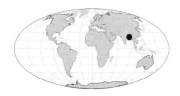

and/or waves; however, these may actually be *safer* than multi-level concrete structures (seen in background) during earthquakes if the concrete structures are not designed or constructed to prevent collapse.

Critical Infrastructure

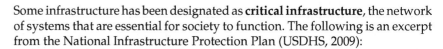
National Infrastructure Protection Plan

Some infrastructure has been designated as **critical infrastructure**, the network of systems that are essential for society to function. The following is an excerpt from the National Infrastructure Protection Plan (USDHS, 2009):

"Protecting and ensuring the continuity of the critical infrastructure and key resources (CIKR) of the United States is essential to the Nation's security, public health and safety, economic vitality, and way of life. CIKR includes systems and assets, whether physical or virtual, so vital to the United States that the incapacitation or destruction of such systems and assets would have a debilitating impact on national security, national economic security, public health or safety, or any combination of those matters. Terrorist attacks on our CIKR, as well as other manmade or natural disasters, could significantly disrupt the functioning of government and business alike and produce cascading effects far beyond the affected CIKR and physical location of the incident. Direct and indirect impacts could result in large-scale human casualties, property destruction, economic disruption, and mission failure, and also significantly damage national morale and public confidence."

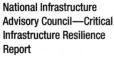
National Infrastructure Advisory Council—Critical Infrastructure Resilience Report

The following are examples of critical infrastructure components/sectors:

- agriculture/food
- national defense
- energy
- national monuments
- dams
- telecommunications
- information technology
- postal
- shipping
- government facilities
- transportation
- drinking water treatment and distribution
- wastewater treatment
- schools/educational facilities
- banking
- emergency services

This list is quite broad, and is drawn from several sources. It lists virtually every aspect of the infrastructure; indeed, it is difficult to think of infrastructure items that are *not* on this list, which emphasizes that infrastructure security is critical to the proper functioning of our country. It also includes some items that you may not have considered, such as national monuments, which are sources of national pride and sense of place. However, within any infrastructure sector, there are certainly some components that are more critical than others.

The Hoover Dam is obviously more critical than a small mill-pond dam. In the field of earthquake engineering, the term **lifeline** is used to differentiate the *most critical* infrastructure components. A municipality may have many bridges, but only a few may be considered "lifelines" because of location or traffic capacity. These bridges should arguably be designed and constructed with less risk of failure than other bridges.

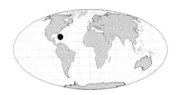

The aftermath of the 2010 earthquake in Haiti demonstrates the critical nature of infrastructure. It is estimated that the earthquake resulted in over 200,000 fatalities, over 300,000 injuries, and more than 1 million people left homeless. The inability to promptly provide relief supplies was widely reported by the media, and largely resulted from ineffective government and a devastated infrastructure. Without properly functioning power supplies, roads, shipping ports, airports, communications, and public safety agencies, supplies could not be effectively distributed. While food, water, and shelter are essential to human survival, the situation demonstrated that, in many ways, restoring the infrastructure is even more important, for without it, necessary supplies cannot be distributed.

In recent years, the link between information technology and the built infrastructure has continued to strengthen, which presents a new set of issues related to security. Many components of the infrastructure are controlled and/or monitored via networked computers. For example, in most large cities and many small municipalities, drinking water treatment processes and water supply components are controlled and monitored by computers. Many roadways, wastewater treatment plants, and power generation and transmission facilities are also controlled or monitored by computer. This integration of computers and infrastructure brings many risks, and the need to protect the information technology infrastructure (**cybersecurity**) goes hand in hand with securing the physical infrastructure.

It is imperative that the components of the critical infrastructure are resilient. The effectiveness of resilient infrastructure depends upon its ability to anticipate, absorb, adapt to, and/or rapidly recover from a potentially disruptive event.

Infrastructure threats include the tremors from an earthquake, the high winds associated with a hurricane, the accidental ramming of a bridge pier by a barge, or the purposeful destruction of that same pier by explosives. Infrastructure components can be made resilient through design, disaster preparedness and mitigation, and through response and recovery activities. Clearly, the flood protection system of New Orleans was not resilient. Nor was virtually every aspect of the infrastructure in Port-au-Prince, Haiti.

One characteristic of infrastructure that improves its resilience is **redundancy**. Most effective engineering designs include redundancy. For example, water distribution networks contain many interconnected pipes arranged in a grid, rather than a branched pattern; one reason for this is redundancy, to ensure that in event of the failure of one pipe, that the vast majority of the remaining system can still operate. In situations where pumping is required,

Redundancy is the duplication of critical components of a system in order to increase reliability of the system, by providing a "back-up" component.

The remarkably intact mountain-top city of Machu

Picchu, located in the Andes Mountains of Peru, was built around 1450 and is considered an engineering marvel and cultural heritage site.

Approximately 3,000 visitors tour the site daily. The site is accessed via train or by a several-day hike ascending thousands of feet along the ancient Inca road.

In January 2010, torrential rains caused substantial damage to the railway (washing out bridges) and the hiking trail, isolating thousands of tourists that had to be evacuated by helicopter and leading to temporary closure of the site. The Peruvian Transport Minister initially stated that rail service would be restored by the end of February 2010, but it was not until April that partial rail service was reinstated. The lack of transportation redundancy caused economic losses to the local tourism industry of $1 million per day during the nine-week shutdown (The Economist, 2010).

two pumps are often mandated where only one is strictly necessary so that in case of a single pump failure, pumping can continue.

The Interstate Highway System also includes redundancies. Consider that if a segment of Interstate highway is closed, traffic can be redirected to a number of possible routes including other Interstates, state highways, and local roads. Such rerouting can lead to many inconveniences, but in many cases does not significantly impact the regional economy. On the other hand, removal of a critical *railroad* bridge from service can have a much greater impact, as the heavy railway system lacks the redundancy of the Interstate Highway System.

Designing for Failure

Although the heading of this section may seem surprising, in some cases, engineering designs are created with failure in mind. For example, a typical stormwater detention pond is designed to handle a 10-year recurrence interval. If a larger (less probable) storm event occurs, the pond is undersized and the water will overtop the pond "walls" (earthen berms). This overtopping could result in erosive scour that could downcut through the berm leading to a catastrophic failure. Many states and/or municipalities require that an emergency overflow be designed to safely pass the larger (but less likely) 100-year storm. This overflow is armored to prevent scour, similar to the design of emergency spillways for dams. Note that for the 100-year storm, the pond will *not* be meeting its primary objective of treating the stormwater, but it will remain intact to effectively treat future smaller storms for which it is designed. Note also that a very large storm event will quite possibly lead to berm failure, but consider that when the 500-year storm occurs, the widespread damage from the flooding will be so excessive that the detention pond failure will be of minor consequence.

In other situations however, failure is an implicit consideration in the design. Buildings can be designed to withstand terrorist attacks, but at prohibitive expense. However, buildings can be *designed to fail* in a way to minimize the loss of life and casualties to occupants by preventing the phenomenon of progressive collapse (e.g., designing a structure to have alternate load paths to support its weight in the event that a primary load bearing element fails). Also, during military operations, semi-permanent bridges may be designed and constructed that can purposefully be destroyed to prevent enemy forces from crossing a river, yet can be readily repaired later.

In the remaining sections of this chapter, we will investigate the security of three major infrastructure sectors: water, energy, and transportation. The framework for ensuring the security of other sectors (e.g., dams, ports) requires similar considerations and methods.

Water Security

Water security relates to drinking water sources and treatment as well as wastewater treatment. If a drinking water treatment facility were damaged by an extreme weather event or by a terrorist attack, daily life could be greatly affected. The scale of potential impacts varies and could include minor service disruptions associated with low water pressure, or could lead to an inability to fight fires or to provide drinking water for several days. At the most damaging extreme, improperly functioning drinking water treatment systems could lead to mass illness and death.

The drinking water industry has always been concerned about the safety of delivered water, and this concern was greatly elevated following the events of 9/11. The following are potential vulnerabilities of drinking water systems.

- Source water protection: Many systems treat water that is first stored in reservoirs, often located many miles from the users. For example, New York City's water supply reservoirs have the capacity to hold 600 billion gallons of water (Figure 18.5) and are located up to 100 miles from the treatment facilities. The concern for source water protection is that terrorists could introduce toxic chemicals or biological agents into reservoirs, rivers, and lakes that treatment plants are not designed to remove. Securing these large, remote water sources is very challenging. For example, securing the perimeter of the reservoirs is very expensive. In some cases, signs (Figure 18.6) are used. Signs are obviously much less expensive than fences, but offer only a marginal level of security.

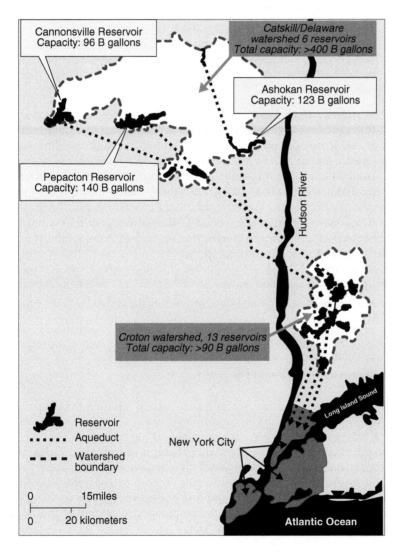

Figure 18.5 New York City's Water Supply. Two major watersheds contribute water to many water supply reservoirs. The largest reservoirs, and their capacities, are labeled. Water withdrawn from the reservoirs is conveyed via three aqueducts (tunnels up to 24 feet in diameter) shown entering the city. The aqueduct system is very complex, crossing under rivers and diverting water into and out of reservoirs.

Source: Redrawn from New York City Department of Environmental Protection.

Figure 18.6 Sign Posted on Property Adjacent to a Drinking Water Reservoir.

Source: M. Penn.

Cost/Benefit Ratios

As of 2006, the U.S. Environmental Protection Agency requires that communities that rely on surface waters for drinking water supply must routinely monitor for Cryptosporidium and upgrade treatment systems if necessary. The EPA justifies this requirement based on estimates that the annualized present value costs associated with these requirements will be approximately $100 million, but that the annualized benefits from avoided illness and loss of productivity range from $300 million to $1 billion, as well as the potential avoidance of the loss of hundreds of human lives (USEPA, 2005). Dividing the annualized benefits by costs gives a **benefit/cost ratio**, in this case ranging from 3 to 10. Benefit/cost ratios greater than 1 indicate that gains outweigh costs, and suggest economic merit.

- Water treatment plants: A treatment plant could also be a terrorist target, resulting in damage to structures or equipment, and potentially releasing dangerous stored chemicals or resulting in an inability to treat the water to appropriate standards.

- Distribution systems: Treated water is transferred from the treatment plant to the consumers via a system of underground water mains, above ground tanks, valves, and hydrants. Injection of toxins into the distribution system, through fire hydrants or other access points, could potentially harm large populations.

- SCADA systems: Modern water treatment systems are typically operated through SCADA (Supervisory Control And Data Acquisition) systems. SCADA provides operators with computerized controls with which to operate and monitor a treatment plant. Hacking into this system could result in improper water treatment that fails to protect the public from toxins and disease, or in shutting down the system completely.

These potential threats have led to the requirement that water treatment facilities must prepare a **vulnerability assessment**. A vulnerability assessment includes the following considerations:

- identify the potential threats
- identify water treatment assets (e.g., elevated storage towers, pumps, chemical storage) to be affected by threats
- estimate probability of each threat occurring
- calculate consequences of losing assets
- evaluate countermeasures

Additional non-terrorism and non-natural disaster threats and hazards exist with water supplies regarding *naturally occurring* pathogens such as Cryptosporidium. In 1993, an estimated 400,000 people in metropolitan Milwaukee (25 percent of the population) became ill and over 100 deaths resulted from an outbreak of Cryptosporidium in the public water supply.

It is estimated that the economic impact of the outbreak was nearly $100 million in medical costs and lost productivity (Corso et al., 2003). After the outbreak, treatment systems were upgraded to prevent further occurrences at a cost of approximately $50 million.

Wastewater systems are also considered to be critical infrastructure, because if they fail, widespread public health impacts as well as environmental damage may result. As we have discussed previously, these effects would in turn have economic impacts. Thus, although you might not think of wastewater treatment facilities as targets of terrorist attacks, they might be. Many of the same steps undertaken to protect against human-induced hazards (e.g., structural hardening/armoring, improved management plans) will also help make a wastewater treatment facility more secure from natural disasters.

Many of the unit processes used in wastewater treatment plants use microorganisms to degrade the wastewater, and these microorganisms could potentially be killed by introducing certain chemicals into the wastewater conveyance system through a toilet or manhole. Some wastewater treatment facilities might also be a target as they, like some drinking water treatment facilities, contain storage areas for potentially dangerous chemicals such as chlorine gas (although this is increasingly rare, for this very reason). Since sewer mains typically travel under roadways, explosions in sewers could make roads unusable. This has happened by mistake, as occurred in Louisville, Kentucky, when a local industry dumped chemicals down the sewer, causing an explosion and subsequent destruction of a street.

 Sewer Explosion Report

Energy Security

The "Northeast Blackout of 2003" affected over 50 million people in the United States and Canada, and provides an example of a system that was not secure. This catastrophe occurred on August 14, 2003, because of a power surge combined with failures in computerized control systems. Initial causes appear to be due to contact between tree branches with power lines. Amazingly, this simple event set off a cascade of events that left people without power from Detroit, Michigan, to Boston, Massachusetts. This left residents unable to perform many routine daily activities, but more importantly it shut down several airports and many mass transit systems that relied on electricity (e.g., street cars), reduced pressure in public water systems, and caused wastewater treatment and conveyance system failures. As illustrated in Figure 18.7, this

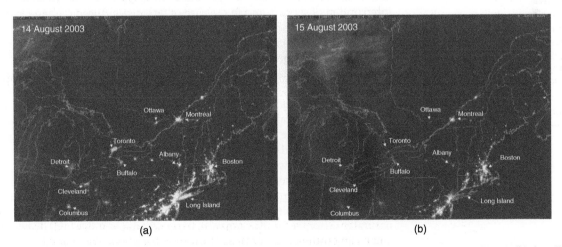

Figure 18.7 **Nighttime Satellite Imagery Before (A) and During (B) the Northeast Blackout.** Note in Figure (b) that: Toronto, Ottawa, and Detroit are completely dark (without electricity); Boston is unchanged from Figure (a); and that the brightness of Long Island and the New York City metropolitan area is much reduced from that of Figure (a).

Source: Redrawn from NOAA/DMSP.

loss of power was evident from space. Note that some areas such as metropolitan New York City had some electricity resources because of local sources or because of supply networks not affected by the out-of-commission system. However, Detroit, Cleveland, Toronto, and Ottawa completely lost power.

In the aftermath of the blackout, Bill Richardson, former Secretary of the Department of Energy, stated that the United States was "a superpower with a third-world electrical grid." This assessment is also reflected in the ASCE 2008 Report Card on Energy:

"The national electric grid currently lacks a significant degree of resilience. Utilities are generally prepared for local and regional responses; however, the national electric grid as a whole lacks a significant degree of resilience should a much broader response be required. Future investments in the system must improve system robustness, redundancy, and rapid recovery. Additionally, new technologies and behavioral changes focused on reduction and increased efficiency are necessary. True system resilience will require a national effort to modernize the electric grid to enhance security and the reliability of the energy infrastructure and facilitate recovery from disruptions to energy supply, from both natural and man-made hazards."

Today there are increased concerns about security at nuclear power plants. Recently the Nuclear Regulatory Commission voted to require all new nuclear power plants to incorporate design features that would ensure that, in the event of a crash by a large commercial aircraft, the reactor core would remain cooled or the reactor containment would remain intact, and radioactive releases would not occur from spent fuel storage pools (Holt and Andrews, 2010). These design features are not required as upgrades to existing reactors, but requirements exist that mitigation strategies be developed in case of such an event. Also, the long-term management of spent nuclear fuel that is currently stored at nuclear power plants is under review.

Transportation Security

The transportation infrastructure includes roadways, waterways, railways, and aviation. Damage to transportation infrastructure can cause loss of life, interruption of service, and in the case of terrorist-induced damage, may lead to demoralization. And like all infrastructure, the quality of the transportation sector affects the economy, and damage to the sector or a component of the sector can have a significant impact on the economy. However, transportation infrastructure (especially the highway system), has an additional function that is very important in times of attacks or natural disasters; in such situations, this system provides the means by which emergency personnel move and by which evacuations take place. The evacuation of Houston, Texas, as Hurricane Rita approached the U.S. Gulf Coast is shown in Figure 18.8. All highways were gridlocked and some motorists were in traffic for more than 24 hours, in what is considered to be the largest evacuation in American history.

The U.S. highway system benefits from a high degree of redundancy, although certain components (e.g., bridges and tunnels) have less redundancy. When a bridge crossing a river is removed from service, whether from accident, natural disaster, or for maintenance, it can add many miles to daily commutes and freight routes. For this reason, bridges and tunnels are sometimes referred to as "choke points."

Consider the case of a bridge that was damaged on Interstate I-40 in Oklahoma in 2002. In a freak accident, the captain of a barge blacked out as the barge approached the bridge. The barge veered sharply off course and struck a pier. The aftermath is seen in Figure 18.9. This accident resulted in

Figure 18.8 Hurricane Rita Evacuation Along I-45.
Traffic on the service road (right) merging with northbound lanes (middle). Note the empty southbound lanes (left), which are closed. After being purged of southbound traffic, these lanes were later reopened and used as additional lanes for northbound evacuation.

Source: Ashish Waghray Photo.

Figure 18.9 I-40 Bridge Collapse.

Source: Courtesy of R. Webster.

Strategic Highway Network

In terms of security, highways serve a very important purpose to the U.S. military. The Strategic Highway Network (STRAHNET) consists of 63,000 miles, of which 47,000 miles are Interstate freeways (Figure 18.10). The purpose of STRAHNET is to provide for movement of personnel and equipment in times of peace and war. Bridges and tunnels on STRAHNET with span lengths greater than 165 feet are defined as "critical" transportation infrastructure. The existence of STRAHNET also emphasizes the point that the interstate highway system, like nearly all components of the infrastructure, has multiple objectives it must meet. Not only must it safely transport cars and vans, tractor trailers, and buses, but it must be able to be effectively used by the military.

Figure 18.10 The Strategic Highway Network.

Source: Federal Highway Administration.

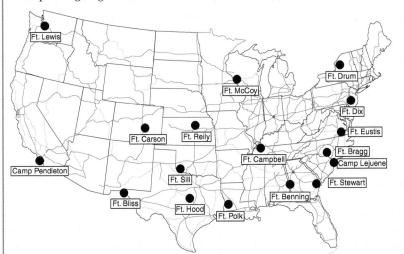

fourteen deaths. Also as a result of this accident, the bridge was closed to traffic for 2 months. (This is a surprisingly short amount of time to design and build a new Interstate bridge.) While the immediate impacts on traffic were significant, thanks to redundancy in the highway system, the 20,000 vehicles that traveled this stretch of highway daily were later able to be rerouted on local roads. Although this accident was significant in terms of loss of life, it did not significantly impact the regional economy thanks to the redundancy. Of course, *local* businesses were significantly affected while traffic was rerouted.

Returning to the principles of Chapter 10, Design Fundamentals, the 1-40 bridge did not "fail" in its design. It simply was not designed to withstand this specific type of impact. So although it was not resilient to this type of accident, it was not a matter of a design failure. You should realize that this pier *could have been designed* to withstand the force of this collision. Larger, stronger piers could have been specified, but at a significantly greater cost.

A significant amount of work has recently been conducted in the transportation security field with respect to the possibility of terrorism. Among the various modes of transportation, transit and rail are the most likely targets. These have less redundancy, are less easily replaced, and provide the potential for more casualties.

Counter-terrorism methods focus on the "4 D's": deterrence, detection, defense, and design. This can be accomplished by a number of methods:

- Key structural elements can be identified and strengthened or shielded.

- Access can be denied to locations where explosives would do the most damage.

- Barriers can be added to increase the "standoff" distance.

Many counter-terrorism strategies are non-technical in nature, and include increased patrolling by security personnel and increased surveillance, but most measures require the expertise of engineers. This type of work is exciting in many respects, as the benefits are clear and the problems require "out of the box" thinking by the engineers. A high level of expertise and a keen understanding of how structures function is required in order to select the key structural members and design how best to protect them.

Risk Considerations

Any activity undertaken to decrease the risk associated with a potential threat or hazard comes with a cost. One of the greatest costs of infrastructure security is money. But there are many other "costs," including loss of privacy, disruption of normal daily activities, and loss of personal freedom. Determining the relationship between cost and benefits (that is, the decreased risk associated with a potential event) is a core concept of **risk management**.

Risk management is the process of identifying, assessing, communicating, and managing risks.

Risk management is especially challenging for infrastructure security given that the threats and hazards have a very low probability of occurring, but also have extremely high consequences. Moreover, large sums of money are required to limit the effects of very low probability events.

This low probability/high consequence relationship is clearly illustrated with the I-40 bridge disaster and makes funding retrofits and other infrastructure hardening projects quite challenging. For the I-40 disaster, the cost to strengthen the piers is immaterial to families and friends of the victims if it would have saved even one of the fourteen lives that were lost. However, cost-benefit analyses typically place a value on human lives (deaths avoided) and casualties. The value varies from approximately $1 million to $10 million per avoided death, depending on the federal agency. For example, the EPA's current "value of a statistical life" is approximately $7 million. Many people have difficulty accepting this concept that a "value" can be placed on a human life. However, it is impossible to do a cost-benefit analysis without such a value. If indeed every life is "priceless," then *unlimited funding* must be available to ensure *zero* risk of loss of life from *any* infrastructure project. In reality, bridge designers and policy makers would have a very difficult time convincing taxpayers, regulators, and lawmakers that *all* piers for *all* bridges should be able to withstand a barge collision (or the strongest conceivable earthquake, or the 1,000,000-year flood).

In the aftermath of the hurricanes that hit the U.S. Gulf Coast in 2005, Congress directed the Army Corps of Engineers to develop a hurricane risk reduction and coastal protection plan. In 2009, the Louisiana Coastal Protection and Restoration (LACPR) Final Technical Report was released. The report provides over 100 alternatives to achieve flood protection from storm surges at the following levels of risk: 100-year, 400-year, and 1,000-year. Cost estimates for the risk reduction alternatives are staggering, ranging from $70 to $115 *billion*. It is important to note that cost is not typically linearly related to risk. For example, to reduce risk by a factor of 10, the cost is not likely to be 5 times greater than to reduce the risk by a factor of 2. The costs of relocating structures, raising structures, and buying land as a function of the recurrence interval of flooding is demonstrated in Figure 18.11

The non-linear function of cost versus recurrence interval is a management challenge that most communities face with issues such as stormwater flooding from common non-catastrophic events. Systems can be designed for a 2-year, a 5-year, or a 10-year recurrence interval, but the cost typically increases sharply in this range of probabilities. This is illustrated in Figure 18.11 where the slope of the cost per increment of risk reduction is steep for smaller return intervals. Many communities design stormwater systems to convey the 10-year storm. This implies that the additional cost to design a larger/stronger/safer system (e.g., 20-year system) is greater than the anticipated damages that would result from the flooding. However, this decision may be based on common practice rather than on a detailed (and expensive) cost-benefit analysis. After substantial damage from flooding *does* occur, certain portions of a stormwater system may be redesigned at a higher level of protection for areas where economic losses are greater.

Selection of a recurrence interval is one portion of a **risk assessment**. Risk assessments are undertaken by communities, businesses, and regulatory

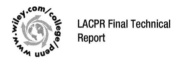

LACPR Final Technical Report

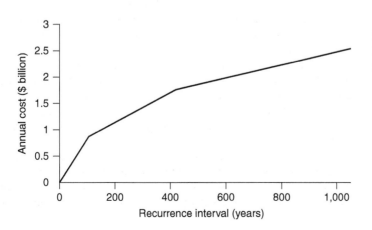

Figure 18.11 Annual Cost of Flood Protection Measures for Various Hurricane Probabilities.

Data source: USACOE, 2009.

agencies in order to quantify and qualify the risks associated with a certain threat or hazard. A risk assessment should answer the following questions:

1. What can go wrong?

2. How can it happen?

3. What is the likelihood?

4. What are the consequences?

In the case of coastal Louisiana, questions 3 and 4 were answered by the LACPR report. Extensive computer modeling was conducted to simulate storm surges with varying probabilities. These storm surges, both in terms of water elevation and time of duration (i.e., the number of hours water will be at a certain elevation) were then utilized to predict flooding from the overtopping of levees. Many potential consequences were evaluated, including not only the direct impacts to people and property, but also the impacts on the coastal ecosystem, regional/national economies, historical places, and traditional cultures.

Outro

Photograph of Fire Truck in Sinkhole

Our infrastructure must be secure to protect the public's health and well-being. Additionally, a secure infrastructure protects the economic vitality of a region or of an entire country and allows appropriate emergency response following a disaster. An example of how the deteriorating infrastructure can have a dramatic effect on the ability of emergency personnel to respond occurred in Los Angeles in 2009; in this case, a sinkhole created by a broken water main "swallowed" the fire truck that was responding to the emergency caused by the broken main! Infrastructure security can be greatly improved through planning, maintenance, analysis, and design.

chapter Eighteen   Homework Problems

Short Answer

18.1   In two or three sentences for each system, explain how redundancy is achieved in:

a. power transmission systems
b. highway systems
c. water treatment and distribution systems
d. source water supply

18.2   How is the selection of a recurrence interval used for design related to risk?

18.3   What are the 4 D's of counter-terrorism?

18.4   What considerations are included in a vulnerability assessment?

Discussion/Open-Ended

18.5   For a local bridge specified by your instructor, discuss how much redundancy is involved. If the bridge were to fail, what would be some social or economic effects?

18.6   Obtain the drinking water distribution system map (or a portion thereof) for a community chosen by your instructor. (Alternatively, if your instructor desires, you can view the map for the community provided on the textbook website at www.wiley.com/college/penn). Explain how redundancy is built into this system.

18.7   Describe the trip you take when traveling from your home/dorm/apartment to a local grocery store. Are there any infrastructure components used for this trip that qualify as a choke point? Explain. Which component is most critical to being able to complete your trip in a timely manner? Explain.

18.8 Recent improvements to security have had a high economic price and society has also had to pay a price in terms of decreased freedom. Give some examples of these losses in freedom. Do you think the steps taken to increase safety and security have been worth the losses in personal freedom? Explain.

Research

Note: For each Research question, limit your report to a single page. Include a bibliography of cited references on a second page.

18.9 Research the basis for the USEPA's value of a statistical life. How was this value determined?

18.10 Research a major infrastructure project for which a cost–benefit analysis is available. Summarize the costs and benefits.

18.11 For a natural hazard of choice (or for a hazard selected by your instructor), report on a recent example from one of the following countries/continents and summarize the impacts on infrastructure.

a. United States
b. South America

c. Asia
d. Europe

18.12 Research the infrastructure impacts of World War II combat/bombing on a major city of your choice (or for a city selected by your instructor).

18.13 Report on the *non-combat* infrastructure impacts associated with the U.S. war effort during World War II on the contiguous United States.

18.14 Summarize the impacts of the 2010 earthquake in Haiti on that nation's:
a. ports
b. roads
c. bridges
d. schools
e. water supply

18.15 Investigate how one of the following has become more secure in the aftermath of the attacks of 9/11.
a. airports
b. ports
c. telecommunications
d. railways
e. food
f. dams

Key Terms

- benefit–cost ratio
- choke point
- critical infrastructure
- cybersecurity
- human-induced hazards

- infrastructure security
- lifeline
- natural hazards
- redundancy
- resilient

- risk assessment
- risk management
- vulnerabilities
- vulnerability assessment

References

"Ruined." *The Economist*. February 13, 2010: 42.

Chen, G., E. C. Witt, D. Hoffman, R. Luna, R., and A. Sevi. 2007. "Analysis of the Interstate 10 Twin Bridge's Collapse During Hurricane Katrina." In: "Science and the Storms—the USGS Response to the Hurricanes of 2005," *U.S. Geological Survey Circular* 1306.

Corso, P. S., M. H. Kramer, K. A. Blair, D. G. Addiss, J. P. Davis, A. C. Haddix. 2003. "Cost of Illness in the 1993 Waterborne *Cryptosporidium* outbreak, Milwaukee, Wisconsin." *Emerging Infectious Diseases*. http://www.cdc.gov/ncidod/EID/vol9no4/02-0417.htm, accessed August 29, 2011.

Holt, M. and A. Andrews. 2010. *Nuclear Power Plant Security and Vulnerabilities*. Congressional Research Service Report for Congress RL34331.

U.S. Army Corps of Engineers (USACOE). 2009. *Louisiana Coastal Protection and Restoration (LACPR) Final Technical Report*. http://lacpr.usace.army.mil/default.aspx?p=LACPR_Final_Technical_Report, accessed August 29, 2011.

U.S. Department of Homeland Security (USDHS). 2009. *National Infrastructure Protection Plan*. http://www.dhs.gov/files/programs/editorial_0827.shtm, accessed August 29, 2011.

U.S. Environmental Protection Agency (USEPA). 2005. *Economic Analysis for the Final Long Term 2 Enhanced Surface Water Treatment Rule*. EPA 815-R-06-001.

chapter Nineteen

Other (No Less Important) Considerations

Legal Considerations contributed by Matthew S. Carstens, Esq.

Learning Objectives

After reading this chapter, you should be able to:

1. Describe why engineers should consider the safety of the public, maintenance personnel, and constructors, and describe how engineers can plan and design for safety.
2. Describe why engineers should consider maintenance needs, and explain how engineers can plan and design to enhance maintenance.
3. Describe why engineers should consider constructability, and how engineers can plan and design to enhance constructability.
4. Describe why engineers should consider aesthetics, and how engineers can plan and design to enhance aesthetics.
5. Describe why engineers should consider political aspects of infrastructure engineering and management.
6. Describe why engineers should consider legal aspects, and how engineers can plan and design to minimize potential for litigation.

Introduction

In this chapter, we will introduce aspects of safety, maintenance, constructability, aesthetics, and political and legal matters. When writing this text, we hesitated "lumping" several additional considerations into a single chapter for fear of an implication of lesser significance. However, the considerations in this chapter are indeed very important to infrastructure engineering and management; for each, there are many engineers that would undoubtedly say "*That* is one of the most important considerations in my line of work!"

Introductory Case Study: Stormwater Management in Milwaukee, Wisconsin

Stormwater management has become a major focus area for municipalities. As increased development occurs within a watershed, increased runoff results. This result is confirmed by the Rational Method (Equation 9.1 of Chapter 9, Analysis Fundamentals), which states that peak runoff flowrate is the product of the land use runoff coefficient, C; rainfall intensity, i; and watershed area, A. Given a constant watershed area and rainfall intensity, the peak flowrate of runoff increases as the runoff coefficient increases; this increase in C occurs with added imperviousness. Without proper management, the runoff flowrates will continue to increase as development progresses in a watershed, as will the costs associated with flood damages.

The Milwaukee, Wisconsin, metropolitan region has invested billions of dollars in recent decades to lessen the impacts of stormwater runoff. One notable project is the "Deep Tunnel," which consists of nearly 30 miles of massive storage tunnels, 14 to 32 feet in diameter, hundreds of feet below the ground surface. This tunnel system can store over 5 million gallons of wastewater and stormwater, the latter from infiltration and inflow and from combined sewers in the older portions of Milwaukee. Other projects include the purchase of over 1,000 acres of undeveloped land to prevent future development and associated runoff; purchase and removal of homes in order to widen urban channels and thus allow more flood storage; the installation of residential rain barrels/cisterns to collect roof runoff; and installation of rain gardens. In order to be effectively implemented, these projects required careful attention to the considerations discussed in Chapters 9 through 18, as well as additional considerations discussed in this chapter.

Safety Considerations

In this chapter, safety will be differentiated from security (Chapter 18). Primary safety considerations of infrastructure projects include protecting the public (users), municipal employees (operators and maintenance personnel), and construction workers (builders) from harm during the construction, start-up, operation, maintenance, and use of the infrastructure components and systems.

PROTECTING THE PUBLIC

Holding public safety paramount is the first Fundamental Canon of the American Society of Civil Engineers (ASCE) Code of Ethics. Protecting the safety of end users of infrastructure is accounted for in many design codes and standards: sight distances for horizontal and vertical curves of roads; lengths of accelerating lanes for highway on-ramps; resilience of structures able to withstand seismic activity, wind loads, and snow loads; maximum

Figure 19.1 **Traffic Lane Closures During Roadwork.**

Source: Courtesy of Caltrans, District 4 Photography/J. Huseby.

The Decreasing Traffic Fatality Rate

The number of fatalities from vehicle accidents has decreased steadily from 1.73 fatalities per 100 million vehicle miles traveled in 1994 to 1.13 fatalities per 100 million vehicle miles traveled in 2009. The decline is attributed to improvements in road designs (including signage and lighting), along with improvements in vehicle safety. However, more than 30,000 traffic fatalities occur per year and efforts to save lives continue. At a recent press conference, U.S. Transportation Secretary Ray LaHood stated, "America's roads are the safest they've ever been, but they can be safer. And we will not rest until they are."

 Zero in Wisconsin: The Campaign to Reduce Traffic Fatalities

allowable contaminant concentrations in drinking water; and extent of buffer zones around airport runways.

Engineers and construction managers are also responsible for protecting the public during construction. In some cases, this can be accomplished by securing a site with adequate fencing and temporarily closing an adjacent sidewalk. However, many projects are not so simple. For example, roadway reconstruction projects often have long stretches of road with reduced lanes (Figure 19.1), entering and exiting construction equipment, reduced visibility from dust, and other factors that can affect public (and worker) safety.

Case in Point: Blasting and the Milwaukee Deep Tunnel Project

During the construction of the Milwaukee Deep Tunnel, concern was raised about the potential for blasting operations to disrupt delicate eye surgeries performed in a nearby medical center. Before construction commenced, an expert was contracted to provide a scientific evaluation demonstrating the lack of impact on procedures at the medical center.

Case in Point: Suicide Prevention

More than twenty recent suicides from jumping off of the Tappan Zee Bridge in New York led state transportation

 Time Magazine Article on Suicide Prevention on the Golden Gate Bridge

officials to the installation of phones with direct connections to suicide prevention hotlines and signs reading "Life is worth living." Minimizing suicide potential with barriers and other measures is a design feature in many new bridges.

PROTECTING OPERATIONS AND MAINTENANCE WORKERS

Practical experience by the design engineer and communications with operations and maintenance personnel during design increase the likelihood that the final design will be safely operated and maintained. When designing, engineers need to consider the needs of the people using their designs. The considerations are practically limitless, but include clearances and access to equipment for repair work, guardrails to protect from falls, environmental conditions to be experienced by workers (e.g., wind, ice, excessive heat), eyewash stations in chemical storage areas, adequate storage areas for tools and equipment, and adequate ventilation.

Many aspects of the infrastructure are located underground and are considered **confined spaces**. Examples include tunnels, vaults, large diameter sewers, and manholes. While many restrictions exist for working in confined spaces, these are sometimes neglected out of ignorance or because they are perceived to be a nuisance, often with tragic outcomes.

> **Confined spaces** are inherently dangerous enclosed areas with limited openings for entry or exit that are not intended for continuous occupancy.

Confined Space Safety

As interns or young engineers working in the field, you may be working around confined spaces. We would like to emphasize that you should NEVER enter a confined space, even for "a minute or two," not even to rescue someone. Indeed, the Centers for Disease Control and Prevention (CDC) estimates that 60 percent of confined space fatalities have involved would-be rescuers. Many hazards may exist in confined spaces, including oxygen-deficiency and flammable or toxic gases. Consider the following sobering case reports from the CDC.

"A 21-year-old worker died inside a waste water holding tank that was four feet in diameter and eight feet deep while attempting to clean and repair a drain line. Sulfuric acid was used to unclog a floor drain. The worker collapsed and fell face down into six inches of water in the bottom of the tank. A second 21-year-old worker attempted a rescue and was also overcome and collapsed. The first worker was pronounced dead at the scene and the second worker died two weeks later. Cause of death was attributed to asphyxiation by methane gas. Sulfuric acid vapors may have also contributed to the cause of death."

"A 27-year-old sewer worker entered an underground pumping station (8' × 8' × 7') via a fixed ladder inside a three foot diameter shaft. Because the work crew was unaware of procedures to isolate the work area and ensure that the pump had been bypassed, the transfer line was still under pressure. Therefore, when the workers removed the bolts from an inspection plate that covered a check valve, the force of the waste water blew the inspection plate off, allowing sewage to flood the chamber, and trapping one of the workers. A co-worker, a supervisor, and a policeman attempted a rescue and died. Two deaths appeared to be due to drowning and two deaths appeared to be due to asphyxiation as a result of inhalation of 'sewer gas.'"

 Additional Confined Spaces Case Studies

 A Guide to Safety in Confined Spaces

Case in Point: Maintenance of Milwaukee Deep Tunnel

Inspections of the Deep Tunnel system are performed to evaluate structural integrity, severity of leaks, and amount of accumulated debris. Extensive measures are taken to ensure worker safety during these inspections, including air quality monitoring to detect poisonous or asphyxiating gases, on-site presence of emergency responders, and redundant communication capabilities.

 Deep Tunnel Inspection Description with Photographs

PROTECTING CONSTRUCTION WORKERS

The Occupational Safety and Health Administration (OSHA) has many requirements in place to protect workers of all industries. Prior to the enactment of these requirements, unsafe working environments were commonplace (Figure 19.2). Today such conditions can be witnessed in developing countries (Figure 19.3). It is not uncommon for construction personnel to complain about, and sometimes ignore, some of the safety requirements, claiming that they reduce efficiency, increase cost, and are unnecessary. But there is no doubt that OSHA requirements have saved countless lives and prevented countless injuries. However, even with these safety requirements, injuries and fatalities do indeed frequently occur at construction sites. Approximately 1,000 fatalities per year occur on construction sites in the United States, a rate

 NAHB-OSHA Jobsite Safety Handbook

Figure 19.2 Construction of the Canada Life Building in 1930. Note that the workers have no hardhats or harnesses for fall protection.

Source: Wikipedia: http://commons.wikimedia. org/wiki/File:Canada_Life_last_stone.jpg.

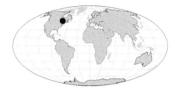

Figure 19.3 Unsafe Construction Methods Currently Used in India.

Source: Courtesy of D. Kraemer.

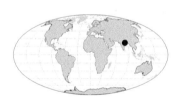

Unsafe Construction Practices Photo Gallery

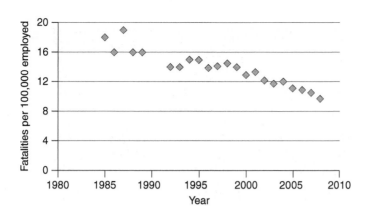

Figure 19.4 Historical Construction Industry Fatality Rates.

Data source: Bureau of Labor Statistics.

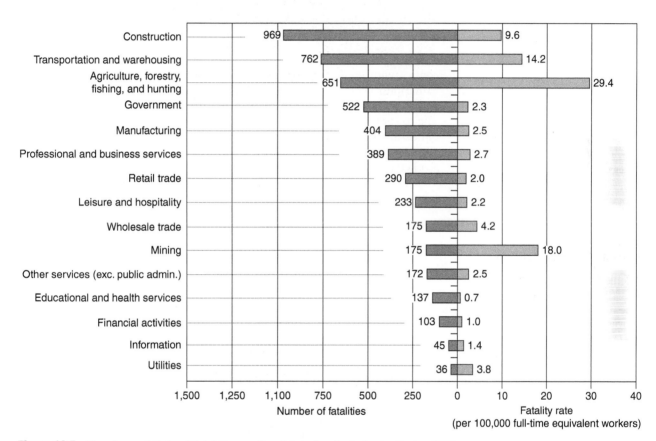

Figure 19.5 **Number and Rate of Fatal Occupational Injuries, by Industry Sector, 2008.**

Source: Bureau of Labor Statistics, 2009.

of approximately 10 deaths per 100,000 workers (Bureau of Labor Statistics, 2006a). More stringent regulations and an increase in the emphasis on safety have resulted in a continually declining rate of construction worker fatalities, which today is nearly one-half the rate of 1985 (Figure 19.4). The work-related injury and illness rate for the construction industry is 2,700 per 100,000, or approximately 3 percent (Bureau of Labor Statistics, 2006b).

Construction workers face the highest risks of any occupation in terms of the number of fatalities and the fourth highest in terms of fatality rate (Figure 19.5). Common causes of deaths, injuries, and illnesses include falls,

Figure 19.6 A Worker Following Safety Regulations on a Bridge Column During Construction.

Source: Courtesy of Caltrans, District 4 Photography/J. Huseby.

Video of Ironworking on Trump Tower in Chicago

contact with equipment or objects, transportation incidents, and exposure to harmful substances or environments. Injuries and fatalities occur for a variety of reasons: unanticipated conditions; not following safe practices due to carelessness or lack of communication; and of course the inherently dangerous nature of the work (Figure 19.6). It is estimated that more than 40 percent of construction fatalities are linked to improper consideration of worker safety during project design (Behm, 2005).

"The regulations are there," said Philip Russotti, a New York lawyer who represents injured workers. "Safety costs money," he said. "If you are motivated by profit, you are motivated to push people to move quickly and cut corners with safety. But you're playing with people's lives."

"Talk About Angels . . ."

The title above was a comment made by a bystander of the November 10, 2009 incident involving a 500 ton crane at a U.S. courthouse construction site in Cedar Rapids, Iowa (Figure 19.7).

The crane fell while more than fifty construction workers were onsite, landing on a truck and two occupied construction trailer offices. Miraculously, no injuries resulted. The courthouse will replace its predecessor, which was inundated during a 500-year flood in 2008.

Figure 19.7 The Tower Crane That Fell in November 2009 at the U.S. Courthouse Construction Site in Cedar Rapids, Iowa.   The photo was taken prior to occurrence of the incident.

Source: Courtesy of P. Fields.

Case in Point: Milwaukee Deep Tunnel Project

The Deep Tunnel is being constructed in three phases, with the third phase currently in progress. During the first phase, which required approximately 1,000 construction workers at the peak of construction, five construction-related deaths occurred. Construction fatalities and disabling injuries in the second and third phases have been successfully prevented by adopting an improved safety plan including independent (third-party) full-time dedicated safety officers.

Maintenance Considerations

The need for maintenance has been highlighted in the ASCE Report Card on the Nation's Infrastructure as well as on national and local news. Common examples of the effects of inadequate maintenance on infrastructure components are listed in Table 19.1.

Inadequate maintenance may occur because of budget limitations, improper communication, or neglect. If not properly maintained, infrastructure components and systems may require more repairs, have shortened periods of service, and pose increased risks to the public. Engineers must be skilled at quantifying and communicating the costs associated with inadequate maintenance to elected officials, public works personnel, and the public. Some cities are going so far as to hire public relations firms to help them educate the public and to gauge citizen response to rate increases that are needed for system upgrades. While this may seem to be an extreme measure, keep in mind that the majority of the public does not understand basic infrastructure, let alone the complexities and needs of infrastructure systems.

The World Without Us

In the bestselling and award-winning nonfiction book, *The World Without Us*, Alan Weisman answers the question "what happens to the Earth when human beings disappear?" It details a long-term environmental "reversal" of our built environment and highlights the immediate initiation of decay resulting from lack of maintenance. We enthusiastically recommend the reading of this book.

Table 19.1

Representative Examples of the Effects of Inadequate Maintenance on Infrastructure Components

Infrastructure Component	Maintenance Required	Result of Inadequate Maintenance
Earthen dams or levees	Mowing to prevent establishment of deeply rooted trees	Tree root penetration weakens structural integrity
Steel bridges	Frequent sanding and repainting	Rust and corrosion weaken structural integrity
Stormwater inlets	Clearing of debris (litter, leaves, grass clippings)	Clogged inlet results in localized flooding
Sanitary sewers	Removal of tree roots that penetrate sewers through cracks in pipes	Blockage of sewage flow and potential backups into basements
Water main valves	Operate each valve annually	Increased likelihood of valves rusting in a fixed position
Street lighting	Trimming branches on nearby trees	Blockage of intended lighting, reduced visibility
Paved roads in cold climates	Sealing cracks in pavement	Water seeped through cracks freezes and expands. After repeated freeze/thaw cycles, potholes develop
Rip-rap armoring of bridge abutments	Replacement of lost material after major floods	Scour of earthen support of abutment compromising structural integrity
Electrical lines	Trimming trees under or in proximity to lines	Power outages and exposure to high voltage due to "flashover" from lines to nearby trees

Constructability Considerations

As noted in a sidebar of Chapter 14, Economic Considerations, a primary benefit of design-build process is that the design *and* construction is carried out by a coordinated team. This approach is often considered superior to the conventional process in which a design firm is selected first followed by competitive bidding to select the construction firm. In such a case, coordination between designers and constructors inherently begins *after* design has been completed. Unfortunately, many design engineers do not have extensive construction experience, and likewise many construction professionals do not have extensive design experience. Construction laborers and managers have often criticized design engineers regarding designs that cannot be readily built.

In an effort to incorporate the construction perspective into the design process, **constructability** reviews are becoming increasingly common. Techniques used to enhance constructability include:

1. Design elements are standardized
2. Module/preassembly designs are prepared to facilitate fabrication, transport, and installation
3. Designs promote construction accessibility of personnel, material, and equipment
4. Designs facilitate construction under adverse weather conditions

In a constructability review, construction professionals (either internal personnel or external consultants) review designs and work with designers to propose solutions to problems that are identified. For complex projects, review teams often include representatives from regulatory agencies, utilities, and material suppliers. It is generally believed that reviews are most effective when conducted periodically *throughout the design process* (e.g., at 30 percent, 60 percent, and 95 percent design completion stages), rather than after design is complete. However, these reviews add expense and extend project timelines. If, however, throughout the course of the reviews, significant problems are averted, overall project cost and time to completion may *decrease*. That is, problems that would have increased project costs and delayed completion may be avoided.

Many studies have shown that constructability reviews provide benefit/cost ratios of 10:1 or greater; that is, the value gained through the reviews is 10 times greater than the cost of the review. The potential benefits include: reduced construction cost because of increased worker productivity; less construction rework; shorter construction schedules; improved construction site safety; increased commitment from design and construction personnel (Rajendran, 2007).

In many respects, constructability reviews are "win-win-win" situations for designers, owners, and contractors. The designers improve their design skills. Owners are more likely to have their project completed on time. And contractors can potentially increase their profit. Consider that the profit gained by project contractors relies on the actual construction cost equaling or being less than the estimated construction cost. Improper consideration of the constructability of a project can lead to cost overruns (resulting in reduced profit, or even worse, a loss).

"Construction friendly," or construction sensitive, considerations have greater potential impact on cost savings when implemented at early project

---

Constructability (or constructibility) was traditionally considered to be the integration of construction knowledge and experience into the engineering and design of a project. Today, constructability is gaining broader application throughout the *entire life* of a project, from planning to decommissioning.

AASHTO Constructability Review Best Practices Guide

The Often Forgotten Benefit

One benefit of constructability reviews that is often overlooked is that design engineers learn more about construction and that construction personnel learn more about design. This increases the likelihood of buildable designs (that rely less on constructability reviews) on future projects. It also helps a constructor understand that some design elements that the constructor views as unnecessarily difficult to construct, redundant, overly conservative, or simply unnecessary, are indeed necessary. Such collaboration can lead to better relationships between the two disciplines that are sometimes at odds with one another.

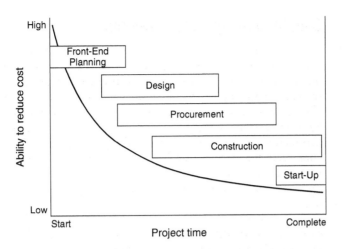

Figure 19.8 Importance of Performing Constructability Reviews Early in the Process. This generalized diagram illustrates the ability to reduce project cost that occurs when constructability improvements are made in earlier stages of a project. Superimposed on this curve is a Gantt chart showing key steps in the project life. Not illustrated on this graph is the fact that once construction is nearly complete, most changes are likely to dramatically *increase* costs.

Source: Adapted from Thomas et al., 2004.

phases (Figure 19.8). The early phases (conceptual planning) are often carried out by architects, and unfortunately the lack of mutual trust, respect, and credibility that sometimes exists between builders and design engineers can also exist between engineers and architects. At the opposite end of the project life cycle, as mentioned earlier in this chapter, operations and maintenance are also important considerations. Extending the scope of constructability reviews to incorporate these post-construction considerations is worthwhile.

"Non-Engineering" Summer Jobs

As academic advisors, the authors of this text meet with undergraduates to plan not only class schedules, but also to discuss career development. During advisory sessions, we typically ask students if they have any engineering related work experience. It is not uncommon for students to underestimate the potential relevance of their work experience, for example by replying "Not really, I only worked on house construction for a few summers." For the very reasons that constructability reviews are performed on many projects, a student's construction experience is *indeed* engineering related. So are experiences such as maintenance or landscaping work. In fact, as we have emphasized throughout this book, because of broad considerations necessary for infrastructure engineering and management, nearly *any* job will provide some experience that will prove to be beneficial in your career, even if for no other reasons than gaining the opportunity to work with other people and developing a sense of responsibility.

Case in Point: Constructability of Milwaukee Deep Tunnel Project

The second phase of the Deep Tunnel project—the 7-mile, 20-foot diameter Northwest Side Relief Sewer—was completed ahead of schedule at a final cost that was $8 million (7 percent) less than the estimated cost of $120 million for a variety of reasons, none less important than the careful consideration of constructability (Figure 19.9). Coordinated efforts between the owner, design engineers, and construction managers resulted in detailed bid documents that allowed bidders to accurately estimate costs with lower contingencies for "surprises."

Another example of constructability is the adoption of a construction method to increase efficiency. When tunneling, the seepage of groundwater into the tunnel can cause delays.

For the Deep Tunnel project, a 2-foot diameter boring was extended 200 feet ahead of the larger diameter tunnel-boring machine. If groundwater seepage was encountered, grouting (highly pressurized placement of concrete) was performed to seal fractures in the rock so that tunneling could proceed with fewer delays.

Milwaukee's Northwest Sewer Relief System—Entry for American Public Works Association 2006 Public Works Project of the Year Award

Figure 19.9 A Large Shaft Under Construction to Connect An Existing Interceptor Sewer (Left) with the Tunnel More Than 100 Feet Below.

Source: Courtesy of Black and Veatch Corporation.

Many recent advances in construction methods and materials allow creative engineers to meet challenges that were considered impossible in the recent past. However, choosing to use a new method carries considerable financial risk, as well as professional risk; that is, there is increased risk of damaging a firm's or a designer's reputation when using an innovative method. The engineer must carefully evaluate whether or not the method or material will work under the range of specific conditions to be expected during construction *and* over the design life of the project.

A Design That a Constructor Wouldn't Have Designed

Structural steel is designed by structural engineers working out of an office, formed and drilled offsite, and transported to the construction site. Figure 19.10 demonstrates a design that would have been corrected by a detailed constructability review. This figure shows a base plate for a vertical support with pre-drilled holes for mounting bolts (circled) placed in a location with poor accessibility (inside the flanges of the I-shaped column) making them difficult to tighten. Also, the mounting bolts should be toward the outer perimeter of the plate for increased stability. Such a connection would not likely have been designed by an engineer with extensive construction experience.

Figure 19.10 A Difficult-to-Construct Design.

Source: Couresy of P. Fields.

Another aspect of constructability is **deconstructability**; that is, designing an infrastructure component so that at the end of its useful life it can be safely and cost-effectively taken out of service. At one extreme, consider the decommissioning of a complex nuclear power facility and how thoughtful design might make the process much less expensive. At the other, simpler extreme, consider road signs that are bolted to a ground base plate rather than being fixed in a borehole filled with concrete.

Case in Point: Speeding Up the Construction Process

 Because of the tremendous disruption to regional transportation in the Minneapolis-St. Paul area, the I-35 W bridge was reconstructed at an accelerated pace. One means of speeding up the construction process was by precasting segments (Figure 19.11 and Figure 19.12) at an offsite location.

Figure 19.11 Precast Segments of the Replacement I-35 W Bridge.

Source: Minnesota Department of Transportation.

Figure 19.12 Placement of Precast Segments of the Replacement I-35 W Bridge.

Source: Minnesota Department of Transportation.

Case in Point: Fighting the Elements

Antarctica may be the most challenging location for construction activities on Earth. The National Science Foundation's 65,000 square feet Amundsen-Scott South Pole Research Station (Figure 19.13) was completed in 2008 after nearly 10 years of construction, at a cost of $150 million. Forty thousand tons of material travelled by ship from California to New Zealand, then by plane to the site in approximately 10-ton increments. The construction season is limited to the "warm" summer period, when temperatures are typically minus 20°F. At these temperatures, workers can be outside for about an hour at a time before going inside to get warm. Construction schedules needed to account for this extra time, as well as the decreased efficiency of workers due to their bulky clothing. Design of the facility also required special attention to constructability.

The design of the station includes many unique features: an adjustable foundation to keep the structure level; the ability to raise the elevation of the entire structure (due to 8 inches of annual snow accumulation that never melts); and an airfoil design that causes the wind to scour the snow out from beneath the structure.

 National Geographic Video of Construction of the Amundsen-Scott South Pole Research Station

 National Science Foundation Special Report on U.S. South Pole Station

Constructability Considerations 367

Case in Point: Fighting the Elements (Continued)

Figure 19.13 The Amundsen-Scott South Pole Station (Foreground) Nearing Completion. The geodesic dome (background) served as the research station from 1975–2003 and has since been removed.

Source: National Science Foundation/M.Conner.

Case in Point: Temporary Infrastructure

A recent \$4 billion expansion to a petroleum refinery in Louisiana created the need for substantial temporary infrastructure during construction. At its peak, nearly 10,000 workers were on site and materials were delivered around the clock. A temporary docking facility on the Mississippi

River and a temporary bridge were constructed. After project completion, the majority of this infrastructure was dismantled.

Building Tomorrow's Landmark Infrastructure

How many of your future engineering projects will people be fighting to preserve in 50 or 100 years? ASCE's Historic Civil Engineering Landmark Program recognizes projects with historical significance, a number of which are outstanding because of their special attention to aesthetics.

 List of "Monuments of the Millennium"

 List and Description of "Seven Wonders of the Modern World"

 List of Historic Civil Engineering Landmarks

Aesthetic Considerations

Great expense was once paid for the architectural grandeur of government buildings (e.g., post offices, city halls, court houses.) This was also true of universities (both public and private). At many older universities, a change in architecture is readily apparent: grand buildings built prior to World War II, sterile ("boring") buildings of the 1960s to the 1980s and a resurgence of architecturally interesting (and often "green") buildings from the last decade.

An emphasis on aesthetics, much like environmental concerns discussed in Chapter 15, is related to the economy. When funding is limited, aesthetics is not given as high of a priority as it might otherwise be given. Additionally, public demand for aesthetically pleasing infrastructure can typically be expected to be proportional to the wealth of a community (or a neighborhood within a community). Virtually everything from road signs to traffic signals to parking ramps and bridges can be aesthetically enhanced at varying degrees of expense. Engineers that seek to enhance the aesthetics of projects should expect to be ensnared in a public debate over whether or not the added expenses are justified. Additionally, engineers must not assume that all clients want the least expensive (and most likely the least aesthetically pleasing) option. Thoughtful communication with the client and the end user should occur to allow a range of alternatives and associated costs and benefits to be considered.

An engineer should be proud of any infrastructure component that he or she designs, regardless of how aesthetically pleasing the final product is. However, there is little doubt that a greater degree of satisfaction is likely if the final product is aesthetically pleasing *in addition* to being a useful contribution to society.

Case in Point: Hart Park Flood Mitigation Project

To decrease flood damage in the city of Wauwatosa (a suburb of Milwaukee) and downstream Milwaukee, the Milwaukee Metropolitan Sewerage District (MMSD, the regional government agency in charge of wastewater, stormwater, and flood management) expanded Hart Park from 20 to 50 acres, added levees, and removed several feet of floodplain soil to increase the floodwater storage volume. Approximately 100 homes and businesses were purchased and removed. Rather than a concrete-lined channel, which was a typical approach in the past (Figure 19.14), the riverbank was designed to create a more natural setting (Figure 19.15).

Figure 19.14 A Concrete-Lined "Stream" in Milwaukee.

Source: M. Penn

Figure 19.15 Hart Park Flood Mitigation Project.

Source: Courtesy of MMSD

Political Considerations

The following is an excerpt from a promotional advertisement for an ASCE professional development seminar titled "Surviving in a Political Environment: Tips for Obtaining, Retaining and Advancing your Job When Working on Public Projects":

> "Consulting and public sector engineers who work on the front lines of local government are frequently exposed to pressures from citizens and elected officials that may be inconsistent with their professional training. Often swirling controversy results and you may find yourself alone standing behind your engineering degree. However, you can improve your resiliency and weather the storms more pleasantly if you have prepared the foundation of trust with local officials. This seminar will focus on what to do and what not to do in order to keep your sanity and your job."

Many civil and environment engineers working on infrastructure projects will admit that politics are one of the least favorite aspects of their jobs. On the other hand, many successful infrastructure engineers actually *embrace* the political process and use it to their professional advantage. There are many aspects of politics that are intertwined with engineering, both rational and irrational. The vast majority of infrastructure projects are publicly funded and thus linked to decisions of public officials and paid for (directly or indirectly) by taxpayers or utility customers.

No discussion of politics can proceed without mentioning the appropriation of funding. At the state level, and especially at the federal level, influential

Figure 19.16 An Example of Public Opposition to a Highway Project.

Source: M. Penn.

legislators are skilled at obtaining funding for their districts (so-called "pork" projects). While there is great disdain by the general public over this practice, it is rampant. (Of course, few people complain when the projects and associated jobs are delivered to *their* districts.) Many infrastructure projects fall into this category of funding.

Infrastructure funding also sometimes suffers from political reappropriation. Transportation funds are a substantial fraction of a state's budget, and it is not uncommon for these funds to be "raided" for money to fund non-infrastructure projects and services.

Elected officials are expected to make educated and ethical decisions on behalf of their constituents. The public is often "split" on decisions of all kinds—rarely does the vast majority agree on anything. Gauging public support is a difficult task. Only in rare cases are public referendums held on municipal funding decisions, and it is impossible for officials to perform statistically valid polls for every decision they face. They *must* make decisions nonetheless. Ignoring a problem is *not* a solution, nor is endless debate without action. President Harry Truman stated that "even a wrong decision is better than no decision at all"; this implies that at some point, regardless of the information available, decisions *must* be made (hopefully with an adaptive management strategy in place as discussed in Chapter 2). After a poor decision is made, it is difficult for elected officials to admit that they made a mistake. It may be equally difficult to defend the "right" decision if the media or public opinion turns against the decision; this is especially true with infrastructure decisions, which are expensive and for which many benefits are often delayed, hidden or misunderstood. Understanding political realities and the pressures under which elected officials perform is important for engineers because infrastructure projects are subject to public and political forces (Figure 19.16).

Saying "No" to Raiding of Transportation Funds

In November 2010, in a statewide referendum, 70 percent of Wisconsin voters approved amending the state constitution to protect the transportation fund.

Case in Point: Public Outcry and the Milwaukee Deep Tunnel

The Milwaukee Deep Tunnel was constructed at great expense (more than $2 billion). By chance, in the first few years after completion, several large (i.e., low probability) storms occurred, resulting in overflows that the tunnel was, in the public's eye, supposed to eliminate. Public outcry ensued, as the uneducated (with regard to engineering design and risk management) public viewed the billions of dollars spent on the project as "wasted" because overflows were still occurring. Additional concern was based on the fact that the project experienced highly publicized delays and cost overruns.

Case in Point: (Continued)

This outcry is highly frustrating for the tunnel owner (MMSD—Milwaukee Metropolitan Sewerage District) and elected officials who championed the project, given that the tunnel has indeed dramatically reduced overflows. The number of overflows has been reduced from an average of about 60 events per year to only a few events per year, and the volume released has also greatly decreased (Figure 19.17). What continues to this day is a classic example of a "battle" between the media, the general public, and politicians on one side and a public works authority (MMSD) on the other side.

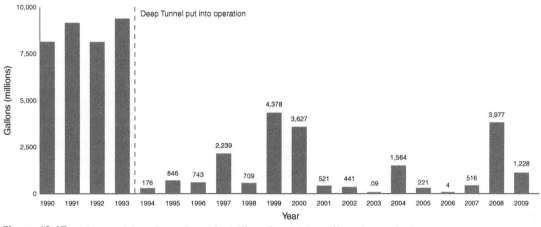

Figure 19.17 **Volume of Overflows from the Milwaukee Metropolitan Sewer System.**

Data source: MMSD.

If It Came To a Vote . . .

The public has a well-documented disdain and distrust for elected officials. Unfortunately, this distrust and disdain is sometimes well deserved. On the other hand, elected officials deserve a bit of sympathy as the choices they make are indeed very difficult, especially those concerned with infrastructure funding.

A public works director for a city once told one of the authors, "If we had a public vote on every infrastructure project, we probably wouldn't have *any* infrastructure." His point is one that we have tried to emphasize throughout this book: in most cities, budgets are tight, and most people feel that their taxes and fees are already too high. If the general public had its choice, they would likely choose against funding what they cannot see and what they do not understand. Of course, their choices would rapidly change once they have sewage backing up in their basements or their roads become "undriveable."

Politics Putting Project Funding on the Shelf

Two counties, several municipalities, and other organizations teamed to write a Transportation Enhancement grant proposal for over $1 million to link the hiking/biking paths of two small communities. The project was funded in 2005. After writing the grant, the membership of the County Board of Supervisors in charge of administering the grant changed due to local elections. The new board was less supportive of the project than the previous board. Implementation of the grant required negotiations between the county and the state environmental regulatory agency regarding the routing of the path. Conflicts soon arose. A new Secretary of the regulatory agency was then appointed, causing further delays. Without significant support from the county administering the funds (even with continued offering of support from the neighboring county), the project has been effectively "shelved." More than $1 million dollars in funding has sat dormant for 5 years.

Politics and the Birth and Death of the Yucca Mountain Nuclear Waste Repository

Approximately $15 billion has been spent on scientific studies and initial construction of the Yucca Mountain Nuclear Waste Repository, 80 miles northwest of Las Vegas, Nevada, for the nation's high-level radioactive waste. The site was chosen by Congress in 1987, in what many consider to be "political bullying"; at that time, Nevada's population was only about 1 million (it is nearly 3 million now) and its political clout was much less than it is today.

The project has been plagued by delays, scientific uncertainty, and opposition, primarily from environmentalists and residents of Nevada. Early in 2010, the Obama Administration decided to eliminate funding and withdraw the license application for the project. The announcement was made without a detailed rationale and ignited intense debate. In the Motion to Withdraw dated March 3, 2010, it is stated "... the Secretary of Energy has decided that a geologic repository at Yucca Mountain is not a workable option for long-term disposition ... It is the Secretary of Energy's judgment that scientific and engineering knowledge on issues relevant to disposition of high-level waste and spent nuclear fuel has advanced dramatically over

the twenty years since the Yucca Mountain project was initiated." Many critics of this decision claim that it was a political move by President Obama as a "favor" to the Senate Majority Leader Harry Reid of Nevada. Supporters of the decision applaud the potential savings of billions of additional dollars to complete the construction of a single long-term storage facility that may ultimately pose a threat to the environment and human health due to inadequate containment at the facility or cross-country rail and truck transportation of radioactive waste (Figure 19.18).

On June 3, 2008, only 2 years prior, the previous administration's Secretary of Energy held a press conference to announce the Yucca Mountain license application to the Nuclear Regulatory Commission (NRC), stating, "We are confident that the NRC's rigorous review process will validate that the Yucca Mountain repository will provide for the safe disposal of spent nuclear fuel and high-level radioactive waste in a way that protects human health and our environment ... On a personal level, let me say that I know some Americans have deeply felt concerns about the Yucca Mountain facility. And I do not seek

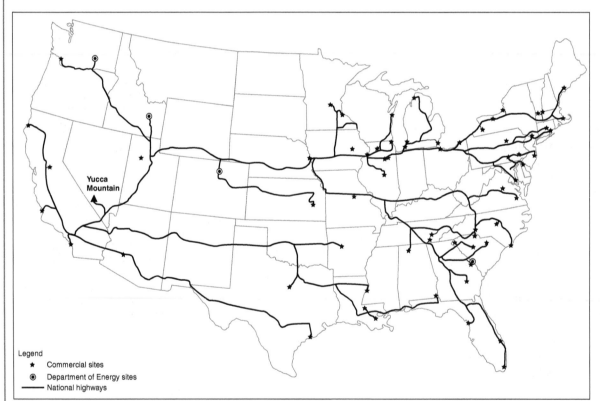

Figure 19.18 Truck Routes for Transport of High-Level Radioactive Waste and Spent Nuclear Fuel to the Yucca Mountain Repository. Rail routes not shown. Long travel routes to a single national facility increase the risk of contamination resulting from accidents or terrorism.

Source: Redrawn from the U.S. Department of Energy, 2002.

to dismiss those concerns nor minimize them; on the contrary, issues of health, safety and security have been paramount during this entire process. They are the driving factors in all decisions we have made and that will continue going forward, as will our commitment to ensuring that this continues to be an open and transparent process. And so I want to speak to all Americans when I say with great confidence that I stand behind this application, as do the men and women who have worked so hard over the past two decades to get us to this point. It is our collective judgment that the Nuclear Regulatory Commission will conclude that this project can and will be completed in a safe and secure manner and, upon completion, that it will contribute to materially improving our nation's energy security."

The Secretary of Energy's "Blue Ribbon Commission on America's Nuclear Future" is currently evaluating options for long-term nuclear waste storage, but it appears that Yucca Mountain is not an option. According to Senator Reid, "one of the things they will *not* discuss is Yucca Mountain, that's off the table." If indeed this is the case, an important question that remains unanswered is "What will come of the years of work and billions of dollars that went into the project?" In business, it is common for a new initiative to include an exit strategy, that is, a plan for discontinuance. Apparently for Yucca Mountain, there was no exit strategy.

Politics, in a more loosely defined manner to mean maneuvering for power or control, can also come into play in the selection of engineering firms for projects. You may find it surprising that the selection of an engineering firm for designing a project is often *not* based on the least expensive or "low-bid" proposal. Using the low-bid process, based solely on cost, has the potential of selecting a firm that is unqualified and which may create a poor quality design. Rather, a process known as **quality based selection (QBS)** is often used. After selection of a "shortlist" of best-qualified firms, interviews and site visits are conducted. The top firm is then selected, and negotiations of fees, terms and conditions commence.

QBS often leads to enhanced project quality because it provides opportunities for cooperation and communication early in the planning phases. QBS may *indirectly* involve cost, however. For example, using QBS allows the project owner to factor in the practice of some firms that estimate low costs and then return repeatedly to the client for more money. On the other hand, QBS allows the project owner to factor in the strong reputations of some firms that often overestimate costs but are known for turning in high-quality designs for less than the original estimate. Another selection process is a modified QBS approach in which firms provide Statements of Qualifications *and* cost bids, with a weighting factor associated with each, for example, 80 percent of evaluation based on qualifications and 20 percent based on cost.

In QBS, a selection board is appointed to review proposals. The nature of the review of qualifications is *inherently subjective*, which is where politics can come into play. While unethical, board members may base selections on favoritism (e.g., choosing the company that "wines and dines" the most, or is managed by a friend or relative). Boards may also make selections in such a manner as to "spread the wealth" to many different companies, even to some that are less qualified.

> **Quality based selection (QBS)** is the selection of an engineering (or architectural) design firm based on several factors *other than cost*, including the qualifications, experience, and reputation of the firm, and the quality of their proposal or **Statement of Qualifications**.

> **Statements of Qualifications** are proposals submitted by firms for QBS that typically include description of the firm, relevant project experience, qualifications and roles of personnel to be assigned to the project, project approach including tasks and plans of execution, and project schedule including a Gantt chart.

A Guide to QBS

Legal Considerations

The collapse of the I-35 W bridge and a traffic fatality in the Fort Point Channel Tunnel in Boston serve as poignant reminders to engineering students and professionals that there are legal duties associated with the work done by engineers and construction professionals. In July 2006, a 3-ton ceiling panel

in Boston's Fort Point Channel Tunnel, a component of the "Big Dig" project, broke loose and fell on a car passing through the tunnel at the time, killing the car's passenger and seriously injuring the driver.

In the aftermath, six employees of the primary concrete supplier were indicted on federal criminal charges for falsifying records over allegedly substandard concrete supplied by their employer. Numerous other civil lawsuits have been filed against other suppliers, including the company that provided the fastener for the concrete slab involved in the fatal collapse.

In the aftermath of the I-35 W bridge collapse, the state of Minnesota and the victims filed civil lawsuits alleging breach of contract and negligence, among other claims, against the successor of the original engineering firm that designed the bridge, along with engineering companies hired by the State of Minnesota to inspect and maintain the bridge. At the time of writing this book, some of those lawsuits are still pending.

The "body of legal rules" that applies to engineers is broad, varies from state to state and jurisdiction to jurisdiction, and emanates from the federal, state, and local governments. The result is a multi-layered system of laws that govern the country as a whole and each state individually, and includes the United States Constitution; each state's constitution; statutes (laws enacted by Congress and state legislatures); federal and state administrative rules and regulations; executive branch orders on the federal and state level; local laws and ordinances; and the decisions of trial and appellate judges in the federal and state courts.

Besides the origins of these laws, responsible engineers should be familiar with the adversarial process in which legal disputes are resolved, that is, the American court system and its alternatives. Like the origins of American law, the court system is multi-faceted, consisting of federal courts and state courts of general jurisdiction, courts with specialized jurisdiction, and appellate courts.

Due to the cost and time involved in resolving a legal dispute in the court system, parties in a legal controversy increasingly rely on **alternative dispute resolution (ADR)**. ADR takes the form of **arbitration** or **mediation**, among other processes, to resolve civil disputes. Arbitration involves the appointment of a single arbitrator or panel of arbitrators, selected because they have more familiarity with the subject of the dispute than a randomly appointed judge may have. The arbitrator(s) hear evidence from both parties in a process that resembles a trial, and in turn issue a decision. A mediation, while similar to an arbitration in terms of the appointment of a neutral third party to hear the parties' positions and arguments, typically involves a facilitation of the parties' positions, encouraging them to agree to a settlement that resolves the matters at issue. The use of alternative dispute resolution is frequently specified in project contracts to avoid the expense and time commitment of court proceedings should a dispute arise.

Against that backdrop of the origins and structure of American law, the following discussion provides a very brief summary of some of the areas of law that affect the work performed by civil and environmental engineers.

TORTS

Generally speaking, torts are civil wrongs. If a person or company is found responsible (or **liable**, in legal terms) for a wrong, the wrong is punished through an award of damages to the party who suffered. In this sense, a tort is similar to a crime. The difference between a tort and a crime, however, is that criminal laws are enforced on behalf of the general public by a prosecutor.

A tort, on the other hand, is enforced by a private party, the **plaintiff**, by seeking redress in civil courts against the wrongdoer, the **defendant**, through litigation.

There are three basic types of torts: (1) intentional torts, (2) negligence, and (3) strict liability. **Intentional torts** are wrongs committed through more than general carelessness. As the name implies, the wrongdoer (or **tortfeasor**, in legal terms) does something intentionally (or with wanton disregard for the consequences) that harms another. Intentional torts include assault, theft, trespass, defamation, false imprisonment and fraud or misrepresentation. Intentional torts are usually not raised in an engineering or construction context, although fraud and misrepresentation can be the basis for claims involving engineers (or other professionals) who are alleged to have over-promised their performance and failed to deliver on such promises to another party's detriment.

Negligence is the failure to use the required amount of care in a given situation. Every activity that poses a potential harm to another person carries with it a corresponding duty of care. For everyday activities such as driving a car, the required duty of care is to act as a "reasonable person." Thus a person who drives a car while texting on a cell phone stands a good chance of being found liable for negligence if he or she causes an accident that results in another person suffering an injury or incurring damages. The tortfeasor's distracted, careless driving does not conform with society's expectation of the standard of conduct, or how a reasonable person should drive, and is therefore held responsible for the harm caused. Note that this example of negligence does not consider the prohibition on texting while driving that many states have enacted, which is a separate legal issue (criminal law).

Engineers, like everyone else, must conform their behavior to a reasonable person standard during their everyday lives. In a professional context however, engineers have legal duties that are specific to the effect the work they perform has on the public, whether directly or indirectly. There is no one uniform standard that engineers must meet in performing their professional duties, but engineers have the legal duty to exercise that degree of care and skill that a *reasonably prudent engineer* would use under the same or similar circumstances. That duty extends to the engineer's client and to third parties, otherwise known as the general public, who work in, drive over, or simply walk by the projects that engineers design and other professionals build. Engineers therefore must be cognizant of their duty to design and build projects that, at a minimum, meet the applicable safety standards, even against the sometimes countervailing pressures of a project owner's budget and deadlines.

The third category of torts is known as **strict liability**. Strict liability torts involve activities that society considers inherently dangerous or hazardous, such as demolition or blasting. With strict liability torts, a person harmed by activities that are viewed as inherently dangerous does *not* need to prove that the tortfeasor was at fault. The general public, as a matter of policy, expects that people engaged in activities like demolition or blasting know what they are doing and do it competently. If there is a failure, a court will not expect the injured party to show a causal connection between the wrongdoer's breach and the resulting injury.

CONTRACTS

In addition to the sources of American law mentioned above, there is an additional source of law that every engineer will encounter throughout their careers. Individuals, companies, and governments all have the ability to make private laws, in the form of **contracts**, between themselves and other entities. At a very basic level, a contract is a promise or a set of promises, which if breached, can be enforced by suing for damages. In a professional engineering context, contracts are used to define the rights and responsibilities of parties on a project and to determine what risks the parties are willing to accept over the course of a project. The owner, architect, engineer, and general contractor *all* attempt to allocate risks that might be encountered on a project *in advance* through contractual arrangements. Some of these risks include:

- adverse weather creating unforeseen delays

- underestimation of costs

- availability of labor that can competently perform the required duties

- delay due to inadequate project plans

- equipment failures or inadequacies

- inadequate analysis of subsurface conditions

- archeological discoveries at the project site

- discovery of hazardous waste

- failure to properly supervise contractors or consultants

- failure to reject deficient work

- failure to report work that does not meet contract specifications or applicable building codes, and

- "Acts of God," a catch-all term used to allocate the risk of otherwise unforeseen circumstances.

Any of these risks can cause overruns and delays, thus increasing costs that have to be borne by one of the contracting parties. The construction contract, therefore, seeks to resolve who is responsible for bearing the risk of loss if any of these circumstances arises over the course of a project, without resorting to expensive and time-consuming litigation to determine the party at fault.

Organizations associated with the construction industry have developed extensive model contracts that address nearly every phase of the construction process. Engineers will likely encounter contracts drafted by the American Institute of Architects (AIA), the Engineers Joint Contract Documents Committee (EJCDC), or other related organizations that are involved in civil engineering projects.

PROPERTY LAW

Property law encompasses the rights and obligations associated with real estate and personal property. To use a practical example from a utility perspective, the construction of an electric transmission line involves, in no small part, numerous property law issues that affect the budget for and the timing of such construction projects. In the construction of a typical utility line (e.g. to convey electricity, natural gas, or water), one of the first concerns to be addressed is whether the entity building the utility line has acquired sufficient

Figure 19.19 Cleared Trees Mark the Right-of-Way for an Electrical Transmission Line.

Source: M. Penn.

rights from property owners, a right-of-way (Figure 19.19), along the proposed route. Utilities typically do not purchase the property needed for the construction of a line. Rather, they acquire **easements** for the purposes of building, maintaining, and updating the line. As part of that process, engineers must consider the size of the easement to be acquired. For example, for an electric transmission line, the size of the proposed easement is determined by the practical concerns of the physical space necessary to build, operate and maintain the transmission infrastructure, along with regulations and standards posed by the Federal Energy Regulatory Commission and the National Electric Safety Code, as adopted in the relevant state's statutes, that determine the amount of lateral and vertical clearance required for the proposed line voltage. Having determined the amount of land required for the easement, engineers will often coordinate with land acquisition agents who carry out negotiations with property owners to purchase the necessary easement rights. Depending on the size of the line to be built, that acquisition process can take months or years, with no guarantee that property owners will ultimately sell sufficient rights along the proposed route, possibly resulting in a revision to the line route or, if necessary, the time-consuming process of condemning land by the process of eminent domain.

An **easement** is a property right that allows the builder to use the property owner's land. The duration of an easement may be perpetual or for a set period of time, for example, for 10 to 50 years.

This example serves as a brief case in point of the property law issues facing engineers. Beyond the relatively simple process of acquiring land rights, civil and environmental engineers will address a myriad of other property law issues, including, but not limited to:

- Nuisance: the right of adjacent landowners to be free from unreasonable interference of their property rights by an engineering project;

- Soil support: the right of land owners adjacent to a project to lateral soil support to prevent collapse of their structures;

- Drainage and surface water rights: the rights of property owners who may be affected by a project's disruption of the natural drainage of water; and

- Local land use controls, or zoning: laws that determine what types of projects may be built in a particular vicinity.

Case in Point: Lawsuits and Milwaukee's Deep Tunnel

In the 1970s, prior to the construction of the Deep Tunnel, the State of Illinois sued the City of Milwaukee claiming that beach closings in Chicago (due to fecal contamination) were the result of combined sewer overflows in Milwaukee. Milwaukee won the lawsuit on appeal before the U.S. Supreme Court. The tunnel was later built nonetheless, driven by federal environmental regulations.

It is common for lawsuits to arise in large-scale projects, and the Deep Tunnel was no exception. During construction, an explosion occurred, killing three men. A consulting engineering company was found liable by the U.S. Department of Labor, but was eventually exonerated by the U.S. Court of Appeals.

Civil wrongful death lawsuits were also filed by the families of the explosion victims and settled out of court, with payments made to the families without admission of responsibility for the accident. The defense attorney for the consulting engineers claimed that there was no evidence of wrongdoing, but a settlement was made because, in the defense attorney's words, "this would have been a real expensive piece of litigation."

In another lawsuit, the same consulting company agreed to pay MMSD (the Deep Tunnel owner) $24 million to settle claims related to cost overruns resulting from subsurface complications of water intrusion and the structural strength of the rock through which the tunnel was bored.

These examples are but a few of the many lawsuits related to the Deep Tunnel project. In such cases, contracts, design documents, communication records, construction field reports, and photographs come under intense scrutiny. The failure by an engineer to appropriately and thoroughly document activities and "agreements" may prove to be costly when lawsuits arise.

Minimizing Legal Action During Construction of Milwaukee's Deep Tunnel

The second phase of the Deep Tunnel, known as the Northwest Side Relief Sewer, is a 7-mile long, 22-foot diameter tunnel to provide additional system storage. This project won the 2006 American Public Works Association's Project of the Year in the Environmental category for large (more than $100 million) projects. There were many innovative construction, planning, and management strategies implemented during the project, including a Disputes Review Board. This Board consisted of three impartial "external" experts jointly selected by the owner and the general contractor. The Board was continually updated on project decisions and progress, and offered non-binding opinions on disputes as a low-cost alternative to mediation, arbitration, or litigation.

A Good Day . . .

The following dialogue opened a casual meeting between one of the authors and a senior project manager for an engineering consulting firm:

Author: *Hi there, how are you doing?*

Project Manager: *Great! I haven't been sued yet today.*

This light-hearted comment demonstrates the relative frequency of lawsuits in civil and environmental engineering.

Outro

Safety, maintenance, constructability, politics, and legal aspects must be considered in infrastructure planning, construction, and management decisions. Proper consideration of these factors will increase the likelihood of project success by saving time and money; increasing the appeal and service life of the project; minimizing injury, frustration, and conflict; and enhancing cooperation among parties involved. These considerations, and those presented in Chapters 11 through 18, comprise the vital "non-design" aspects of engineering that many engineers will claim are more important than design itself.

Short Answer

19.1 What is the difference between mediation and arbitration?

19.2 Why is it generally better to initiate constructability reviews in early project phases?

19.3 Why do standardized "model" contracts exist?

19.4 What is the standard of care for an engineer in performing his or her professional duties?

19.5 Summarize a confined space case study (available at www.wiley.com/college/penn), identifying hazards and how the situation could have been avoided.

19.6 Identify the worker safety hazards in Figure 19.3.

19.7 List the five employment sectors with the highest fatality rates.

Discussion/Open-Ended

19.8 For an active construction site of your choice (or for one chosen by your instructor) on your campus, identify public safety measures.

19.9 For an active construction site of your choice (or for one chosen by your instructor) in your community, identify public safety measures.

19.10 Why might lawsuits arise on a project even though very detailed contracts have been signed?

19.11 Which of the Seven Wonders of the Modern World do you find most interesting? Why?

19.12 Why is it often advantageous to have external consultants (rather than internal personnel) involved in constructability reviews?

19.13 Identify buildings on your campus that are designed with and without aesthetic considerations.

19.14 After a company institutes a policy of performing constructability reviews on all projects, why might the benefit–cost ratio of the reviews increase on future projects? After several years of completing constructability reviews, why might the benefit–cost ratio on projects eventually decrease?

19.15 Identify an infrastructure component on your campus that would benefit from an aesthetic upgrade. Explain the modifications necessary to achieve this upgrade.

19.16 Identify an infrastructure component on your campus that was designed with aesthetics in mind. Identify a less aesthetic option. In your opinion, was the additional expense necessary to improve aesthetics worthwhile?

19.17 Identify an infrastructure component in a community that would benefit from an aesthetic upgrade. Explain the modifications necessary to achieve this upgrade.

19.18 Identify an infrastructure component in a community that was designed with aesthetics in mind. Identify a less aesthetic option. In your opinion, was the original additional expense worthwhile?

Research

19.19 Browse ASCE's Historic Civil Engineering Landmark Program (available at www.wiley.com/college/penn). Identify three landmarks that paid special attention to aesthetics when they were designed and built.

19.20 As suspension bridges age, their cables are often weakened through corrosion. One method to combat corrosion is by dehumidifying the cables, as explained in the paper entitled "Prevention of main cable corrosion by dehumidification," available at www.wiley.com/college/penn. Briefly explain the process by which bridge cables are dehumidified, and compare its effectiveness to painting.

19.21 Research the construction of the Amundsen-Scott South Pole Station. Prepare a one-page summary of interesting aspects of this project.

Key Terms

- alternative dispute resolution (ADR)
- arbitration
- confined spaces
- constructability
- contracts
- deconstructability
- easements
- intentional torts
- liable
- mediation
- negligence
- quality based selection (QBS)
- statement of qualifications
- strict liability

References

Behm, M. 2005. "Linking Construction Fatalities to the Design for Construction Safety Concept." *Safety Science*, 43(8): 589–611.

Bureau of Labor Statistics. 2006a. "Fatal Workplace Injuries in 2006: A Collection of Data and Analysis." http://www.bls.gov/iif/cfoibulletin2006.htm, accessed August 29, 2011.

Bureau of Labor Statistics. 2006b. "Occupational Injuries and Illnesses: Counts, Rates, and Characteristics, 2006." http://www.bls.gov/iif/oshbulletin2006.htm, accessed August 29, 2011.

Bureau of Labor Statistics. 2009. "National Census of Fatal Occupational Injuries in 2008." http://www.bls.gov/news.release/pdf/cfoi.pdf, accessed August 5, 2010.

Rajendran, S. 2007. "Constructability Review Process—A Summary of Literature." In: *Constructability Concepts and Practice*, J. Gambatese et al. (Ed.). Reston, VA: ASCE.

Thomas, S. R., J. R. Sylvie, and C. L. Macken. 2004. "Best Practices for Project Security." NIST GCR 04-865.

U.S. Department of Energy. 2002. "Final Environmental Impact Statement for a Geologic Repository for the Disposal of Spent Nuclear Fuel and High-Level Radioactive Waste at Yucca Mountain, Nye County, Nevada." DOE/EIS-0250.

chapter Twenty Analysis II

Learning Objectives

After reading this chapter, you should be able to:

1. Apply the Critical Path Method.
2. Estimate the ground-level concentration of a contaminant emitted by a stack.
3. Calculate the forces on a levee and relate these forces to levee failure.
4. Estimate the amount of deflection in a beam.
5. List items that must be considered when undertaking a Transportation Impact Study.

Introduction

In Chapter 9, we presented the analysis process and provided examples of how it is used in the subdisciplines of civil and environmental engineering. In the intervening chapters, we have presented many considerations that are necessary in "real-world" analyses. In this chapter, we provide you with five analysis applications that take into account the additional considerations we have presented in Chapters 11 through 19.

Construction Engineering: Critical Path Method

The managing of construction projects requires several analysis tasks. In Chapter 9, we introduced Gantt charts as a means of organizing project activities. In this section, we present the Critical Path Method, or CPM, as a more powerful means of analyzing project schedules. Like Gantt charts, CPM requires the construction manager to break down the construction project into its component parts, and allows him or her to analyze the resulting plan in order to increase efficiency of the project.

CPM helps you identify the **critical activities**. The delay of non-critical activities, to a certain extent, will not delay the finish date of the project. Understanding which activities are critical can also help the project manager focus on the critical activities; this helps the manager to stay on schedule or even to speed up the project so it is completed ahead of schedule. Often, infrastructure owners provide incentives to the builders for early completion, on the order of thousands to hundreds of thousands of dollars per day.

The **critical path** is defined as the path on which the critical tasks lie. To identify the critical path, a CPM diagram is often drawn. The steps to completing a CPM analysis are:

1. Create a list of the activities, or tasks.
2. Estimate the duration of each activity.
3. For each activity, identify predecessor activities, if any.
4. Create a diagram based on the sequence and duration of the activities.

For example, consider the activities needed to design and construct a stormwater detention pond. We have discussed these ponds in Chapter 5, Environmental Infrastructure. For this example, the first step to designing the pond is to complete a topographic survey of the area. Following this, a preliminary engineering plan is created. This information is then sent to the regulatory agency (e.g., a state environmental protection agency) for its approval. The engineering design will continue to be refined while waiting for regulatory approval. Once approval is obtained, earthwork can commence, by which the depression for the pond is excavated and embankments are formed. As this construction is proceeding, appropriate erosion control measures should be implemented. In this example, we will assume that once approximately half of the pond is excavated, the outlet device can be constructed. Following completion of the excavation, the site will be seeded with an appropriate seed mixture and a fence will be constructed around the perimeter.

The activities necessary to design and construct the detention pond are summarized in Table 20.1. The predecessor activities for each activity can readily be obtained from the information provided in the previous paragraph. The duration of each activity has also been added to the table.

Once this information has been collected and tabulated, it can be visualized with a CPM diagram. We will use the following conventions when creating CPM diagrams:

1. Start and end the CPM diagram with "Start" and "End" nodes, which do not have a duration associated with them.

2. Assign each activity a letter and draw each activity as a letter enclosed by a box.

3. Above each boxed activity, label the length of time to complete that activity, using consistent time units.

4. Draw arrows from node to node based on the sequence dictated by the list of predecessor activities.

Critical activities are those activities, that if delayed, will delay the completion of the entire project.

Construction Incentives

In Chapter 1, we discussed the collapse of the I-35 W bridge in  Minneapolis, Minnesota. A total of $27 million in incentives were provided for the construction of the replacement bridge. The incentives included a $200,000 *per day* bonus for each day that the bridge was open before December 18, up to a maximum of 100 days early. Indeed, the bridge opened exactly 100 days before the required opening date. The amount of the incentive was calculated based on the roadway user cost impacts of the collapsed bridge. Roadway user costs include the additional gasoline consumed and lost productivity associated with traffic that had to be rerouted.

Table 20.1

	Activity Label	Activity Description	Duration (days)	Predecessors
Example CPM Activities for a Stormwater Detention Pond	A	Site survey	2	None
	B	Preliminary design	3	A
	C	Regulatory approval	30	B
	D	Refine design	1	B
	E	First half of earthwork	6	C, D
	F	Second half of earthwork	6	E
	G	Outlet device	5	E
	H	Final seeding	1	F, G
	I	Fence	2	F, G

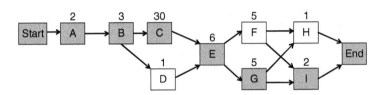

Figure 20.1 **Sample CPM Diagram.** The critical path is shaded.

The information in Table 20.1 is plotted as a CPM diagram in Figure 20.1. The CPM diagram provides valuable insight on the project. For example, the CPM helps us to readily determine that the earliest that Activity E (beginning of earthwork) can begin is 35 days (= 2 days + 3 days + 30 days) after the start date. Even though the design may be finalized after 6 days (Start → A → B → D), the earthwork cannot begin until Activity C (the obtaining of regulatory approval) has been completed. By the same logic, we see that in general, the *minimum* time to complete a project is equal to the length of the *longest* path from start to end. Thus, the least amount of time in which the stormwater detention pond can be designed and constructed is 48 days by following path Start → A → B → C → E → G → I → End(= 2 days + 3 days + 30 days + 6 days + 5 days + 2 days). This path is termed the critical path and the activities on that path are the critical activities.

A CPM diagram highlights the fact that activities that can be conducted at the same time should be conducted at the same time. In the case of this simplified example, there is no need to wait until seeding is completed before installing the fence. Indeed, waiting for the seeding to be completed before installing the fence would add 2 days to the length of project completion.

CPM helps emphasize that great care should be taken in ensuring that critical activities are completed on time. For example, adding 1 week to the approval process (Activity C) will add 1 week to the time of project completion, whereas adding 1 week to the "refine design" step will not necessarily add any time to the duration of the project.

Environmental Engineering: Air Pollution

Smokestacks (more correctly referred to as "stacks") are a means of dispersing air pollutants at high elevations. The stack emissions form a **plume** downwind of the stack, which mixes with cleaner air to dilute the pollutant concentration (Figure 20.2; Figure 2.22 of Chapter 2).

A **plume** is a "cone-shaped" region of contaminated air originating at a single point source, caused by the mixture of stack emissions and "clean" air. Note that plumes of contamination also form in groundwater (e.g., gasoline from a leaky storage tank mixing with groundwater) and in surface waters (e.g., wastewater effluent mixing with the river into which it was discharged.)

Figure 20.2 A Plume Emitted from a Coal-Fired Heating Plant. The visible plume in this case is mostly condensed water vapor. Pollutants other than particulate matter are "invisible" gaseous emissions.

Source: M. Penn.

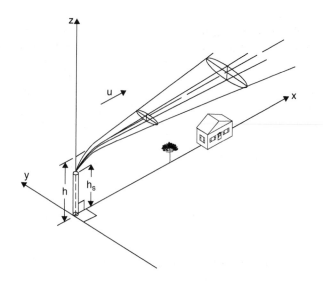

Figure 20.3 Plume Dispersion in Three Dimensions. The effective stack height, h, is often greater than the actual stack height, h_s due to plume rise at the stack exit.

The concentration of pollutants varies within the plume and is assumed to be negligible outside of the plume. The dispersion (spreading) of the plume occurs in three dimensions, and eventually at some distance downwind, the plume reaches ground level (Figure 20.3). The concentration of pollutants at ground level is a human health concern.

Environmental regulations, including those associated with air pollution, are ever-changing. For example, if evidence from health studies suggests that the "safe" concentration of an air pollutant is lower than current regulations permit, new regulations may be enacted. An existing industry may then be required to take steps to lower the concentration of pollutants at ground level. This can be accomplished by implementing one (or more) of the following options.

1. One option is to make the stack taller, thus raising the stack outlet elevation which results in more dilution and lower ground level concentrations. However, note that the same *mass* of pollutant is emitted from the stack despite the increase in elevation. This approach may not be favored by regulatory agencies seeking to decrease the amount of air pollutants discharged.

2. A better approach, although not always feasible, is to try to reduce the amount of pollutants being generated. This may be accomplished by changing the processes; e.g. by using a cleaner fuel source or modernizing equipment.

3. A common approach is to install air pollution control devices in order to reduce the amount of pollution entering (and consequently exiting) the stack. The question for the engineer is thus, "How much pollution must be removed by the air pollution control equipment so that the *existing* stack will meet *new* air requirements?"

The Gaussian dispersion equation (Equation 20.1) is a mathematical model used to determine the concentration of pollutants in air resulting from stack emissions. It provides concentrations of pollutants in units of micrograms of pollutant per cubic meter of air at any location downwind of the stack having coordinates x, y, and z. This model can be used to determine the stack height necessary for a given emission rate to ensure safe concentrations at ground level. The model can be used to provide necessary *design* information (i.e. stack height requirement) or to *analyze* pollution problems.

$$C(x,y,z) = \frac{Q}{2\pi u \sigma_y \sigma_z} \left[\exp\left(-\frac{1}{2}\left(\frac{z-h}{\sigma_z}\right)^2\right) \right.$$

$$\left. + \exp\left(-\frac{1}{2}\left(\frac{z+h}{\sigma_z}\right)^2\right) \right] \exp\left(-\frac{1}{2}\left(\frac{y}{\sigma_y}\right)^2\right) \qquad \textbf{(20.1)}$$

$C =$ concentration of pollutant in air at any point x, y, z (μg/m^3, note: $\mu = 10^{-6}$)

$x =$ distance downwind (m)

$y =$ horizontal distance from centerline of stack in the direction of the wind (m)

$z =$ elevation above ground level (m)

$h =$ effective stack height (m)

$Q =$ pollutant emission rate (μg/s)

$u =$ wind speed at stack elevation (m/s), *not* at ground-level

$\sigma_y =$ lateral coefficient of dispersion (m), and

$\sigma_z =$ vertical coefficient of dispersion (m).

The coefficients of dispersion, σ_y and σ_z, are case specific and vary with distance downwind and atmospheric stability (which is a function of wind speed and sunlight). A detailed analysis of historical meteorological conditions is needed to accurately determine appropriate coefficient values and wind speeds.

The maximum concentration will be at the centerline of the plume; thus the last term in the equation can be omitted if $y = 0$ (the centerline). Also, at ground level, $z = 0$. By substituting this relationship into Equation 20.1, the ground-level concentration at the centerline of the plume can be expressed by Equation 20.2.

$$C(x) = \frac{Q}{\pi u \sigma_y \sigma_z} \left\{ \exp\left(-\frac{1}{2}\left(\frac{h}{\sigma_z}\right)^2\right) \right\} \qquad \textbf{(20.2)}$$

The Gaussian model is available in spreadsheet form on the textbook website. A worksheet is available in the spreadsheet for each of three different types of atmospheric stability: unstable, moderately unstable, and stable.

This application demonstrates how some tools can be used for design *or* analysis. It also suggests the need for engineers to stay current with changing regulations.

Gaussian Plume
Spreadsheet Model

numerical example 20.1 Ground Level Concentration of Nitrogen Dioxide Emitted from Stack

An industry emits $20,000,000\,\mu$g/s (20 g/s) of nitrogen dioxide through a 50-meter high stack (with an effective stack height of 55 meters). The wind speed is 5 m/s in a moderately unstable atmosphere. What is the maximum ground-level concentration of nitrogen dioxide, in units of μg/m^3? If the regulatory limit for ground-level concentration of nitrogen dioxide is $100\,\mu$g/m^3, how much nitrogen dioxide may be emitted?

numerical example 20.1 — Ground Level Concentration of Nitrogen Dioxide Emitted from Stack (Continued)

solution

To solve the first part of the problem, the input values are entered into the Gaussian Plume Spreadsheet Model located on the textbook website. This yields a maximum ground-level concentration of $178\,\mu g/m^3$ at a distance 600 meters downwind.

If the regulatory limit for ground-level concentrations is $100\,\mu g/m^3$, the corresponding emission rate is calculated using a trial-and-error approach with the spreadsheet model to be $11{,}200{,}000\,\mu g/s$. This equates to a 44 percent reduction in emissions for the same stack height to lower the ground-level concentrations to this "safe" level.

The next step for an engineer would be to investigate available technologies for nitrogen dioxide removal to determine the most cost-effective approach to reach this level of reduction.

Geotechnical Engineering: Katrina Floodwalls

You might think that in the aftermath of Hurricane Katrina, the major focus of engineering efforts was to rebuild the breached floodwalls and levees. To the contrary, the major focus was to analyze why the failures occurred, and to incorporate this knowledge into improved designs with a lower likelihood of future failure. This work represents a type of forensic engineering, which was discussed in a sidebar in Chapter 7, Infrastructure Systems.

The analysis of New Orleans floodwalls performed in the aftermath of Hurricane Katrina requires geotechnical engineering knowledge beyond the scope of this book. As such, we will present here the concepts of the analyses, rather than the technical details.

In simplest terms, floodwalls and levees must act to counter the forces created by high water elevations. Water behind the floodwall or levee exerts a hydrostatic force. (Recall that hydrostatic pressure was defined in Chapter 3, Structural Infrastructure). The hydrostatic force on a vertical planar surface may be calculated using Equation 20.3, and values are provided in SI units that will yield a force with units of newtons ($1\,N = 1\,kg \cdot m/s^2$).

$$F = \frac{\rho g h A}{2} \qquad (20.3)$$

where $\rho =$ density of water ($1{,}000\,kg/m^3$)
  $g =$ acceleration due to gravity ($9.8\,m/s^2$)
  $h =$ depth of water (m)
  $A =$ area of floodwall against which water is in contact (m^2)

One of the major oversights implicated by the forensic investigation of the Katrina floodwalls was the failure to account for deflection (movement) of the floodwalls due to increased forces caused by high water elevations. This deflection is illustrated in Figure 20.4, and served to separate the floodwall from the compacted soil at its base. A gap formed behind the wall, and water filled the gap. The increased water depth ($h_2 > h_1$ in Figure 20.4) increased hydrostatic pressure acting on the wall because the pressure is a function of the *depth* of water.

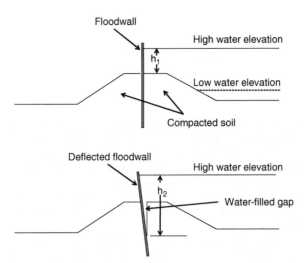

Figure 20.4 A Floodwall Integrated into a Levee (Top) and the Same Floodwall Deflecting Due to Forces from High Water (Bottom). The deflection causes separation of the wall from the soil and the creation of a water-filled gap.

The water-filled gap ran along the length of the levee (where length is the dimension "into" the page in Figure 20.4), but was only a few inches wide. You may wonder how such a small gap can have such a large impact. To understand, consider Figure 20.5, which shows two completely filled tanks of equal height, and thus equal water depth. The end walls (the dark-shaded areas) are identical in size (and thus area), but the two tanks have different widths. The force on the end walls of the two tanks is identical, since the hydrostatic pressure and the area on which it is acting are identical. *Thus the hydrostatic force is independent of the width of water.* In the case of the deflected floodwall in Figure 20.4, the *depth* of the water-filled gap is critical; the width of the gap is irrelevant.

Equation 20.3 demonstrates that the hydrostatic force on a vertical planar surface increases in proportion to the depth of water and the "wetted" area of the wall. As shown in Figure 20.4, the *depth* of water has increased significantly as a result of the formation of the water-filled gap ($h_2 > h_1$). Also, the effective area of the wall with which water is in contact has increased. Thus we expect the force on the floodwall to be greatly affected by the water-filled gap. The water-filled gap also caused other types of levee failures that are outlined in the *What Went Wrong and Why* report (ASCE, 2007), and are beyond the scope of this book.

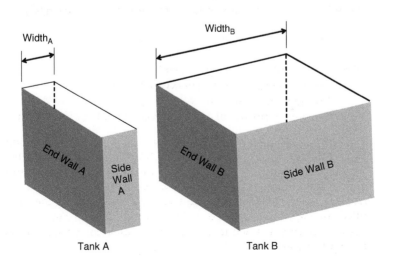

Figure 20.5 A Hydrostatic Paradox. The two completely filled tanks are the same height and length, but have different widths. As such, the hydrostatic force on End Wall A is equal to the hydrostatic force on End Wall B. The hydrostatic force on Side Wall B is, however, greater than that acting on Side Wall A due to the larger wall area.

numerical example 20.2 Effect of Water-Filled Gap

Consider the two floodwalls in Figure 20.4. If $h_1 = 2$ m and $h_2 = 4$ m, how much greater is the force on the floodwall behind which the water-filled gap has formed?

solution

Without deflection, the force due to a depth of water of 2 meters (h_1) can be calculated from Equation 20.3, substituting height (h) multiplied by length (L) for area (A). We will calculate the force per 1 m length, L, of floodwall; thus, the answer can be multiplied by the floodwall length to obtain the actual force acting on the wall.

$$F = \frac{\rho g h(hL)}{2} = \frac{\left(1,000 \frac{\text{kg}}{\text{m}^3}\right)\left(9.8 \frac{\text{m}}{\text{s}^2}\right)(2\,\text{m})(2\,\text{m} \cdot 1\,\text{m})}{2}$$
$$= 20,000\,\text{N per 1 m of floodwall length}$$

With deflection, the height of water now equals 4 m (h_2) and the force per 1 m of floodwall can be calculated corresponding to the wetted area.

$$F = \frac{\rho g h(hL)}{2} = \frac{\left(1,000 \frac{\text{kg}}{\text{m}^3}\right)\left(9.8 \frac{\text{m}}{\text{s}^2}\right)(4\,\text{m})(4\,\text{m} \cdot 1\,\text{m})}{2}$$
$$= 80,000\,N \text{ per 1 m of floodwall length}$$

Thus the force increases four-fold!

Structural Engineering: Deflection of Beams

Structures are designed to withstand **ultimate loads** (maximum expected loads) without failing (e.g., collapse). Designing a structure to withstand ultimate loads considers any anticipated combination of snow loads; wind loads; stopped bumper-to-bumper traffic; pressure from blasts; forces from impacts; or any other probable extreme event that the structure may be subjected to during its design life.

Structures are also designed to react to **service loads** (routine loads) in ways that do not substantially damage the structure or cause users to be uncomfortable (e.g. swaying). Hardy Cross, a famous structural engineer, stated that "strength is essential, but otherwise unimportant." His point was that designing for ultimate loads is essential, but if a structure sways, vibrates, or deflects too much, the structure has not met its intended purpose for use. **Serviceability** refers to the ability of a structure or structural member to handle service loads. It is important to note that structures may deflect, crack, or vibrate within allowable limits and remain serviceable and structurally sound. The primary goal of serviceability is to address the comfort of the structure's users. Too much deflection can unnerve building occupants and may crack finishes such as paint or sheetrock. If a high-rise building sways too much, occupants may become sickened. Too much sway in a bridge can unnerve stopped drivers. Lack of serviceability does *not* necessarily imply lack of structural integrity—that is, a swaying skyscraper may not be at risk of collapse.

For the remainder of this section, you will learn how to analyze a beam in order to estimate the amount of **deflection**, which is one aspect of serviceability. If a weight such as a bottle of water is placed on the ruler in Figure 20.7, you

Minimizing Skyscraper Sway

High wind speeds can cause skyscrapers to sway. In order to minimize this sway, the structures must either be designed to resist the wind loads or be designed to be more aerodynamic.

Engineering the Impossible: Built to Sway (Video)

Deflection is the distance a structural member moves when experiencing a load as compared to when there is no load applied.

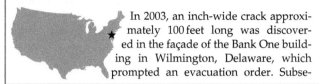

Case in Point: Serviceability of Bank One Building in Wilmington, Delaware

In 2003, an inch-wide crack approximately 100 feet long was discovered in the façade of the Bank One building in Wilmington, Delaware, which prompted an evacuation order. Subsequent structural analysis determined that the building was *not* in danger of failure (collapsing) however. This example demonstrates the fact that even though a building can carry loads without failing structurally, its utility is compromised if occupants are caused to be uncertain or fearful.

Case in Point: The "Wobbly Bridge"

The Millennium Bridge (Figure 20.6) in London is a pedestrian bridge, built in part to celebrate the new millennium. When the bridge first opened, as many as 2,000 people crossed the bridge at one time. The motion of their walking set up vibrations in the bridge that caused it to sway laterally (thus the nickname "Wobbly Bridge"). Many people walking on the bridge were unnerved by the experience. After only 3 days of being in service, the bridge was closed down for over 2 years. Dampers were built into the bridge to stop the swaying at a cost of more than $7 million.

Video of Wobbly Bridge

Figure 20.6 **Millennium Bridge, London.**

Source: Copyright © Simon Bradfield/iStockphoto.

can envision the ruler bending, or deflecting. The ruler, supports, and bottle of water are redrawn as a schematic in Figure 20.8.

In Figure 20.8, the beam is deflecting a distance Δ as a result of being loaded with a force F. Figure 20.8 demonstrates a "simply supported" beam loaded at the center. In future classes, you will consider the deflection of such a beam under different loading conditions and the deflection of many other types of structures. Also recall that in Chapter 10, Design Fundamentals, we introduced the West Point Bridge Designer software, which calculated deflection in the trusses of a bridge. The deflection at the center of the bridge varied as

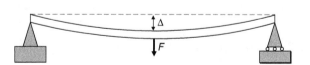

Figure 20.8 **Deflection in a Simply Supported Beam.** The beam is termed "simply supported" because it is supported on each end. The support at the left is fixed and prevents the left end of the beam from moving side to side, while the support on the right end, which is sitting on rollers, is able to move side to side. In future structural engineering courses, you will learn the implications of such types of supports.

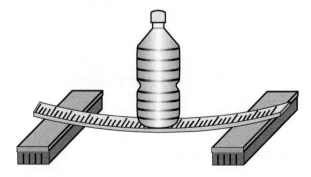

Figure 20.7 **Water Bottle Load Applied to a Ruler Supported Between Two Erasers.**

the truck moved across the bridge (i.e., the point of application of the load changed along the span), and you could see this happen in "real" time.

The amount of deflection that a beam will undergo depends on several factors. The magnitude of the load will greatly affect the deflection, as well as the location of the load. Some loads may be idealized to act as **point loads** (that is, they act at a single point, such as the water bottle on the ruler) while others may be distributed along the beam (e.g., several water bottles placed side by side along the ruler). The length of the beam will also affect the deflection. All other things being equal, a longer span will deflect more than a shorter span. The cross-sectional shape of the span is also very important. The cross-sectional shape may be rectangular (solid or hollow), circular (solid or hollow), I-shaped, T-shaped, etc. Finally, the type of material used will also affect the amount of deflection. All other things being equal, steel deflects less than aluminum for example. In "real life," as we have discussed, the choice of material would also include aesthetics, longevity, reliability, maintenance needs (see sidebar), life-cycle costs, weight, environmental impacts, and availability (see Case in Point: Petronas Towers).

All structures deflect. You may not have considered this, but a highway bridge deflects due to the loading of only a single vehicle. Such deflection is practically immeasurable, but it does occur. A rule of thumb for the allowable deflection in a beam is that the deflection should be no greater than 1/360th of the beam length. Suspension bridges, due to their typically long spans, are able to undergo noticeable and significant deflections. For example, the Golden Gate Bridge is designed to deflect as much as 10 feet at its center span! However, this is allowable given that the middle span length is 4,200 feet, and thus a deflection of up to 11.7 feet (4,200 feet/360) would be considered acceptable according to our rule of thumb.

Weathering Steel

Exposed steel requires regular painting to avoid rust and corrosion. One nearly maintenance-free type of steel exists termed **weathering steel**. This is a special alloy of steel that, when exposed to rain, forms a protective rust-like coating. Normally we think of rust as a "bad" thing, but in this type of steel, it creates a protective layer that prevents interior penetration of rust. The truss bridge in Figure 3.10 of Chapter 3 was constructed of weathering steel, and indeed has a unique appearance.

Case in Point: Petronas Towers

At nearly 1,500 feet tall, the Petronas Towers (Figure 20.9) were the tallest buildings in the world until they were surpassed by the Tapei 101 skyscraper in 2004. Unlike many skyscrapers, the Petronas Towers were constructed of reinforced concrete. Steel is commonly used for skyscrapers given its much higher strength-to-weight ratio; a concrete skyscraper may weigh nearly twice as much as a corresponding steel skyscraper, which has significant implications for foundation design and the design of lower-story columns. In this case, concrete was chosen because steel was prohibitively expensive to import and concrete was readily available. Additionally, recent advances in concrete technology have increased its strength-to-weight ratio.

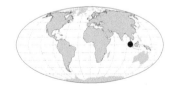

Figure 20.9 Petronas Towers in Kuala Lumpur, Malaysia. Construction began in 1991 and was completed in 1998.

Source: Copyright © Ales Kramer/iStockphoto.

For simply supported beams of rectangular cross-section subjected to a single concentrated load at mid-span, the deflection can be calculated using Equation 20.4.

$$\Delta = \frac{FL^3}{4\,Ebh^3} \qquad \text{(20.4)}$$

where F = load
L = length of span
E = modulus of elasticity
b = beam width
h = beam height—see Figure 20.10 for beam dimension orientation.

The modulus of elasticity, E, is a material property. Values for E for some common materials are provided in Table 20.2.

It is important to remember that Equation 20.4 applies to a specific set of conditions: a simply supported beam of rectangular solid cross-section that is loaded by a single load at the center of the beam. Different equations will be required for deviations from these conditions: if the beam is cantilevered rather than simply supported (on both ends); if the beam is an I-beam rather than one having a rectangular cross-section; if a distributed load acts on the beam rather than a single point load; or if a point load acts anywhere other than at the center. In future civil engineering courses, you will derive relationships for these and other scenarios.

This simple equation also demonstrates the importance of the orientation of the beam. That is, consider a wooden plank with dimensions 1 inch by 8 inches. Most people's intuition tells them that the amount of deflection will vary significantly if that board, or beam, is oriented with the long edge vertical as opposed to horizontal; this intuition is confirmed by Equation 20.4, as demonstrated in Numerical Example 20.3.

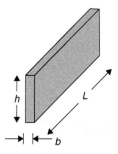

Figure 20.10 Beam Dimension Orientation. *H* is the vertical dimension, regardless of whether the beam is "on edge" (as shown) or "flat."

Table 20.2

Modulus of Elasticity for Common Materials

Material	$E(\text{lb/in}^2)$
Aluminum	$10.2 \cdot 10^6$
Carbon steel	$29.0 \cdot 10^6$
Cast iron	$14.5 \cdot 10^6$
Maple wood	$1.5 \cdot 10^6$
Poplar wood	$0.74 \cdot 10^6$
Stainless steel	$27.6 \cdot 10^6$

numerical example 20.3 Deflection of a Plank

A 1 inch by 8 inch piece of maple lumber, 12 feet in length, is supported at both ends. A 200-pound person stands in the middle of the plank. How much does the plank deflect if the board is laid "flat" as compared to if it is laid "on edge?" That is, how much does the plank deflect if $b = 8.0$ inches and $h = 1.0$ inch as compared to the case where $b = 1.0$ inch and $h = 8.0$ inches?

solution

We can solve this problem using Equation 20.4. First, we will examine the case where the plank is laid "flat" on two supports, with $b = 8$ inches (or $8/12$ feet) and $h = 1$ inch (or $1/12$ feet).

$$\Delta = \frac{FL^3}{4Ebh^3} = \frac{200\text{lb} \cdot (12\,\text{ft})^3}{4 \cdot \left(1.5 \cdot 10^6\, \frac{\text{lb}}{\text{in}^2}\right) \cdot \left(\frac{144\,\text{in}^2}{1\,\text{ft}^2}\right) \cdot \left(\frac{8.0}{12}\,\text{ft}\right) \cdot \left(\frac{1.0}{12}\,\text{ft}\right)^3} = 1\,\text{ft or 12 in.}$$

For the case where the beam is laid on edge, we can substitute $b = 1/12$ feet and $h = 8/12$ feet into Equation 20.1. This calculation yields a deflection of only 0.19 inches, 1/60th as much as when the beam is laid flat! If you have ever tried to stand on a piece of lumber placed on edge, you would know that there is another structural condition to consider—that of stability. Beams are typically designed on edge to minimize deflection, but require bracing to provide horizontal support that lessens lateral movement and twisting (torsion).

The type of material has a large effect on the amount of deflection. If the maple board of Numerical Example 20.3 is replaced by a steel bar of the same dimensions, but all other conditions remain the same, how much will the deflection be?

solution

First, we will consider the steel bar laid "flat." Equation 20.4 is used, and the same data is inserted into the equation as for Numerical Example 20.3, except that the modulus of elasticity, E, is equal to $29 \cdot 10^6 \text{ lb/in}^2$.

$$\Delta = \frac{FL^3}{4Ebh^3} = \frac{200\,\text{lb} \cdot (12\,\text{ft})^3}{4 \cdot \left(29 \cdot 10^6 \dfrac{\text{lb}}{\text{in}^2}\right) \cdot \left(\dfrac{144\,\text{in}^2}{1\,\text{ft}^2}\right) \cdot \left(\dfrac{8.0}{12}\,\text{ft}\right) \cdot \left(\dfrac{1.0}{12}\,\text{ft}\right)^3} = 0.64\,\text{inches}$$

This is significantly less than the 12-inch deflection calculated for the maple plank of the same dimensions. Also, on your own, you can verify that when laid on edge, the deflection of the steel beam is 0.01 inches, which would be undetectable without a precise measuring device.

Using the rule of thumb that deflection should be less than 1/360th of the span, the maximum deflection is 0.4 inches for the 12-foot beam considered in this example. Thus, the steel bar laid flat will deflect "too much," but the beam oriented on edge will deflect an acceptable amount.

Transportation Engineering: Transportation Impact Study

A traffic impact study (TIS), sometimes called a traffic impact analysis (TIA), is conducted to determine the effect that proposed development or redevelopment will have on existing traffic patterns. For many developments, public opposition is often directed at the increased traffic associated with the development. A TIS is typically undertaken by a consultant on the behalf of the developer. Development activities such as expanding an existing shopping center, building a grocery store on a brownfield, or building a residential subdivision on a greenfield, will most likely generate additional traffic on nearby roads, and how this new traffic will interact with the remainder of the transportation *system* must be analyzed.

To determine whether a TIS is warranted, a variety of quantitative threshold values are used such as the number of **trips** generated per day or during the peak hour, amount of acreage to be rezoned, number of dwelling units added, or the amount of square footage to be occupied by the new development. Estimates of the threshold values are commonly based on trip generation tables.

Trip generation can be estimated based on the land use (e.g, office building, residential housing, shopping mall). An example of a trip generation table with typical rates is shown in Table 20.3. The most commonly used tables are published by ITE (Institute of Transportation Engineers), but individual municipalities or regions may have their own trip generation tables. Depending on the type of land use, trips may be expressed based on the number of users; the size of the parcel of land; the size of the building; or the number of dwelling units.

By using Table 20.3, an estimate of the average trips per day can be obtained. Depending on the land use and how busy the existing roads are, the TIS may

In transportation terminology, a **trip** represents travel from one location to another. Thus, taking "a trip to the store" actually counts as two trips: travel from your residence to the store and return travel from the store.

Table 20.3

Sample Trip Generation Data	Land Use	Daily Trip Rate
	University	2.5 trips/student; or 100 trips/acre
	Bank	200 trips/1,000 square feet
	Post office	90 trips/1,000 square feet
	Medical office	50 trips/1,000 square feet
	Bowling alley	30 trips/lane
	Mobile home park	5 trips/dwelling unit
	Apartment complex	8 trips/dwelling unit
	Single family residential	9 trips/dwelling unit
	Shopping center	1,200 trips/acre

also need to consider traffic during *peak* ("rush") hours in the morning and evening. Traffic generation tables will often include peak trip information.

Once it is determined that a TIS is required, data about the existing roadway system and any other previously approved developments and planned transportation improvements within the study area are gathered. Data gathered for the transportation system include roadway geometric features (curves, lane widths, and grades), parking regulations, adjacent land uses, posted speed limits, travel times, accident history, sidewalk and crosswalk locations, driveway locations, traffic signs and pavement markings, traffic signal operations with phasing and timing, and peak period traffic counts for roadway segments and intersections. The data gathered are used for analyzing the present and projected operational characteristics of the transportation system.

In essence, the purpose of a TIS is to characterize the current supply and demand of **roadway capacity** and to forecast how the development will affect the supply and demand. The existing demand for roadway capacity can be estimated using traffic counts and assessing the **level of service (LOS)** of the roadway. Traffic counts can be obtained by one or more persons tabulating the number of cars and other vehicles, or by use of an automated traffic counter.

The LOS for a roadway is characterized based on travel times, drivers' freedom to maneuver, the number of traffic interruptions, and other factors. The level of service ranges from LOS A (least congested) to LOS F (most congested). You may think that all roads are designed for LOS A; however, the LOS must be balanced with funding realities, land availability, and other factors.

The level of service concept is applied to two-lane roads, multi-lane roads, freeways, intersections, bike lanes, and sidewalks. For each of these, the LOS is related to quantitative information. For example, for certain types of two-lane roads, the LOS depends on how fast vehicles are traveling and how much time vehicles spend following other vehicles. This is illustrated in Figure 20.11. According to this figure, if vehicles can travel at an average speed of 60 mph and do not spend much time (less than 35 percent of the time) behind other vehicles, that road would be characterized as LOS A.

To estimate the future demand of roadway capacity, transportation engineers identify any other developments to occur in the study area during the forecast period and use trip generation tables such as Table 20.3 to estimate trips to be generated by the developments.

Once a study has been completed that characterizes the current supply of the roadway capacity and the current and proposed demands on the roadway

Driver Comfort

Transportation engineers often use the term "driver comfort." They are *not* referring to the plush leather seats of a luxury sedan. Rather, they are referring to the ease with which drivers can get from point A to point B. Traffic congestion creates discomfort and stress because of delays, difficulty in switching lanes, etc.

Roadway capacity refers to the amount of traffic a roadway can carry. *Demand* for roadway capacity is characterized by how much traffic the roadway carries. *Supply* is the actual roadway capacity (i.e., how much traffic the roadway can potentially carry). The existing supply of roadway capacity can be characterized by the geometric features of the roadway (number and width of lanes; grades; existence of turning lanes) and the number and effectiveness of traffic signals.

Level of service, or **LOS**, is a qualitative measure of the roadway's operating conditions and the perception of those conditions by motorists and/or passengers.

Figure 20.11 Level of Service (LOS) for Two-Lane Highways.

Source: From *Highway Capacity Manual 2000.* Copyright, National Academy of Sciences, Washington, D.C., Exhibit 20-3, p. 20-4. Reproduced with permission of the Transportation Research Board.

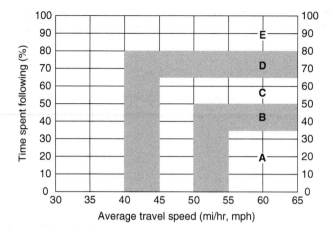

capacity, the TIS report can be written. Typically, the contents of a TIS report include:

- background description of the site, including a map
- plan of the proposed development
- analysis of existing conditions (LOS, speed study, traffic counts, crash history)
- trip generation predictions
- future LOS with and without development
- recommended mitigation steps, or improvements, if any.

There are many possible mitigation steps that can be taken if necessary, which may either address increasing the future supply of roadway capacity or decreasing future demand. Mitigation efforts that address the supply of roadway capacity include adding additional travel lanes or turning lanes, building a new highway interchange, or widening lanes. Alternatively, the capacity of the existing roadway can be increased by making the roadway more efficient, perhaps by enhancing existing signs and pavement markings, or adding traffic signals. To decrease the demand for future roadway capacity, the proposed development can take measures to encourage non-motorized transportation by designing walkways with pedestrians in mind, by integrating the development carefully into existing bike trails, or by integrating with existing mass transit. Recall that reducing the demand for motorized transportation is one of the characteristics of green buildings. Also, the design of multi-use developments, in which people can work, shop and live, can greatly reduce the demand for the existing roadways.

The regulatory authority that reviews the TIS has several options at its disposal. It can accept the TIS and any mitigation efforts proposed. It can request additional mitigation efforts. Or, it can prohibit the development from occurring.

Impact Fees

The costs of widening roads, adding traffic signals, or some other mitigation activity is often borne by the developer. The municipality may oversee the construction of such infrastructure projects and recover the costs by assessing an impact fee to the developer.

Outro

The analyses presented in this chapter serve as examples of the types of work performed by civil and environmental engineers. Such analyses are common, as are many other more complex analyses requiring advanced technical knowledge. These analyses typically require a *systems* viewpoint and the consideration of non-technical aspects to ensure that all perspectives are incorporated, without which applicability and acceptance are likely to be limited.

chapter Twenty Homework Problems

Calculation

20.1 The activities needed to build a home are outlined in the following table. Based on this information:

a. create a CPM diagram;
b. highlight the critical path;
c. note which tasks lie on the critical path;
d. determine the minimum length of time to complete the project.
e. if the roofing takes 3 days longer than predicted, by how much will the length of time to complete the entire project increase?
f. if the wiring takes 3 days longer than predicted, by how much will the length of time to complete the entire project increase?

Activity Label	Activity Summary	Duration (days)	Predecessors
A	Excavating	5	
B	Pouring foundation	3	A
C	Framing	12	B
D	Plumbing	7	C
E	Wiring	8	C
F	Roofing	4	C
G	Finishing wall exteriors	7	C
H	Plumbing and wiring inspection	1	D, E
I	Covering interior walls	6	F, H
J	Finishing interior	8	I
K	Landscaping	3	G

20.2 Some representative activities necessary to develop a greenfield site into a small shopping center are listed in the following table along with their duration and predecessors.

a. create a CPM diagram;
b. highlight the critical path;

c. note which tasks lie on the critical path;
d. determine the minimum length of time to complete the project;
e. which would delay the project completion more: a 3-week delay in planning council approval or a 1-week delay in the stormwater permit approval?

Activity Label	Activity	Duration (days)	Predecessor
A	Site survey	3	
B	Site design	5	A
C	Planning council approval	5	B
D	Stormwater permit approval	20	B
E	Water main extension approval	15	B
F	Site grading/ earthmoving	15	C, D, E
G	Underground utility installation	20	F
H	Stormwater pond construction	15	F
I	Road construction	20	G
J	Parking lot paving	5	G
K	Landscaping	5	H, I, J

20.3 Derive Equation 20.2 from Equation 20.1, given $z = 0$, and $y = 0$.

20.4 For a pollutant emission rate of 3 lbs/day from a 50-foot tall stack with a wind speed of 10 mph, use the Gaussian Plume Spreadsheet Model to prepare a plot of ground-level pollutant concentration along the centerline as function of distance downwind from the stack. Your x-axis should vary from 0 meters to 2,000 meters.
Determine your solution for the following atmospheric conditions:

a. unstable
b. moderately stable
c. stable.

20.5 Determine the stack height necessary to reduce the maximum downwind ground-level concentration in Homework Problem 20.4 by 50 percent if all other variables remain constant.

20.6 Prepare a graph of the deflection of a simply supported beam as a function of load at the center. The x-axis should vary from 0 to 500 pounds. The beam is 10-foot-long, 2-inch × 6-inch in cross-section, and oriented with the 2-inch side horizontal. Show separate series on a single graph, one for each of the following materials:

a. aluminum

b. cast iron

c. stainless steel

20.7 Will a 10-foot-long aluminum beam (dimensions 2-inch × 4-inch) bend "too much" if loaded in the center by a 500-pound load if:

a. the beam is oriented "on edge" with the 2-inch side horizontal;

b. the beam is oriented "flat" with the 4-inch side horizontal.

20.8 What is the maximum load that a 20-foot long maple beam (4-inch × 4-inch) can carry at its center if it is not to deflect more than 0.25 inches?

20.9 A windowless room with a footprint of 10-foot by 20-foot and an 8-foot ceiling has a tight fitting door. A water pipe in the ceiling of the room breaks and the room begins to fill with water. How does the force on the 3-foot by 7-foot door vary with water depth? Answer the question by preparing a graph (using a spreadsheet) of the hydrostatic force (in Newtons) on the door as a function of water depth. Does the size of room footprint matter?

20.10 Redo Numerical Example 20.2 for seawater rather than freshwater. Use a value of $1,030\,kg/m^3$ for the density of seawater. By what percentage does the force on the floodwall increase as compared to freshwater?

Short Answer

20.11 Define CPM. Why is CPM useful to project managers?

20.12 Define "critical path."

Discussion/Open-Ended

20.13 Under what conditions might "traffic calming" be recommended as a mitigation effort for a TIS?

20.14 Do the gates on the Panama Canal have to hold back the entire "force" of the Pacific Ocean? Explain.

Research

20.15 On the textbook website (www.wiley.com/college/penn) are links to three example TIS reports. Read one of the TIS reports, as assigned by your course instructor, and summarize it in a single page. Do you think the proposed mitigation efforts/recommendations are reasonable?

20.16 Research a LEED-certified building. Find out ways that this building has minimized demands on the existing motorized transportation infrastructure.

Key Terms

- critical activities
- critical path
- deflection
- level of service (LOS)
- plume
- point loads
- roadway capacity
- service loads
- serviceability
- trips
- ultimate loads
- weathering steel

References

American Society of Civil Engineers Hurricane Katrina External Review Panel. 2007. *The New Orleans Hurricane Protection System: What Went Wrong and Why*. American Society of Civil Engineers, Reston, VA.

Transportation Research Board. 2000. *Highway Capacity Manual 2000*. Washington, DC: Transportation Research Board.

chapter Twenty One Design II

Learning Objectives

After reading this chapter, you should be able to:

1. Design basic wood formwork for a concrete wall.
2. Determine the dimensions of a primary clarifier for wastewater treatment.
3. Explain the design principles for floodwalls.
4. Design a beam to not exceed allowable stresses.
5. Design horizontal curves of a roadway.

Introduction

In Chapter 10, we presented the design process and provided examples of how it is used in the subdisciplines of civil and environmental engineering. In the intervening chapters, we have presented many aspects such as economic, social, and sustainability considerations that are necessary in "real-world" design. In this chapter, we provide you with five design applications that take into account the additional considerations we have presented in Chapters 11 through 19.

Construction Engineering: Concrete Mix and Formwork

Concrete Admixtures

Concrete may be enhanced by adding chemical or mineral admixtures. Admixtures can be used to:

- accelerate, or in some cases, retard curing times

- increase workability, durability, or strength

- add color for aesthetics

- inhibit corrosion of rebar

Waste products such as coal fly ash (fine particulates removed by air pollution control devices from coal-fired power plants) and ground blast furnace slag (from steel production) may be recycled into concrete as substitutes for a substantial fraction of the cement.

Concrete is often incorrectly referred to as "cement." Cement is the binding agent in concrete, and is made by heating limestone and other materials at very high temperatures in a cement kiln. Concrete is a mixture of cement, water, and aggregate (sand and gravel, Figure 21.1). Concrete is typically about 10 to 15 percent cement, 60 to 75 percent aggregate, and 15 to 20 percent water on a volume basis. Important characteristics of concrete include its workability when wet, its final strength, and its durability after curing. Ready-mix concrete is concrete that is mixed to project specifications at a batch plant, either on-site or at a remote location.

The amount of water in a concrete mix is critical to performance. A minimum amount of water must be added to initiate the chemical reactions that bind the cement to the aggregate. Additional water increases the workability of the concrete, so that it is not too "stiff." However, as shown in Figure 21.2, the strength of concrete decreases as more water is added. Consequently, there is a tradeoff between strength and workability. Additionally, concrete strength increases with curing time (Figure 21.2).

Concrete is often poured, or placed, into forms (molds) to create columns, walls, and footings. Steel reinforcing bars ("rebars" or "rerods") are often added to provide additional tensile strength. The forms are kept in place during initial curing.

Forms are typically removed as soon as possible once the concrete has cured to an adequate strength. The concrete will continue to cure and strengthen after the forms are removed. Temperature is very important to the rate of curing. At concrete temperatures below 70°F, curing slows (virtually stopping at 40°F). At higher temperatures, excessive evaporation of water from concrete can be problematic.

Figure 21.1 A Broken Concrete Step in Which Aggregate is Visible.

Source: M. Penn.

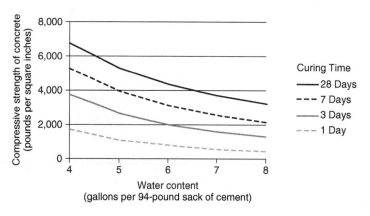

Figure 21.2 Ranges of Compressive Strength of Concrete as a Function of Water Content and Curing Time at 70°F.

Source: Adapted from Love, 1973.

Concrete Curing Temperatures

As concrete cures, chemical reactions increase the temperature. For the recently completed $240 million Colorado River Bridge bypassing the Hoover Dam, it was was determined that extreme desert heat (air temperatures exceeding 110°F) would result in curing temperatures that exceeded concrete design specifications. One approach to reducing curing temperatures is the circulation of water through pipes embedded in the concrete. However, this option was determined to be infeasible due to site-specific constraints. As a result, the seemingly extreme option of using liquid nitrogen to pre-cool the concrete before placement was used (Figure 21.3).

Figure 21.3 Liquid Nitrogen Being Injected into a Ready-Mix Truck to Cool Concrete Before Placement in the Construction of the Colorado River Bridge.

Source: Federal Highway Administration.

Concrete Canoes

The American Society of Civil Engineers (ASCE) sponsors a concrete canoe competition annually. Students design lightweight concrete mixes, many of which are lighter than water, from which canoes are constructed (Figure 21.4). Although the races are the most exciting portion of the competition for the teams and spectators, 75 percent of the overall scoring is based on a design report, an oral presentation, aesthetics, and durability.

ASCE National Concrete Canoe Competition Rules and Regulations

Figure 21.4 A Concrete Canoe on Display at the 2010 National Competition Hosted by the California Polytechnic State University-San Luis Obispo. This canoe placed third nationally in the Final Product category, and fourth overall. The display stand is a model of the Mackinac Bridge, the third longest suspension bridge in the world, which spans the Mackinac Strait between Lake Michigan and Lake Huron.

Source: Courtesy of Michigan Technological University, ASCE 2009–2010 Concrete Canoe Team.

Figure 21.5 Forms for a Concrete Wall.

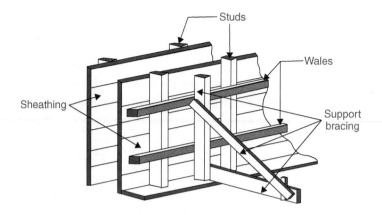

Studs

Wales

Sheathing

Support bracing

The design of formwork is necessary for efficient construction and to ensure that the final product meets design specifications for shape, surface finish, and strength. Examples of forms for walls and columns are shown in Figure 21.5 and Figure 21.6 respectively. Formwork at a construction site is shown in Figure 21.7.

The design of formwork involves the temporary structural support for the sheathing that acts as the mold for concrete. The formwork must support the hydrostatic pressure of the freshly placed concrete, the weight of

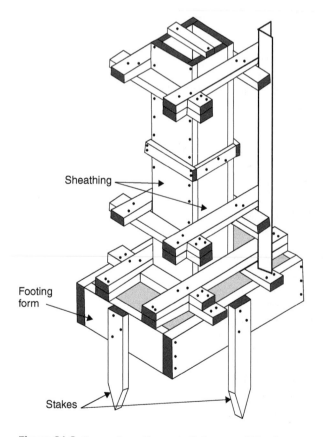

Sheathing

Footing form

Stakes

Figure 21.6 Forms for a Concrete Column and Footing.

Figure 21.7 Formwork on Second-Story Columns (Right), and Cured Columns After Form Removal (Left, Background, and Below). Note the rebar projecting from the tops of the columns.

Source: M. Penn.

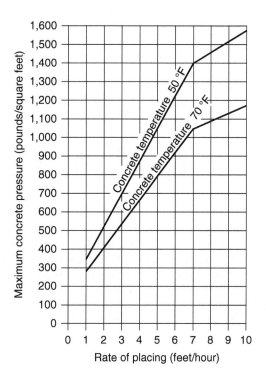

Figure 21.8 **Concrete Pressure on Form Sheathing as a Function of Concrete Placement Depth Rate.** Note that for a given depth, the hydrostatic pressure of freshly poured concrete is approximately 2.5 times greater than that of water since the unit weight of concrete is approximately 2.5 times greater than that of water (150 lbs/ft^3 for concrete, 62 lbs/ft^3 for water).

Source: Adapted from Love, 1973.

construction workers, the weight of equipment, and the impact of vibration if used to consolidate the concrete. Formwork may be of wood, aluminum, steel, fiberglass, or foam board.

Simplified wood formwork design will be presented here for construction of a concrete wall. First, consider that the hydrostatic pressure of the placed concrete on the form sheathing is a function of both temperature and the rate (i.e., depth per time) of concrete placement in the form. This is demonstrated in Figure 21.8.

The rate of concrete placement is important because as concrete begins to cure, or "set," the hydrostatic pressure decreases. When fully cured, concrete exerts no hydrostatic pressure on the forms. Consequently, higher placement rates will lead to a greater depth of liquid concrete, which will exert higher hydrostatic pressure as shown in Figure 21.8. Figure 21.8 also demonstrates that, all other things being equal, higher temperatures decrease the maximum concrete pressure; this is due to the fact that higher temperatures lead to faster curing.

Higher pressures require stronger formwork to prevent deflection of the sheathing and potential seepage of wet concrete between sheathing panels, termed blowout. Thus, there is a tradeoff between rate of concrete placement and the cost and time of erection of the formwork. Construction engineers must balance this tradeoff to decide what is best for a given project depending on budgets and timelines.

Once the desired placement rate and resultant concrete pressure is determined, the stud spacing to support the sheathing can be determined based on the type of sheathing (Figure 21.9). Note that as the sheathing thickness increases, so does the stud spacing. Thicker sheathing provides additional resistance to the concrete pressure, requiring less stud support. For example, at a concrete pressure of 800 pounds per square feet, 1-inch plywood sheathing requires 14-inch stud spacing compared to 8-inch spacing for 1/2-inch plywood sheathing.

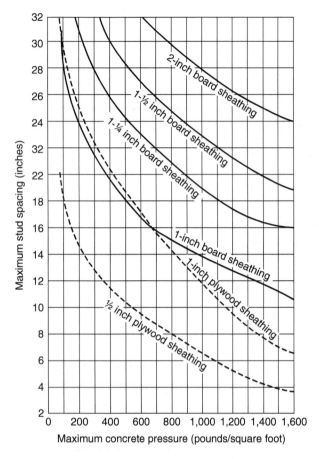

Figure 21.9 Maximum Stud Spacing as a Function of Concrete Pressure.

Source: Adapted from Love, 1973.

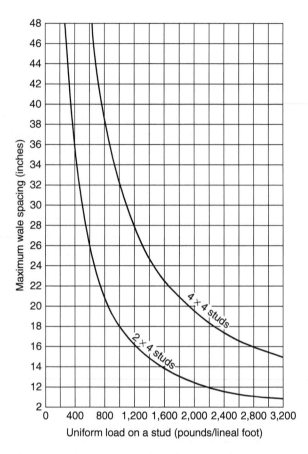

Figure 21.10 Wale Spacing as a Function of Stud Size and Stud Loading.

Source: Adapted from Love, 1973.

The stud loading (pounds/lineal foot) is equal to the product of concrete pressure pounds/square feet and stud spacing (feet). The vertical studs are cross-supported, horizontally, by wales (refer to Figure 21.5). The wale spacing is determined from Figure 21.10 based on the calculated stud loading and the stud dimensions. Note that wale spacing increases with stud size, as larger studs require less support. For example, a load of 1,000 pounds/lineal foot requires 18-inch spacing for 2-inch × 4-inch studs as compared to 32-inch spacing for 4-inch × 4-inch studs. The construction engineer again is faced with a tradeoff, this time between stud size (and cost) and the number of wales needed for support.

numerical example 21.1 Form Design

Determine the 2-inch × 4-inch stud and wale spacing to support 1-inch plywood forms for a concrete wall that is to be placed at a rate of 5 feet per hour with a concrete temperature of 70°F.

solution

According to Figure 21.8, the concrete pressure on the forms will be 800 pounds/square feet.

Next, the stud spacing required is found to be approximately 14.2 inches by using Figure 21.9. This spacing is rounded down to 14 inches for ease of measurement in the field. Note that this also makes the design slightly conservative; by rounding down, rather than up, we are ensuring that the stud spacing is more than sufficient to support the concrete pressure.

The stud loading is determined by:

$$\text{Stud loading (lbs/lineal ft)} = \text{concrete pressure (lbs/ft}^2) \times \text{stud spacing (ft)}$$
$$= 800 \text{ lbs/ft}^2 \times 14 \text{ inches} \times 1 \text{ ft/12 inches}$$
$$= 930 \text{ lbs/lineal ft}$$

Using this stud loading and Figure 21.10, the wale spacing for 2-inch × 4-inch studs is found to be approximately 18.5 inches. Again, this is rounded *down* to 18 inches to incorporate a modest safety factor.

Because of cost savings, modular form systems made of aluminum or steel (Figure 21.11) are replacing wood forms in practice. Modular forms last longer than wood, and require less time to install. The modular sections are pinned or clipped together, and wood or aluminum wales are fastened horizontally to provide deflection support as required.

There are many types of concrete that have particular advantages with regard to constructability, design strength, durability, and aesthetics. You will learn about these, as well as how to design concrete mixes (relative amounts of water, cement, and aggregate, as well as the size of aggregate), in future civil engineering courses.

Variations of Formwork

Not all formwork is temporary (i.e., removed after curing). Permanent insulated concrete forms are becoming increasingly popular as energy prices rise.

Not all formwork is used to cast concrete in-place. Precast concrete panels, or tilt-up concrete walls are formed and cured in a controlled environment either onsite, or offsite and transported to a construction site for placement.

Information on Insulated Concrete Forms

Time-Lapse Video of Insulated Concrete Form Installation

Tilt-Up Concrete Panels Product Information

Figure 21.11 **Wooden Forms (Left) and Modular Forms (Right) for Reinforced Concrete Foundation for a High-Rise Building in Downtown Chicago.**

Source: P. Parker.

Remember that full-sized versions of textbook photos can be viewed in color at www.wiley.com/college/penn.

Environmental Engineering: Primary Treatment of Wastewater

Wastewater treatment processes were introduced in Chapter 5. Here we will provide you with the fundamental design concepts of primary treatment. Primary treatment follows preliminary treatment that removes grit and debris. The objective of primary treatment is to remove floatable solids (fats, oils, and greases) and readily settling solids (organic and inorganic particles). The settling of particles leads to "clearer" water and thus settling tanks are termed clarifiers. Stokes' Law (Equation 21.1) is often used to estimate settling velocities.

$$V_S = g \cdot d^2 \cdot \frac{\rho_p - \rho_w}{18 \cdot \mu} \tag{21.1}$$

where

$V_s =$ settling velocity (m/s)
$g =$ gravitational constant (9.8 m/s^2)
$d =$ particle diameter (m)
$\rho_p =$ particle density (kg/m^3)
$\rho_w =$ wastewater density, temperature dependent (kg/m^3)
$\mu =$ viscosity[1] of wastewater, temperature dependent (kg/ms)

A pair of primary clarifiers are shown in Figure 21.12. Floatable solids are skimmed from the water surface of the tank and collected in a sump before being pumped to a digester for further treatment. After settling to the tank bottom, settled solids are scraped to an accumulation hopper (Figure 21.13) and then pumped to a digester.

In reality, wastewater contains particles of all shapes, sizes, and densities, thus limiting the use of Stokes' Law. It is common in engineering to develop empirical relationships based on collected data for use as the basis for design. One such relationship will be used here.

Figure 21.12 Twin (Parallel) Primary Clarifiers in Platteville, Wisconsin. Wastewater enters the tanks in background and exits in foreground. 1: Accumulated floatables on water surface. 2: Skimmer blades push floatables to the accumulation area for removal. The twin tanks provide redundancy, to allow partial treatment in one tank while the other is being repaired or undergoing maintenance. Under normal conditions, both clarifiers are used in parallel, each treating one-half of the influent.

Source: M. Penn.

[1] Viscosity is the resistance of a fluid to flow. Motor oil and molasses have high viscosities while water has relatively low viscosity.

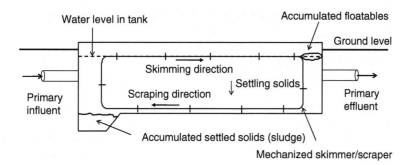

Figure 21.13 Side View of a Simplified Primary Clarifier.

First, consider that for primary clarifiers, the percentage of particles (also termed suspended solids, SS) removed increases as a function of the time that the wastewater is in the clarifier (**detention time**, τ).

$$\tau = \frac{V}{Q} \qquad (21.2)$$

where
$\tau =$ detention time (hr)
$V =$ volume of tank (m^3)
$Q =$ flowrate of wastewater through the tank (m^3/hr)

The relationship between detention time and removal efficiency is shown in Figure 21.14 for both SS (which includes inorganic *and* organic solids) and **BOD**, or **biochemical oxygen demand**. BOD is the target contaminant for removal in the subsequent secondary treatment step.

Note from Figure 21.14 that the removal efficiency dramatically increases with detention times up to about 2 hours, and begins to plateau after approximately 4 hours. Particles that are not removed are very small and settle *very* slowly. In Stokes' Law, the settling velocity is a function of the diameter *squared*; thus, a 0.01-mm diameter particle settles 100 times slower than a 0.1-mm particle of the same density. The typical design detention time is 2 hours for primary clarifiers. Longer times require larger and more expensive tanks, which provide only a minimal increase in treatment efficiency, and thus are not cost-effective.

Using a specified detention time and a known design flowrate of wastewater, the required volume of the tank can be determined by solving Equation 21.2 for volume. If rectangular in shape, the length (L), width (W), and depth (D) must be determined based on the volume. Most states have design codes that stipulate the minimum depth of primary tanks. Typical depths of primary tanks are 4 meters. What remains to be determined is the length and width of

BOD, or **biochemical oxygen demand**, is a measure of the degradable organic content of the wastewater.

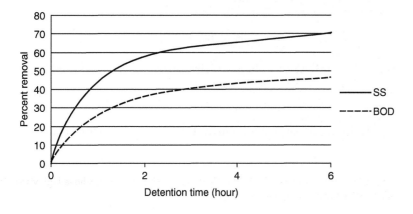

Figure 21.14 Removal of SS and BOD in Primary Clarifiers as a Function of Detention Time.

Source: Adapted from Steel & McGhee, 1979.

the tank. A long and "skinny" tank is impractical from a construction excavation standpoint. A short and "wide" tank is subject to inefficient treatment as the wastewater may "short circuit" the tank (exiting before its designed detention time). Typical length-to-width ratios are 5:1.

numerical example 21.2 Preliminary Sizing of Primary Clarifier

For a wastewater flowrate of 150 cubic meters/hour and a specified detention time of 2 hours, calculate the dimensions of a primary clarifier. The tank depth is 4 meters and the length-to-width ratio is 5.

solution

In order to determine the dimensions of primary clarifiers for site layout and structural design purposes, we can solve Equation 21.2 for tank volume:

$$\text{Tank volume} = \tau \cdot Q$$

The volume of the tank is the product of the tank's length, width, and depth, and this relationship can be substituted into the previous equation.

$$\tau \cdot Q = L \cdot W \cdot D$$

Thus, we can rearrange for tank width:

$$W = \frac{\tau \cdot Q}{L \cdot D}$$

Substituting the given length-to-width ratio yields the following:

$$W = \frac{\tau \cdot Q}{(5\,W) \cdot D}$$

Solving for W yields the following:

$$W = \sqrt{\frac{\tau \cdot Q}{20}}$$

For a wastewater flowrate Q of 150 cubic meters/hour and a specified detention time τ of 2.0 hours, the tank dimensions would be ($L \times W \times D$): 19 meters by 3.9 meters by 4.0 meters. If two parallel redundant tanks were to be used, the surface area, SA, of each tank would be halved:

$$SA = L \cdot W = 19\,\text{m} \times 3.9\,\text{m}/2 = 37\,\text{m}^2.$$

The $L{:}W$ ratio would remain 5; thus the dimensions can be determined:

$$L \cdot W = 5W^2 = 37\,\text{m}^2.$$

The width of each tank would be 2.7 meters and the length would be 13.6 meters.

Additional design requirements for primary clarifiers include:

- Selection of skimmer/scraper to remove settled solids and floatables (these are designed by manufacturers and selected or specified by environmental engineers.

- Design of sumps for collected solids (sludge) at the bottom of the tank and floatables.

- Hydraulic design of influent and effluent flow control structures to ensure mixing at the entrance and to minimize resuspension of settled solids at the entrance and exit by reducing wastewater velocities.

- Hydraulic design of pumps and piping to transport removed floatables and settled solids to the digester.

- Structural design of the reinforced concrete for the tank walls and base.

- Selection of tank covers (in some instances) to control odors.

Geotechnical Engineering: Floodwalls

Cantilevered floodwalls are often referred to as "T-walls" because the cross-section has the shape of an inverted "T" (Figure 21.15). T-walls are designed to prevent three common types of floodwall failure: sliding, overturning, and exceeding soil-bearing capacity. Sliding occurs when lateral forces on the water side of the floodwall exceed resisting forces of the supporting soil (Figure 21.16). Overturning occurs when a wall rotates along the base of the toe (Figure 21.17). Bearing capacity failure occurs when the pressure on the soil at the base of the wall exceeds the bearing capacity of the soil (Figure 21.18).

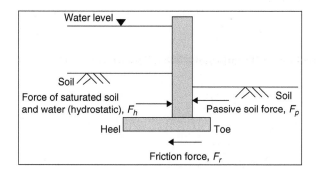

Figure 21.15 A Cantilevered Floodwall, or T-Wall. Horizontal forces acting on the wall are shown.

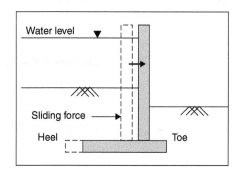

Figure 21.16 Failure of a T-Wall by Sliding.

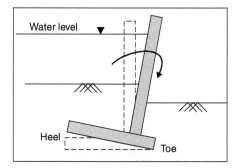

Figure 21.17 Failure of a T-Wall by Overturning. Note the rotation occurs at the toe.

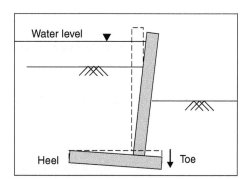

Figure 21.18 Failure of a T-Wall Due to Excessive Soil Pressure and Settlement. Note that the base has settled, primarily at the toe.

Figure 21.19 Failure of an I-Wall Due to Scour of Supporting Soil Due to Overtopping of the Floodwall. The base of a T-wall (not shown) provides additional protection from scour due to overtopping water.

Source: Adapted from ASCE, 2007.

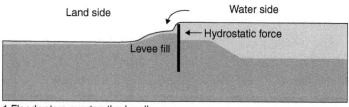

Land side   Water side
Hydrostatic force
Levee fill

1.Floodwaters overtop the I-wall.

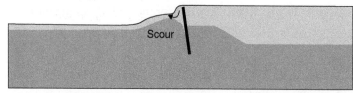

Scour

2.The water scours soil from the land-side of the I-wall and washes it away.

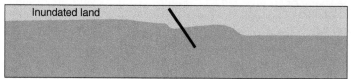

Inundated land

3.I-wall fails due to lack of foundation support.

Preventing bearing capacity (settlement) failure in foundations is the purpose of footings, which you learned to design in Chapter 10. T-walls also minimize the risk of failures from excessive scour of soil on the land side of the wall due to overtopping of the floodwall by high water (Figure 21.19), which occurred at some I-walls (a vertical wall without a horizontal base) during Hurricane Katrina. In this section, we will discuss floodwall design to prevent sliding failure.

In order to prevent sliding failure, the resistive lateral forces of the soil must be greater than the cumulative lateral sliding force, F_H, of the water (hydrostatic) and of the saturated soil on the high water side of the wall (Figure 21.15). A factor of safety of 1.5 is typically used in designs, such that: Sum of resistive forces/Sum of sliding forces $\geq$ 1.5.

The resistive forces are the friction force, F_R, at the base of the footing and the passive soil force, F_p, above the base of the toe (Figure 21.15). The passive soil force is a function of the soil depth above the base and the passive soil pressure coefficient, K_p, which is a measure of the lateral pressure that a soil type can withstand without movement. In simple terms, imagine the passive soil force as the force resisting the blade of a bulldozer. The bulldozer must overcome this force in order to move the soil. The greater the K_p of the soil, or the greater the soil depth on the blade, the more force required by the bulldozer.

The second lateral resisting force, the friction force, is the product of the soil coefficient of friction, C_f, and the net (downward) vertical force acting on the base. The friction force acts laterally along the base of the T-wall. In simple terms, again using the bulldozer example, the friction force is equal to the force required by a bulldozer to push (slide) an object along a soil surface. The net vertical force, and thus the friction force, increases as the base width, B, increases (increasing the weight of the concrete base) and as the height of soil above the heel or toe, D_h or D_t increases (increasing the weight of soil acting on the base).

Design values for T-wall base widths are provided in Table 21.1 and Table 21.2 for two scenarios:

Table 21.1

Base Width Design for Scenario 21.1 Common parameters are shaded.		Column A	Column B
	Soil type	Clean sand	Silty sand
	Coefficient of friction, C_f	0.55	0.45
	Wall height, H (feet)	6	6
	Depth of soil at heel, D_h (feet)	3	3
	Depth of soil at toe, D_t (feet)	3	3
	Base width, B (feet)	6.5	8

Source: Adapted from FEMA, 2001.

Scenario 21.1: Different soil *types* with all other parameters being equal (Table 21.1). The soil types are clean sand (without fines) and silty sand (with fines), each with different values of K_p and C_f.

Scenario 21.2: Different soil *depths* (over both toe and heel of the base) with all other parameters being equal (Table 21.2). The soil depths considered are 3 feet and 4 feet.

By comparing columns A and B in Table 21.1, we can see that for different soil types, the base width increases from 6.5 feet to 8 feet because the silty sand has a lower value of K_p and C_f than clean sand. By increasing the base width in silty sand, the net vertical force increases, which results in more weight of concrete in the base and more weight of soil above the enlarged base. This increased net vertical force counteracts the lower C_f, and increases the resistive friction force. In this case, changing the base width does not affect the passive soil force, because the depth of soil remains constant.

By comparing columns A and B in Table 21.2, we can see that when the soil depth increases from 3 feet to 4 feet over the base (for the same soil type), the base width decreases from 8 feet to 5.5 feet. The increased soil depth results in an increased net vertical force and thus an increased frictional resistive force for any given base width. Also, the passive soil force increases with soil depth for a constant value of K_p.

This simplified design approach does not incorporate all of the forces that may be acting on a floodwall. Actual design may require accounting for additional soil and hydrodynamic forces that you will learn about in a geotechnical engineering course (in which you will also gain the technical knowledge to design floodwalls in order to prevent overturning and pressure failures not covered in this section). For example, in "soft" soils with low bearing capacities, the foundation is extended to greater depths (Figure 21.20).

Table 21.2

Base Width Design for Scenario 21.2 Common parameters are shaded.		Column A	Column B
	Soil type	Silty sand	Silty sand
	Passive soil pressure coefficient, K_p	3.0	3.0
	Coefficient of friction, C_f	0.45	0.45
	Wall height, H (feet)	6	6
	Depth of soil at heel, D_h (feet)	3	4
	Depth of soil at toe, D_t (feet)	3	4
	Base width, B (feet)	8	5.5

Source: Adapted from FEMA, 2001.

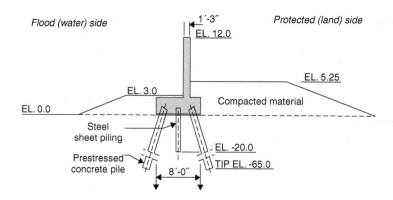

Figure 21.20 Extended Foundation of a T-Wall in Soft Soil. Typical elevations (abbreviated as "EL." in diagram) and dimensions shown. Note concrete piles extend 65 feet below the base of the T-wall.

Source: USACOE.

Labels in figure: Flood (water) side; Protected (land) side; 1'-3"; EL. 12.0; EL. 5.25; EL. 3.0; Compacted material; EL. 0.0; Steel sheet piling; EL. -20.0; Prestressed concrete pile; TIP EL. -65.0; 8'-0"

Structural Engineering: Design of Structural Members

As we discussed in Chapter 20, a structure must be able to support ultimate loads without failing, and must also support service loads without deflecting or vibrating excessively. In this section, we will discuss the concept of designing a beam to accommodate an internal stress such that it will not fail when subjected to ultimate loads.

In structural engineering, **stress** is defined as a force per area. To understand the concept of stress, consider the cantilevered member of Figure 21.21. The beam is being subjected to an axial force and is in tension. This force is carried throughout the entire member. When analyzing the interior cross-section of the member as shown in Figure 21.22, this force F is distributed across the cross-sectional area A. The stress, σ, may be defined with the variables of Figure 21.21 and Figure 21.22 using Equation 21.3.

$$\sigma = \frac{F}{A} = \frac{F}{b \cdot h}$$

(21.3)

where

$F =$ force applied
$A =$ cross-sectional area
$b =$ structural member width
$h =$ structural member height

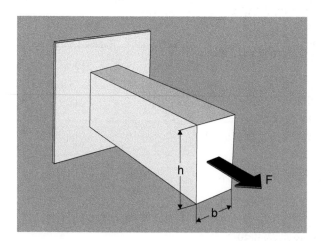

Figure 21.21 A Cantilevered Structural Member in Tension.

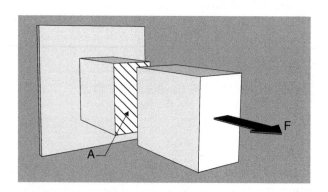

Figure 21.22 Internal Cross-Sectional Area (A, Hatched) of Structural Member Across Which the Tensile Force is Distributed.

For common SI units of newtons for force and m^2 for cross-sectional area, stress has units of N/m^2, which is equal to one pascal (Pa). The stresses involved in structural engineering are relatively large, and are often expressed as MPa (megapascals, or 10^6 Pa). For English units, common units are lb (force), in^2 (area), and psi or pounds per square inch (stress).

You can readily imagine that the member in Figure 21.21 and Figure 21.22 could withstand a certain force before it failed (i.e., "broke"). Certain materials are "strong" in the sense that a large force per unit area is needed before they fail as compared to "weak" materials. Said another way, stronger materials can withstand larger stress.

The engineer designing a beam needs to ensure that it does not fail when subjected to ultimate loads. One method to do so is to calculate the stress carried by a beam and compare it to the beam's **failure stress**. For this textbook, we will term this method of analysis and design the **maximum permissible stress method**. While this exact method is not used by practicing engineers, it is based on the same concepts used in practice, and is simplified for use by first- and second-year students.

> **Failure stress** is the stress that a structural member can carry before it fails.

The calculation of stress for a simply supported beam is somewhat more complex than for a cantilevered structural member under pure tension. For a simply supported beam loaded at its center, the maximum permissible stress carried by the beam is given by Equation 21.4, where the subscript "ss" refers to the simply supported beam. In future structural engineering courses, you will learn how to derive this equation and similar equations for other beam arrangements (e.g., cantilevered beam, simply supported beam carrying a distributed load).

$$\sigma_{ss} = \frac{3 \cdot F \cdot L}{2 \cdot b \cdot h^2} \qquad (21.4)$$

where

$L =$ beam length

$F =$ magnitude of point load at center of simply supported beam

$\sigma_{ss} =$ maximum permissible stress.

To *analyze* a beam to determine whether it could carry its ultimate loads, Equation 21.4 would be solved for the maximum permissible stress, given F (value for the ultimate load) and the beam dimensions (width, height, and length). The maximum permissible stress would then be compared to the ultimate stresses corresponding to the material type using Table 21.3. Theoretically, if the stress calculated using Equation 21.4 is less than the stress in Table 21.3, then the beam will not fail. However, in reality, σ_{ss} should be less than the ultimate stress in Table 21.3 by a considerable amount as determined by the factor of safety (*FS*), defined by Equation 21.5. A typical value for *FS* is 1.5; thus, the value of the maximum permissible stress calculated using Equation 21.4 should be at least 1.5 times less than the ultimate stress of the material.

Table 21.3

Ultimate Stresses for Various Materials

Material	Failure Stress (psi)
Aluminum	60,000
Concrete	500
Pine wood	6,000
Stainless steel	75,000
Steel	36,000

$$FS = \frac{\text{ultimate stress}}{\sigma_{ss}} \qquad (21.5)$$

To *design* a beam, a similar process would be followed. The designer can manipulate several variables, including the beam material and cross-sectional dimensions. A factor of safety will likely be specified by industry design standards.

numerical example 21.3 Preliminary Sizing of a Beam

Consider a simply supported steel beam that needs to carry a load at its center of 2,000 pounds. The beam length is 15 feet (180 inches). If the beam height is to be twice its width and a factor of safety of 1.5 is to be used, determine the beam's height and width using the maximum permissible stress method.

solution

From Equation 21.5, σ_{ss} can be calculated by knowing the factor of safety and the ultimate stress of steel, obtained from Table 21.3.

$$\sigma_{SS} = \frac{\text{ultimate stress}}{FS} = \frac{36,000\,\text{psi}}{1.5} = 24,000\,\text{psi}$$

Thus, we must ensure that the beam cross-section is large enough that the stress developed is less than 24,000 psi. We can now use Equation 21.4 to calculate the dimensions that can effectively carry this stress. Given that the height is to be twice as large as the width, we can substitute into Equation 21.4 the expression $h = 2b$.

$$\sigma_{SS} = \frac{3 \cdot F \cdot L}{2 \cdot b \cdot h^2} = \frac{3 \cdot F \cdot L}{2 \cdot b \cdot (2b)^2} = \frac{3 \cdot F \cdot L}{8b^3}$$

This equation can be solved for b, knowing values of σ_{ss}, F, and L. L will be inserted using units of inches to keep units consistent.

$$b = \left(\frac{3FL}{8\sigma_{SS}}\right)^{\frac{1}{3}} = \left(\frac{3 \cdot (2 \cdot 10^3\,\text{lb}) \cdot 180\,\text{in}}{8 \cdot 2.4 \cdot 10^4\,\text{psi}}\right)^{\frac{1}{3}} = 1.8\,\text{inches}$$

Given this value of b, we conclude that the height h is 3.6 inches.

---

numerical example 21.4 Redesign of a Beam, Considering Beam Deflection

A beam of the dimensions calculated in Numerical Example 21.3 should structurally withstand the specified load. Reflecting on Chapter 20, perhaps you are asking, "I wonder if the beam will deflect too much?" Using the rule of thumb that the deflection should be less than 1/360 of the beam length (0.042 foot or 0.5 inch in this case), we can investigate whether this beam will deflect excessively.

solution

To calculate the deflection, we will use Equation 20.4 from Chapter 20. To keep units consistent, the beam length L will be inserted with units of inches.

$$\Delta = \frac{FL^3}{4Ebh^3} = \frac{2,000\,\text{lb} \cdot (180\,\text{in})^3}{4 \cdot (29 \cdot 10^6\,\text{psi}) \cdot 1.8\,\text{in} \cdot (3.6\,\text{in})^3} = 1.3\,\text{inches}$$

We can conclude that although the beam will not break when a force of 2,000 lb is exerted, it will deflect too much. Consequently, the beam needs to be re-designed to ensure that deflection is within allowable limits. We will determine the necessary cross section dimensions by solving Equation 20.4 from Chapter 20 for b (knowing that $h = 2 \cdot b$ and $\Delta = 0.5$ inches).

$$b = \left(\frac{F \cdot L^3}{32 \cdot E \cdot \Delta}\right)^{\frac{1}{4}} = \left(\frac{2,000\,\text{lb} \cdot (180\,\text{in})^3}{32 \cdot (29 \cdot 10^6\,\text{psi}) \cdot 0.5\,\text{in}}\right)^{\frac{1}{4}} = 2.2\,\text{inches}$$

Given this value of b and the specified height-to-width ratio, the beam height, h, equals 4.4 inches. A beam with these dimensions will not deflect "too much." Since these dimensions are larger than the original beam we designed, we can be sure that the stress in the beam will be much lower than the ultimate stress. This can be confirmed by inserting the dimensions into Equation 21.4; the result demonstrates that the ultimate stress in the simply supported beam will be less than the previously determined design criteria for stress of 24,000 psi.

$$\sigma_{ss} = \frac{3 \cdot F \cdot L}{2 \cdot b \cdot h^2} = \frac{3 \cdot 2000\,\text{lb} \cdot 180\,\text{in}}{2 \cdot 22\,\text{in} \cdot (4.4\,\text{in})^2} = 19{,}000\,\text{psi}$$

This design results in a FS value of 1.9 (36,000 psi/19,000 psi).

Transportation Engineering: Horizontal Curves

In plan view, roads are composed of straight sections, termed tangents, and horizontal curves. A horizontal curve provides a transition between two tangents and facilitates the movement of traffic between the two tangents. The tangents meet at an imaginary point of intersection (POI) as shown in Figure 21.23.

The radius of the horizontal curve and the amount of **superelevation** dictate the speed at which a vehicle can move through a horizontal curve. The cross slope of superelevation is often expressed as a percent, where a superelevation of 5 percent would describe a road cross-section that has a slope of 0.05 feet of rise (vertical) per 1 foot of run (horizontal). This is similar to our previous discussion on road crowns in Chapter 4, Transportation Infrastructure, in which the normal cross section of a road has a cross slope of 2 percent downward from the centerline crest. Figure 21.24 provides a scaled drawing of a 40-foot wide roadway superelevated at 5 percent.

Intuitively, most people understand that as the radius of a curve increases, a vehicle can move through that curve more rapidly. Tight curves (i.e., curves with small radii) require low speeds while large radius curves allow a higher

Superelevation refers to the extent that a curve is banked or tilted in cross-section. It is also known as camber or cross slope.

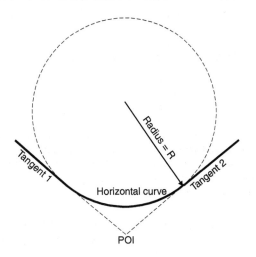

Figure 21.23 A Horizontal Curve Between Two Tangents and the Point of Intersection (POI) of the Tangents.

Figure 21.24 **Roadway with Superelevation of Five Percent.**

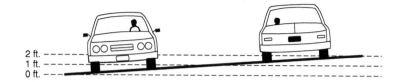

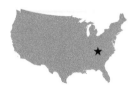

travel speed. In addition, as the rate of superelevation increases, the rate at which vehicles can safely move through the curve increases. This is the reason that automobile race tracks are superelevated at very high slopes; for example, the Bristol Motor Speedway (Figure 21.25) has a superelevation of 10 percent, which corresponds to 36° and is the steepest superelevation of any NASCAR track. To put this in perspective, 36° is the approximate slope of an Olympic ski jump or a "black diamond" downhill ski run.

Equation 21.6 is a governing equation for the design of horizontal curves, and agrees with intuition as it takes into account the rate of superelevation and the curve radius. In addition, the friction between the vehicle's tires and the roadway surface will affect the design radius. Note that Equation 21.6 is a non-homogeneous equation; you must use the units specified in order for the radius to be calculated in units of feet.

$$R = \frac{v^2}{15(e+f)} \tag{21.6}$$

where

$R =$ radius of curve (ft)
$v =$ velocity of vehicle in curve (mi/hr)
$e =$ rate of roadway superelevation (ft/ft)
$f =$ side friction factor, expressed as a decimal

Figure 21.25 **The Resurfacing of Bristol Motor Speedway.** This photograph illustrates the superelevation of a NASCAR racetrack.

Source: Courtesy of Baker Concrete Construction.

Remember that full-sized versions of textbook photos can be viewed in color at www.wiley.com/college/penn.

numerical example 21.5 Horizontal Curve Design

A rural road has been designated for resurfacing and realignment. Currently, it has a design speed of 55 mph, but a horizontal curve between two perpendicular tangents is only designed for 25 mph.

a. Calculate the radius of the existing curve if the curve is not superelevated and the side friction factor is 0.1.

b. When realigning the road, engineers wish to use a design speed of 45 mph for the curve. How might they achieve this?

numerical example 21.5 Horizontal Curve Design (Continued)

solution

a. We address this problem by using Equation 21.6, given that $e = 0$ and $f = 0.1$. These values can now be substituted into Equation 21.6.

$$R = \frac{v^2}{15(e+f)} = \frac{(25\,\text{mph})^2}{15(0+0.1)} = 420\,\text{ft}$$

b. We have two variables to manipulate in Equation 21.6, e and R, and thus there are an infinite number of solutions. For example, if the curve is superelevated to 4 percent ($e = 0.04$), using Equation 21.6 with a velocity of 45 mph and side friction factor of 0.1 yields a radius of 960 feet. Alternatively, a superelevation rate of 6 percent decreases the radius to 840 feet. Choosing the appropriate superelevation rate and radius would be accomplished by referring to appropriate codes and standards, as well as considering land use issues and construction costs.

As we have emphasized in this textbook, engineering consists of much more than inserting values into equations and solving those equations. This is certainly true in the case of horizontal curve design. For example, as the designed radius of a curve increases, the amount of land required will also increase. As a result, more property will be affected; this affected property might be very expensive to purchase, or may include wetlands or other environmentally sensitive areas. If the land is privately owned, the potential for conflict with landowners will increase. Also, safety is paramount in the design of horizontal curves. When the radius requires a lower speed limit than that of the incoming tangent, the potential for accidents increases. Small-radius curves, by making vehicles deccelerate and then accelerate, are less fuel efficient than large-radius curves. Moreover, designers must take into account transitions

 Flash Animation of Roadway Transitions

"Dead Man's Curve"

One of the few 90-degree curves in the Interstate Highway System exists along I-90 in Cleveland, Ohio (Figure 21.26). It was first opened to traffic in 1962 with a 50-mph speed limit. Many accidents, including truck rollovers, occurred. In 1965, the posted speed limit was lowered to 35 mph, but the accidents continued. In 1969, the curve was superelevated and the median guardrail was replaced with a concrete barrier. Recently, larger signs, better night-time lighting, and flashing warning lights have been added.

Mahmoud Al-Lozi, a highway system planner for the Northeast Ohio Areawide Coordinating Agency, commented on why the accident rate remains high, despite all of the efforts to increase driver safety, "They get used to traveling curves posted at 45 mph at 55 mph with no problems, so when they encounter a bad one, they're not ready for it. Dead Man's Curve is one of those where they say 35 mph and they mean 35 mph" (Sweeney, 2001).

Figure 21.26 A Sign Alerting Drivers to "Dead Man's Curve" on I-90 in Cleveland, Ohio. Note that this is *not* an off-ramp.

Source: P. Parker.

Figure 21.27 **Existing and Proposed Alignment.**

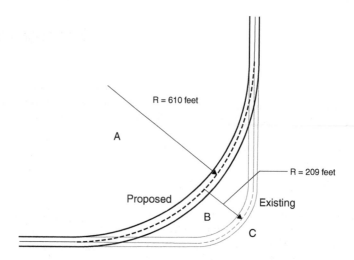

between crowned cross-sections of tangents and superelevated cross-section of curves.

We can see the effect of curve radius on consumption of land by considering a hypothetical case. Figure 21.27 illustrates an existing and proposed alignment. The proposed alignment is presented as dark lines and the existing alignment as light lines. Offsets have been drawn 15 feet to the side of each centerline to represent the edge of pavement. For each case, the dashed line is the circular arc segment that comprises the horizontal curve.

The length of the proposed redesigned horizontal curve is 960 feet, which multiplied by a pavement width of 30 feet, yields a road area of 29,000 square feet, or 0.67 acres. Additional land will need to be purchased beyond the edge of pavement for the right-of-way. If the right-of-way width is 40 feet (20 feet from the centerline), the total land area to be acquired is 0.88 acres. Depending on property values in the area, the cost of this land may amount to a significant fraction of the total project cost. In addition, the design engineers must address this question: what should be done with the land between the proposed and existing alignment (denoted with a 'B' in Figure 21.27)? This land has an area of 0.93 acres, and most likely belonged to property owner A before construction began, but may have little or no value to the owner as it is no longer contiguous with his or her property. Moreover, property owner C may have no use for the land either. In such a case, the road owner (e.g., the county, state, or township) may need to purchase that land also from property owner A and be responsible for maintaining that land in the future.

The values reported in the previous paragraph were obtained from the computer aided drafting (CAD) program with which Figure 21.27 was drawn. You can easily replicate this work in any CAD program; alternatively, answers with satisfactory precision can be obtained by use of a compass, protractor, and engineering scale.

Outro

In future engineering courses, you will be able to analyze and design much more complicated infrastructure components than we have presented in this book. We hope that in your future courses, where the tendency may be to focus on technical topics, that you continue to view infrastructure components and sectors in light of the many non-technical considerations that we have presented in this textbook.

chapter Twenty One   Homework Problems

Calculation

21.1   Reconsider Numerical Example 21.5. What is the radius of the redesigned curve if there is no superelevation? How does your answer compare to the answers provided in the example?

21.2   Reconsider Numerical Example 21.5. One way to increase the allowable traffic speed around the curve is to increase the side friction factor f. If the pavement is constructed of concrete, it can be diamond-ground, effectively increasing f from 0.1 to 0.18. Alternatively, a type of asphalt overlay termed "open-graded friction course" can increase f from 0.1 to 0.15.

a. For the existing curve (i.e. radius is 420 feet and no superelevation), how much will these two rehabilitation techniques increase the design speed of the existing curve?

b. The asphalt overlay can be applied in such a way to increase superelevation. How much will increasing e to 0.4 and 0.6 increase the design speed for the existing curve?

21.3   Use a spreadsheet to create a graph of minimum curve radius as a function of vehicle speed, given that $e = 0.0$ and $f = 0.15$. Show vehicle speed on the x-axis, varying from 0 to 60 mph.

21.4   Two horizontal tangents meet at a $90°$ angle. For a 55 mph design speed, design two "safe" curves, specifying values of e, f, and R for each of your designs. Also, use a CAD program or a careful hand drawing to estimate the horizontal curve length of each of your designs.

21.5   A community of 510,000 people generates wastewater at a rate of 105 gallons/capita/day. The community is growing at a rate of 0.50 percent per year. When designing a new WWTF, design flows are typically projected 15 years into the future.

a. Using the typical detention time, calculate the volume of primary clarification needed (in cubic feet).

b. If ten parallel rectangular clarifiers are to be designed, provide the dimensions (in feet) of each tank. Use typical depths and length-to-width ratios.

21.6   What is the settling velocity of a 0.1-mm diameter sand grain in wastewater if the particle density is 2.5 times that of water, the wastewater density is $1,000\,kg/m^3$, and the wastewater viscosity is $0.891 \times 10^{-3}\,kg/m \cdot s$? How long will it take this particle to settle to the bottom of a 4 meter deep tank? Given this settling time and the typical detention time of a primary clarifier, would the particle be removed?

21.7   If concrete is placed at $70°F$ at a rate of 2 feet/hour, what is the required stud spacing for 1-1/4-inch board

sheathing? Draw a dimensioned diagram of the studs for a 4-foot high by 8-foot wide sheet of board sheathing. Include a stud at both ends of the sheathing.

21.8   For Homework Problem 21.7, diagram the wale spacing for 2-inch × 4-inch wales. Include a wale at the base of the form, but not at the top.

21.9   Consider a simply supported steel beam that needs to carry a load at its center of 10,000 pounds. The beam length is 20 feet. If the beam height is to be twice its width and a factor of safety of 1.5 is to be used, determine the beam's height and width using the maximum permissible stress method.

21.10   Consider a simply supported aluminum beam that needs to carry a load at its center of 10,000 pounds. The beam length is 20 feet. If the beam has dimensions of $h =$ 5-inch and $b =$ 10-inch and a factor of safety of 1.5 is to be used, determine whether the beam will fail according to the maximum permissible stress method described in this chapter. Also, for this beam, calculate the deflection at the center of the span.

21.11   A beam made of stainless steel is 50 feet long and is to carry a load of 10 tons at its center. Using a factor of safety of 1.5, you are to specify the dimensions (b and h) of a rectangular cross-section beam such that it will not deflect too much nor carry more than its failure stress. Propose three sets of dimensions that will work.

Short Answer

21.12   Why isn't Stokes' Law typically used to design primary clarifiers?

21.13   If a WWTF designer fails to account for BOD removal in a primary clarifier, what impact will this have on the design of secondary treatment?

21.14   Why can forms be removed before the concrete has completely cured?

21.15   By doubling the detention time in a primary clarifier from 2 hours to 4 hours, the percent removal of suspended solids does not double. Why not?

21.16   Compare the relative performance of an I-wall to a T-wall with respect to sliding failure.

Discussion/Open-Ended

21.17   One method that would avoid needing to realign the roadway of Figure 21.27 would be to lower the speed limit for the entire road. What are the pros and cons of such an approach?

21.18 Why might both wood and modular forms be used at the same site, as seen in Figure 21.11?

Research

21.19 Some soils in New Orleans are such that they require extended foundation depths (see Figure 21.20) for floodwalls. Research these soils and write a one-paragraph summary of your findings.

21.20 Explain how extended foundations affect the likelihood of T-wall failures due to:

a. sliding

b. soil bearing capacity.

Key Terms

- detention time
- failure stress
- floodwall failure
- formwork
- maximum permissible stress method
- stress
- superelevation
- T-wall

References

American Society of Civil Engineers (ASCE) Hurricane Katrina External Review Panel. 2007. *The New Orleans Hurricane Protection System: What Went Wrong and Why*. American Society of Civil Engineers, Reston, VA.

Federal Emergency Management Agency (FEMA). 2001. *Engineering Principles and Practices of Retrofitting Flood-Prone Residential Structures*. FEMA 259, edition 2.

Love, T. W. 1973. *Construction Manual: Concrete and Formwork*. Carlsbad, CA: Craftsman Book Company.

Steel, E., and J. J. McGhee. 1979. *Water Supply and Sewerage*, 5th Edition. New York: McGraw-Hill.

Sweeney, J. F. 2001. "Dead Man's Curve Could Be Worse—In Fact, It Was." *Cleveland Plain Dealer*, April 22, 2001.

Chapter Opener Photo Credit

Chapter 1: M. Penn.

Chapter 2: Wikipedia: http://commons.wikimedia.org/wiki/File:Widnes_Smoke.jpg.

Chapter 3: Copyright © oversnap/iStockphoto.

Chapter 4: M. Penn.

Chapter 5: M. Penn.

Chapter 6: Federal Emergency Management Agency/M. Rieger.

Chapter 7: U.S. Army Corps of Engineers/Digital Visual Library.

Chapter 8: Copyright © Alan Crawford/iStockphoto.

Chapter 9: Courtesy of Ayres Associates.

Chapter 10: Copyright © OSTILL/iStockphoto.

Chapter 11: Courtesy of the Massachusetts Department of Transportation.

Chapter 12: P. Parker.

Chapter 13: M. Penn.

Chapter 14: Prill Mediendesign & Fotografie/iStockphoto.

Chapter 15: M. Penn.

Chapter 16: Prill Mediendesign & Fotografie/iStockphoto.

Chapter 17: Courtesy of S. Owusu-Ababio.

Chapter 18: Courtesy of R. Webster.

Chapter 19: Courtesy of Caltrans, District 4 Photography/J. Huseby.

Chapter 20: M. Penn.

Chapter 21: Andy McNeill.

Index